STEM CELLS – LABORATORY AND CLINICAL RESEARCH

STEM CELLS ORIGINS OF CYTOKINE INTERACTIONS

STEM CELLS – LABORATORY AND CLINICAL RESEARCH

Additional books and e-books in this series can be found on Nova's website under the Series tab.

STEM CELLS – LABORATORY AND CLINICAL RESEARCH

STEM CELLS ORIGINS OF CYTOKINE INTERACTIONS

PRASAD S. KOKA
EDITOR

nova
Medicine & Health
New York

Library of Congress Cataloging-in-Publication Data

ISBN: 978-1-53615-302-6

Published by Nova Science Publishers, Inc. † New York

CONTENTS

PREFACE

CD34+ pluripotent hematopoietic stem cells (HSCs) are generally involved in all aspects of the human system as well as in the stem cell therapeutic strategies, as in bone marrow transplantation (BMT). This book deals with several original research articles on HSC proliferation and containment. Lineage directed differentiation of HSCs and use of easily available herbal extracts for greatly increased proliferation into erythrocytes to treat thrombocytopenia are herein reported.

Chapter 1 - HSCs respond to the growth factors and the herbal medicine is a welcome substitute with lack of known side effects for treatment purposes. However molecular mechanisms as in signal transduction are more easily deciphered than those derived from the herbs which may well be taken up for future studies. Mutations in the signal transduction genes such as JAK-2 have been shown to cause dysregulation of the CD34+ cell differentiation causing erythrocytosis, which is the reverse of thrombocytopenia, both of which are undesirable clinical consequences! Erythrocytes have been shown to date back to ancient tissues from the fossils.

Chapter 2 - Objectives: Determine complete and partial transection rates (CTR/PTR) of follicular units in follicular unit extraction (FUE) and the techniques employed to minimise the follicular transfection rates (FTR). *Methods*: FTR values were evaluated for 50 cases. The "permanent" hair around the scalp are extracted from the follicular roots and grafted through small circular incisions around a follicular unit on the top areas (occipital part) of the scalp where loss of hair in a natural manner that has occurred. Classification of the FTR was performed using Bernstein and Rassman FOX test into five grades. *Results*: Complete FTR - Mean ± SD = 8.0 8 ± 6.98 and partial FTR - Mean ± SD = 16.78 ± 12.01. This includes graft quality of all the 5 grades of FOX test for the 50 cases of individuals who have undergone follicular transfection or dissection. *Discussion*: Follicular transfection or dissection could be affected by the individual hair and skin characteristics and the procedure is generally desired for cosmetic reasons. Baldness which is restricted to occipital area of the scalp is a major cause to undergo hair transplants which involve autologous follicular removal procedures from the lateral parts. *Conclusion*: Low FTR is very essential for good harvest of grafts in FUE and better outcome. Newer methods and techniques should be employed for minimising transfections. More studies are required to validate this study.

Chapter 3 - Increasing evidence highlights the multipotent potential of mesenchymal stromal cells (MSCs), produced in peripheral tissues, bone marrow etc. to promote tissue regeneration and hence cell based therapies. In this regard, the human placenta has attracted attention for translational research due to its immunomodulatory properties, large stem cell number in tissue, and ease of availability for clinical experimentation. The placenta plays the role of an endocrine organ in providing the foetus nutrition and also exerts some control over maternal hormones during gestation through expression of several hormones including the growth hormone releasing hormone (GHRH), placental growth hormone (GH) and insulin like growth factor (IGF-1). While on one hand, the role of GHRH in the placenta is unclear and there are only speculations of a paracrine role for GHRH in the placenta; on the other hand, GHRH and its analogues have been used to pre-condition bone marrow MSCs for cell based therapies in treatment of angiogenesis and GHRH antagonists are being considered for cancer therapy, due to its role in regulating somatic growth. Given this scenario, the authors studied the effect of GHRH and IGF-1 on the proliferation and differentiation of human placenta derived MSCs. GHRH and IGF-1 by themselves and in combination stimulated faster proliferation of MSCs compared to controls. The combination of GHRH and IGF-1 also enhanced the rate of osetocytic and chondrocytic differentiation. GHRH was particularly effective in inducing chondrogenesis in p-MSCs. The authors' study opens the possibility for hitherto unidentified roles for GHRH in the placenta.

Chapter 4 - Skeletal remains from ancient populations are usually the only material left for archaeologists to examine. Multidisciplinary research has revealed structures consistent with the same shape and size as modern erythrocytes in ancient bone marrow and skeletal tissue. These findings have been questioned because postmortem fungi contamination has a similar morphology to erythrocytes. Due to the risk of fungi contamination, identification of erythrocytes by specific proteins using immunohistochemical staining has been considered a requirement for accuracy. However, immunohistochemical staining only demonstrates that the target protein is present on the cell surface and if the protein is degraded, the researcher may misclassify an erythrocyte as a contaminant and loose information. The authors provide an alternative approach using dyes that can verify erythrocytes, hemosiderin, and fungi but does not require specific cell-surface proteins. This method requires less time and is not as costly as immunohistochemical studies. Staining with dyes can provide important information of the quality of the sample before advanced procedures are employed. Paraffin-embedded demineralized bone sections from archaeological excavations of two Bronze Age burials and commercially available control *Aspergillus* slides were stained using haematoxylin and eosin (H&E), PAS staining with a Schiff reagent to detect fungi (carbohydrates), and an iron-specific stain that detects hemosiderin in degraded erythrocytes. The authors' research demonstrates that it is possible to distinguish between erythrocytes and fungi using a combination of staining techniques.

Chapter 5 - In this report the authors will discuss the early formation of the cancer stem cells (CSCs) both in terms of human genetic evolution and also in the role of stem cell differentiation from totipotency via pluri- and multi-potency to terminal differentiation. In addition, the authors will address the issues of therapies that target specifically CSCs, even within the conglomeration of cancer cells as a whole and the onset and development of this disease. The authors will address the possibility that interactions between normal hematopoietic stem cells (HCSCs) and cancer cells leads to the formation of cancer stem cells to monitor emergence of malignant transformation that give rise to or from cancer stem cells. If and whether indeed the HCSCs lose their stem cell characteristics over time during the course of the disease progression, would be a matter of great interest in cancer etiology. But this does not seem to happen in leukemias. The reason as to why HCSCs in preference to other types of stem cells would play a role in the formation of cancer stem cells is that the HCSCs are the components of blood formation, even if these cells are not solely involved only in the formation of primarily leukemias *in vivo,* due to their imperative circulation in the body. As the authors have previously reported, it is plausible for the cell biologically engineered generation and characterization of hematopoietic cancer stem cells, *in vivo*. The authors had observed evidence for non-leukemic malignancies such as tumorigenesis in these humanized mice. These would be expected to lead to potential anti-cancer therapies due to the generally perceived implication that the CSCs are therapy-resistant as opposed to their non- CSC counterparts. Presumably these malignancies exist in latency and cause a relapse and metastasis despite the use of currently available anticancer therapies which may yet kill CSCs in a selective manner, or a replenishment of CSCs will occur post-treatment. A drug dependent susceptibility possibly partially and transiently may also be likely. The anticancer treatment resistance could potentially be phenotype dependent of the CSC *in toto*. This will not only help in understanding the (stem cell) origin of cancer, but could also be used in translational and clinical treatments that may ultimately benefit many patients with cancer.

Chapter 6 - *Background:* Hypertension (HTN) is a highly prevalent disease and a leading cause of morbidity and mortality worldwide. It leads to several long term complications including cognitive impairment. Yoga is a form of alternative and complementary therapy known to be effective in improving cognitive function. There is however, a lack evidence of the role of yoga on cognitive function in hypertensive patients. *Aim:* To study the impact of short term residential yoga-based lifestyle intervention on cognitive functions in hypertensive patients. *Methodology:* Sixty hypertensive subjects, within age range 30 to 60 years (Mean ± SD; 46.92 ± 9.2 yrs) were enrolled in this study. Thirty subjects (IAYT) followed 6 days of residential yoga program, consisting of 5 hrs of different yoga sessions including physical practices, meditation, devotional songs, a low salt & calorie diet and counseling sessions at SVYASA University campus in Bengaluru. Thirty subjects (Non-IAYT) followed their daily routines. Subjects with diabetes, psychiatric problem, stroke, head injury, neuro-deficits, history of recent surgery, and previous exposure to yoga in last one year were excluded from the study. All the subjects were administered Digit Letter Substitution Test and Digit Forward and Backward Test at the baseline and after 6 days. *Results:* For the IAYT group, there was a significant improvement in Digit Forward and Backward Test ($p>0.001$) scores and in DLST test ($p>0.001$) after 6 days of intervention compared to baseline. Whereas, non-IAYT group showed, no significant improvement in any of the variables assessed. Therefore, compared to the Non-IAYT group, the IAYT group showed significantly better improve in DF ($p < 0.001$), DB ($p < 0.001$) and DLST scores ($p < 0.001$). *Conclusion*: Short term residential yoga practice helps in enhancing cognitive function in hypertensive patients.

Chapter 7 - Background: In bone tissue engineering, biphasic calcium phosphate (BCP) with different hydroxyapatite (HA)/β-tricalcium phosphate (β-TCP) ratios can be combined with regeneration-competent stem cells, and bioscaffolds like fibrin that trigger neovascularization. The authors have shown recently that human adipose stem cells (hASCs) on BCP with a HA/β-TCP ratio of 20/80 (BCP20/80) incorporated in fibrin gel show enhanced osteogenic and/or vasculogenic differentiation potential compared to BCP with a HA/β-TCP ratio of 60/40 (BCP60/40) *in vitro*. A hypoxic microenvironment for the cells might occur in these hASC-seeded BCP-based composites after implantation *in vivo*. This limits oxygen supply which is essential for successful bone tissue engineering. Objectives: Here the authors aimed to test the effect of hypoxia on osteogenic and/or vasculogenic differentiation of hASC-seeded BCP60/40 or BCP20/80 scaffolds in fibrin gels as well as fibrin gel degradation. Methods: BCP60/40-based composites and BCP20/80-based composites were cultured under normoxia (20% O2) or hypoxia (1% O2) for 12 days, and total DNA content, alkaline phosphatase activity, nitric oxide production, vascular endothelial growth factor production, and fibrin gel degradation were assessed during culture. Results: Hypoxia similarly decreased total DNA content at day 7 (1.3-fold) and 12 (1.2-fold) in both composites. It also decreased alkaline phosphatase activity at day 5 (5.0 and 4.7-fold), day 7 (4.5 and 6-fold), and 12 (7.1 and 6.8-fold) in BCP60/40-based composites and BCP20/80-based composites. Hypoxia similarly increased vascular endothelial growth factor production at day 5 (3.9-fold) and day 7 (2.6-fold) by both BCP-based composites. Hypoxia did not affect nitric oxide production and fibrin gel degradation in both BCP-based composites at all-time points. Conclusion: Hypoxia decreased osteogenic differentiation but enhanced vasculogenic differentiation of cultured BCP-based composites. These *in vitro* results implicate that cell-seeded BCP-based composites implanted *in vivo*, might enhance vascular endothelial growth factor production resulting in sprouting and tube formation leading to increased survival of cells in these composites, and stimulation of bone formation.

Chapter 8 - The osteogenesis is a complex process that involves an accurate control of bone development, growth as well as remodeling during postnatal life. In particular, osteoblasts arise from mesenchymal stem cells through a process regulated in several steps that leads to the osteogenic differentiation. Class I homeobox genes (HOX in human and Hox in mice), are 39 transcription factors, mainly involved in the regulation of embryonic development. It has been shown that HOXA10 was able to activate several osteoblast related genes like Osx. The HOXA10 gene was considered a key factor able to control osteogenesis. In the last decade, several non-coding RNAs (ncRNAs), have been identified like crucial factors during osteogenesis. HOTAIR is an lncRNA related to Osteosarcoma and able to induce the Epithelial Mesenchymal Transition. The aim of this paper was understood the role of HOTAIR and HOXC locus in the control of Osteoblasts development in MSCs treated with adipose tissue obtained with Lipogems method.

Chapter 9 - The differentiation capabilities of human $CD34^+$ stem cells can be exploited in understanding the development of erythrocytes and its related erythroid lineage disorders. Hence, in the present study, human $CD34^+$ stem cells were isolated from a donor and cultured in DMEM medium. Immunocytochemistry (ICC), Scanning electron microscopy (SEM) and flow cytometry were used to assess the purity and characteristic features of $CD34^+$ stem cells. This pure culture was induced with erythrocyte differentiating medium (EDM) containing erythropoietin (EPO), stem cell factor (SCF), interleukin 3 (IL-3) and Granulocyte-macrophage colony-stimulating factor (GM-CSF) for 21 days, the morphological changes of erythrocytes were assessed with Leishman's stain and hematopoietic colony forming unit (CFU) assay. The expression of erythrocytes-specific markers ζ-globin, ε-globin, β-globin, *Glycophorin A* (GPA) and *BGP1* genes were evaluated by quantitative Real-time PCR. The ICC and FACS results showed (12×10^5 cells/mL) purity of the culture expressing CD34 marker. These pure cells on culturing with EPO, SCF, IL-3 and GM-CSF for 21 days showed conspicuous biconcave morphology when stained with the Leishman's stain. The results of the CFU-E assay indicated majority of the differentiated cells are erythroid progenitors. The qRT-PCR analysis showed distinct expression with ζ-globin, ε-globin, $\boldsymbol{\beta}$-globin, *GPA* and *BGP1* genes in differentiated erythrocytes compared to $CD34^+$ cells. Further, GPA^+ expression was also confirmed by ICC. All these results conclusively elucidate that the $CD34^+$ stem cells were successfully differentiated into erythrocytes and these differentiated cells may be useful in the treatment of various erythrocytosis and erythroid lineage disorders.

Chapter 10 - The Myeloproliferative disease that arises due to variation in the haematopoietic stem cell (HSC), and usually proliferates and dominates over normal HSC causing formation of enormous amount of erythrocytes (absolute erythrocytosis) in the blood is called Polycythemia vera (PV). Idiopathic erythrocytosis (IE) is absolute erythrocytosis of undefined etiology. In the present study, 54 patients with primary absolute erythrocytosis were thoroughly worked up including Endogenous erythroid colony forming assay (EEC), followed up for 2-4 years and analyzed retrospectively. The diagnoses were based on BCSH guidelines (2007) and WHO revision criteria (2016). The authors diagnosed 33% (18/54) patients as Polycythemia vera (PV) and 66.6% (36/54) patients as absolute erythrocytosis most probably Idiopathic Erythrocytosis (IE). Overall, hypertension (74%), dyslipidemia (55%), overweight (18%), consumption of alcohol (66.6%), smoking (55%) were seen and all these could have contributed to the pathology. Splenomegaly was seen in 14/18 (77.7%) in PV, 6/36 (16.6%) in IE. Serum EPO levels were below normal range 12/18 (66.6%) in PV, 14/36 (38.8%) in IE patients. EEC formation was observed 10/12 (83.3%) in PV, 14/22 (63.6%) in IE patients. Differences in WBC, platelet counts, EPO levels between PV and IE were statistically significant. In PV patients, *JAK-2* exon 14 mutation (V617F) was seen in 17/18 patients (94.4%). Only one patient was *JAK-2* (exon 12, 14) Negative and none were JAK2 exon 12 positive. IE was twice as common as PV in the authors' study group and interestingly, most of the IE patients were EEC positive. The number of EEC colonies was significantly high ($p < 0.001$) in PV patients when compared with IE patients and normal positive controls.

Chapter 11 - Multiple sclerosis is an autoimmune disease characterized by inflammatory demyelination of the nerves in the central nervous system (brain and spinal cord). Depending on the attacked sites, the disease manifestations vary from limited sensory and or motoric deficits to wide spread neurological and cognitive deficits. Similar to other autoimmune neurological and non-neurological diseases, the available therapeutic strategies rely on the pharmacological modulation of the immunity, which could be associated with significant side effects. In this chapter, the light will be shed on another innovative therapeutic strategy that relies on the biological modulation of the immune response, together with the inhibition of the recruitment of the immune cells into the central nervous system after distortion of the blood brain barrier, as well as the enhancement of the regenerative ability towards the restoration of efficient blood brain barrier and the regenerative recovery of the neurological deficits.

Chapter 12 - Biological investigation was conducted to study *in vitro* changes in morphological properties and metabolic activity of human continuous cell lines after treatment with the morphogenetic field hologram information (the field). The authors used human tumor cell lines, including cervical (HeLa) and lung cancer (A549), as well as one normal fetal lung fibroblasts cell line (MRC-5) to investigate the effect of such treatment. Metabolic activity of investigated cell lines was evaluated by MTT assay. Results showed that MRC-5 cells exposed 3 h to the field had 50% increases in metabolic activity 48 h after treatment in comparison to untreated control cells. Similar but lower effect was observed in both tested tumor cell lines. Changes in cells depended also on the exposure time. Morphology of the cells was examined by visual observations using light microscopy. Metabolic activity was evaluated by MTT assay. Results of this study showed that application of the field has measurable influence on biological systems.

Chapter 13 - Although, the studies about the genotoxic and cytotoxic effects of methacrylate resins have increased considerably, the understanding of molecular regulatory basis is still poorly understood. HOTAIR is a non-coding RNA related to several diseases; originally, it was found in colorectal, gastric cancer and hepatoma. Moreover, HOTAIR is one of the most important regulator of the Epithelial-Mesenchymal Transition process. The aim of this paper was to investigate the ability of three commercial acid resins (Unifast, JetKit and Duralay) to control the HOTAIR gene expression. The cytotoxic activity in HGFs, measured by MTT assay, was always conditioned by high concentrations of the resins. On the other hand, the proliferation rate does not change in HGFs treated with Unifast and Duralay, while Jet Kit is a negative regulator of the cell proliferation. The HOTAIR gene expression was regulated in HGFs treated with Unifast and Duralay, while JetKit did not induce change in charge of HOTAIR expression. In this study, the ability of different commercial polyacrylic acid resins to regulate lncHOTAIR gene has been demonstrated for the first time. Moreover, no data have previously been reported about the HOTAIR gene expression in HGFs. Several experimental evidences identify HOTAIR as a crucial regulator of EMT; pathologically, any alteration of EMT process leads to several diseases. Particularly, overgrowth of gingival epithelium is due to the deregulation of EMT specific markers. In conclusion, the authors' results could lead to better understanding the role of HOTAIR in different gingival pathologies related to the EMT deregulation.

Chapter 14 - Cerebro vascular insult is commonly known as stroke which is caused by interruption of blood supply and rupture of blood vessels or due to the abnormal vascular structure. This leads to focal tissue loss and death of neural cell population. Even though stem cell transplantation is considered as likely treatment choice, it will be associated with certain unfavorable outcomes such as tumor formation, immune mediated graft rejections. Studies have shown that regenerative effects promoted by stem cells in larger part through their secreted bio-factors and exosomes. Neural stem cells are present in some parts of the adult brain they participate in regeneration events. As the result ischemic shocks, capability of synthesizing trophic factors and signaling molecules are severely compromised in injured areas of Central Nervous System. Indeed, several experimental studies shed lights on practical applicability of neurotrophic factors (NTFs), MSCs secretome and exosomes in inducing regenerative repair in the nervous system. In the case of stroke, the major concern is to protect the penumbra by limiting consequential cell death, further enhancing the regeneration and repair. Therefore, providing neuroprotective and regenerative support through exogenous NTFs, stem cell secretome and exosomes or by stimulating their endogenous secretion can help to compensate tissue loss, regain neural plasticity and functional recovery after stroke. Intranasal administration of NTFs and stem cells derived neuro therapeutics ensures efficient, safe and direct delivery to central nervous system which bypasses blood brain barrier. This cell-free therapeutic strategy appears to be safe and effective also eliminates the invasive transplantations and its associated adverse side effects.

Chapter 15 - *Background:* Hypertension significantly affects the mental health of the patient. Elevated anxiety and depression are commonly observed in hypertensive subjects compared to normotensives. Previous studies have demonstrated the positive impact of IAYT on mental health in many chronic health problems. *Objectives:* To study the effects of Integrated Approach of Yoga Therapy (IAYT) on anxiety and depression in hypertensive subjects with elevated anxiety and depression. *Methodology:* The authors screened 131 hypertensive patients within age range 40-60 years for anxiety and depression (using Hospital Anxiety Depression Scale (HADS)), 92 subjects found to have either elevated anxiety or depression or both. These potential subjects underwent one week residential yoga program at Holistic Health Care Centre, S-VYASA University, Bangalore-India. Subjects who were on antipsychotic medication, had recent surgery, previous exposure to any form of yoga in past one year, secondary hypertension, resistant hypertension were excluded from the study. Heart rate, systolic & diastolic blood pressure, anxiety, and depression were assessed at baseline and after one week of IAYT intervention. Data was analyzed using SPSS version 10. *Result:* At the baseline, of 92 potential subjects, 47 (51.08%) had clinical anxiety, 26 (28.26%) had clinical depression, and 19 had both (20.65%). After one week of IAYT intervention significant decrease in pulse rate ($p < 0.001$), systolic ($p < 0.001$), & diastolic blood pressure ($p < 0.001$), anxiety ($p < 0.001$), and depression ($p < 0.001$) was observed compared to baseline. *Conclusion:* Present study indicates the potential role of IAYT in improving mental health & cardiac variables in hypertension subjects. However, further randomized control trial should be performed to confirm the findings of the present study, which will be implemented in future studies.

Chapter 16 - *Background:* The signals received by Hematopoietic stem/ progenitor cells (HSCs) for proliferation and differentiation within the bone marrow environment are critical for understanding and characterization of their mechanisms, which likely to decipher coordinated expression of related functional genes & associated multiple pathways. This study was aimed to identify potential biomarkers of transient effect on Umbilical Cord Blood (UCB) derived CD34+ cells in response to the treatment of a particular combination of cytokines Interleukin-3 (IL-3), Fms-Like Tyrosine Kinase-3 Ligand (FLT-3), and Stem Cell Factor (SCF). *Method:* The microarray data with accession number GSE3003 was extracted from NCBI to identify global signatures and to construct & analyze Protein-Protein Interaction (PPI) networks eliciting different biochemical responses in molecular signaling. *Results:* In this study, a total of 22,283 gene transcripts were identified. The most significant genes that are found to be responsible for HSC responses were 144 & 111-proliferation, 98 & 58-differentiation and 10 & 28 -self- renewal among the group of up-regulated and down-regulated genes respectively. Further, nine notable genes (RBM3, DHX32, PSMD9, SAR1A, SMYD3, ST6GAL1, CKAP4, CTR9) were identified having a crucial role in molecular signaling. Strikingly, among the gene-interacting partner associations, few combinations like RRAGC & PSMC2; CKAP4 & MANF on proliferation and CTR9 & CNTNAP2 on self-renewal properties have been found unexplored previously. *Conclusion:* The novel combinations-RRAGC & PSMC2; CKAP4 & MANF and CTR9 & CNTNAP2 that have been found unexplored previously would be a stepping stone to unlock their long-touted therapeutic & synergistic gene therapy potential through siRNA-mediated knock down techniques.

Chapter 17 - Stem cells from human exfoliated deciduous teeth (SHED) are originally referred to as fibroblastoid-colony-forming-cells because one of their characteristics is related to proliferation and differentiation potential. In this study, the authors *developed an efficient protocol for colony-forming-unit (CFU) assay for SHED.* The culture conditions including culturing time, seeding cell density and modifications of media supplements were optimized. The CFUs were determined on the proportion of colony types, sizes and the numbers of colony. Isolated SHED expressed CD44, CD73, CD105 and showed multipotent ability. Higher yield was obtained when SHED was cultured at the density of 500 cells per well of a 6-well plate for 14 days in Dulbecco's modified Eagle medium (DMEM) with 10% fetal bovine serum (FBS). The number and size of CFU was increased with the increase of FBS concentration in which the transit-amplifying cell (TAC) colonies were the most abundant, followed by the stem cell (SC) and differentiated cell (DC) colonies, respectively. In DMEM with 20% FBS and 100 μM L-ascorbic acid-2-phosphate, the CFU rate was highest with increasing the proportion of SC colonies. The improvement of the culture conditions described herein could be adopted to easily acquire more functional SHED for applying in cell therapy.

Chapter 18 - *Andrographis paniculata* (Burm.f.) Nees (Nelavembu) a natural herb extract largely used to increase the platelet count in thrombocytopenia condition. Normally, Hemato-poietic stem cells (HSCs) are key cell in the formation of platelets and depletion of platelets results in thrombocyte-penia. The present study is aimed to understand the effect of *Andrographis paniculata* (Burm.f.) Nees extract (APE) on proliferation of HSCs that eventually exhibit anti-thrombocytopenic effect. Andrograph-olide a prominent compound in APE is known to stimulate thrombopoietin from liver which showed greater affinity towards thrombo-poietin receptor compared to eltrombopag and pegacaristim (compounds used for generation of platelets). The exertion of APE on HSCs proliferation showed a significant effect on cell count with increasing concentrations from 1 μM to 2 mM. The results were concurrent with the in vitro IL-3 and G-CSF treated HSCs. These findings explain that administration of APE results in the HSCs proliferation and eventual differentiation into platelets.

Chapter 19 - The balance of stem cell self-renewal and differentiation maintains tissue homeostasis in different organs. Excess of stem cell self-renewal may lead to tissue hyperplasia and/ or tumorigenesis; while excess of differentiation may lead to tissue degeneration and/ or tissue aging. Asymmetric stem cell division, producing one stem cell and one differentiating cell, is a simple way to maintain the balance between stem cell and differentiated cell populations. Several recent studies suggest that the biology and behavior of lung stem and progenitor cells is controlled by both extrinsic signals and intrinsic factors. The intrinsic aspects of stem cell control provide the context in which extrinsic signals are interpreted.

Chapter 20 - Stem cells have been classified as drugs and so will necessitate compliance with regulations. Cell therapy initially started in the form of a bone marrow transplant for hematopoietic reconstitution to produce blood, a complex liquid tissue. Several years later, this experience has culminated in the administration of specialized cells with one or more intended therapeutic effects; this is a hallmark of cellular therapy. The authors are currently witnessing the emergence of this 'new age medicine' which uses a variety of stem cells for several unmet medical needs. Stem cells function by one or more mechanisms to regenerate the damaged tissue / organs by stimulating tissue-resident stem cells and by secreting different cytokines and growth factors. They also possess the potential to recruit themselves in the damaged part of the body and play a pivotal role to reinstate degenerated area through neoangiogenesis and neovascularization. While it is no secret that cells have been used as drugs for decades in one or the other forms, testing a specific therapeutic effect of cells objectively in the form of clinical trials is relatively recent. Lessons in the establishment of quality systems have been learnt as a part of establishment of blood banking and subsequently, cord blood banking, with emphasis on key components in the assurance of the safety and potency of cell therapy preparations. As more stakeholders in the academic medical centers and private companies are vying to be part of this new age medicine, there is a compelling need to get to the granular details of the development of the infrastructure, method developments, research requirements, analytical tools to assess and quantify efficacy, before it becomes part of main stream medicine. In this article, the authors discuss the various challenges specific to this industry, with reference to establishment of a quality management program.

In: Stem Cells Origins of Cytokine Interactions
Editor: Prasad S. Koka
ISBN: 978-1-53615-302-6

Chapter 1

STEM CELLS ORIGINS OF CYTOKINE INTERACTIONS

Prasad S. Koka*, PhD

CD34+ pluripotent hematopoietic stem cells (HSCs) are generally involved in all aspects of the human system as well as in the stem cell therapeutic strategies, as in bone marrow transplantation (BMT). This book deals with several original research articles on HSC proliferation and containment. Lineage directed differentiation of HSCs and use of easily available herbal extracts for greatly increased proliferation into erythrocytes to treat thrombocytopenia are herein reported. HSCs respond to the growth factors and the herbal medicine is a welcome substitute with lack of known side effects for treatment purposes. However molecular mechanisms as in signal transduction are more easily deciphered than those derived from the herbs which may well be taken up for future studies. Mutations in the signal transduction genes such as *JAK-2* have been shown to cause dysregulation of the CD34+ cell differentiation causing erythrocytosis, which is the reverse of thrombocytopenia, both of which are undesirable clinical consequences! Erythrocytes have been shown to date back to ancient tissues from the fossils.

Manipulation of the multipotent mesenchymal stem cells (MSCs) continues to be "exploited" by scientists from basic science uses, for the clinics and commercial external applications as in alopecia. Hormone or cytokine treated proliferations are some aspects whereas MSCs are generally capable of providing some essential cytokine or growth factor related self-secretions into their culture media. Conditioned medium (CM) from different cultures are a well-known byproducts of cell cultures. These from adipose tissue have been shown to regulate signal transduction genes in the MSCs, suggesting that despite the stromal MSCs secreting many factors into their CM, still are mechanistically susceptible to interactions, as expected, with the other generally systemic stem-derived

* Corresponding Author Email: kokaprasad005@gmail.com

cells which possess the stem-ness origins. The beneficial and deleterious effects of the secreted components in the CM and including platelet rich plasma (PRP), for clinical and cosmetic applications, is as reported require quality management of the stem cells based products. Besides, regulatory agencies such as FDA do need to be in a position, if not undertake to reproduce these experimental investigations, evaluate the produced side effects data to be of as authentic as of relevance, for not only efficacy but safety as well, to be devoid of side-effects.

The yet "unresolved" mystery of cancer stem cells (CSCs) persists with various theories and hypotheses in vogue or float among the stem cells scientific community, seeking leadership in acceptance of their own research based revelations. Whether the CSCs are ultimately responsible for relapse of malignancies, or also their persistence as in leukemias, have been of great interest. One of the articles in this book deals with CSCs in cancer disease evolution and its etiology.

CSCs are more likely to be of HSC origin than of any other stem cell types. Exactly at which type of stem cell-stage of differentiation upstream does this occur is at this point still a debatable issue/topic of interest? Diversity of growth factors from diverse stem cell types, possess their own characteristics at the molecular and hence functional levels. Interactions between different types of cells expectedly involve their secreted factors to help prevent or fight disease in the best case scenario. On the contrary, worst that can happen would be onset of detrimental causes for the living beings of which humans are most necessarily implicated in the acquirement of diseases, whether non-communicable, or infectious in nature.

Alternative herbal and complementary herbal/yoga therapies are here to decrease the negative symptoms but not serve as complete a cure. Neither is allopathy in its entirety as we continue to battle cancer, multiple sclerosis, AIDS, etc. Vitamins and supplements that have been in use for health benefits and these are also derived from biological sources. These therapies are a counter to the deleterious environmental, morphogenetic, or even biological systemic consequences.

In: Stem Cells Origins of Cytokine Interactions
Editor: Prasad S. Koka
ISBN: 978-1-53615-302-6

Chapter 2

FOLLICULAR TRANSECTION RATES IN FOLLICULAR UNIT EXTRACTION METHOD

Sandeep Mahapatra[1], Vignesh Subramanian[2], Rani James[2], Kaushik D. Deb[2] and Prasad S. Koka[2,*]

[1]Derma Solutions Hair Transplant and Skin Clinic, Bangalore, India
[2]Merisis Therapeutics - DiponEd BioIntelligence LLP, Bangalore, India

ABSTRACT

Objectives: Determine complete and partial transection rates (CTR/PTR) of follicular units in follicular unit extraction (FUE) and the techniques employed to minimise the follicular transfection rates (FTR).

Methods: FTR values were evaluated for 50 cases. The "permanent" hair around the scalp are extracted from the follicular roots and grafted through small circular incisions around a follicular unit on the top areas (occipital part) of the scalp where loss of hair in a natural manner that has occurred. Classification of the FTR was performed using Bernstein and Rassman FOX test into five grades.

Results: Complete FTR - Mean ± SD = 8.0 8 ± 6.98 and partial FTR - Mean ± SD = 16.78 ± 12.01. This includes graft quality of all the 5 grades of FOX test for the 50 cases of individuals who have undergone follicular transfection or dissection.

Discussion: Follicular transfection or dissection could be affected by the individual hair and skin characteristics and the procedure is generally desired for cosmetic reasons. Baldness which is restricted to occipital area of the scalp is a major cause to undergo hair transplants which involve autologous follicular removal procedures from the lateral parts.

Conclusion: Low FTR is very essential for good harvest of grafts in FUE and better outcome. Newer methods and techniques should be employed for minimising transfections. More studies are required to validate this study.

[*] Corresponding Author Email: kokaprasad005@gmail.com.

Keywords: hair follicle, transfection rate, extraction, scalp hair and skin, characteristics, transplants

INTRODUCTION

Alopecia androgenetica and *Alopecia areata*, and *Alopecia capitis totalis* are hereditary, patchy and total hair loss, respectively [1, 2, 3]. *Alopecia universalis* is total body hair loss. Stem cells treatment for vitiligo may address this autoimmune disorder that also comprises of *Alopecia areata* [4]. Keratinocyte-derived cytokines including keratinocyte growth factor are influenced positively in vitiligo and associated epidermal microenvironment [5, 6, 7, 8, 9]. Platelet rich plasma (PRP) or platelet rich follicular medium (PRFM) helps promote expansion of human hair follicle stem cells *in vitro* and also proposed for *Androgenic alopecia in vivo* [10, 11, 12]. It would be interesting and potentially beneficial if PRP/PRFM were to be administered in conjunction with hair follicular transplants.

FUE is a method of extracting, or "harvesting," donor hair in a follicular unit hair transplant procedure [13, 14, 15, 16, 17, 18]. Small circular incision in the skin around a follicular unit from the lateral part of the scalp is made, separating it from the surrounding tissue. It is then transplanted into the bald region of the scalp. Such a procedure results in the regeneration of permanent hair from the engrafted hair follicles.

Determination of FTR is useful for realizing the effectiveness of perfection to the maximum extent of FUE for subsequent grafting procedure on the same individual [19]. Successful FTR depends not only upon the professionals performing the procedure, but also on the physiological condition of the permanent hair follicle on the donor scalp area of the individual who is subjected to the FUE procedure. However those hair transplant team professionals who have significant and considerable experience and have achieved near or full perfection in undertaking these procedures can bypass the estimation of FTR as described herein and visually through microscopic examination perform the completion of the hair transplants. The reasons for bypassing the quantitative estimation of FTR by this time consuming analysis, by the more experienced professionals, is to avoid discomfort to the individuals undergoing these procedures and the repeated administration of the localized anaesthesia during the targeted completion of the entire procedure within the day (~12 hours). The time duration of the total procedure from follicular extraction to their grafting would depend upon the follicle yield that would include an idea of the FTR equivalent through visualization comprising also the microscopic examination, even if not estimated as described in this article, besides the number of team members from the technicians to the physicians.

The multiple causes for the cosmetic undertaking of follicular transfection/dissection include the individual hair and skin characteristics.

In this chapter, we describe the FTR during FUE that generally would set the stage for the subsequent autologous hair follicular engraftment mostly for those individuals who choose to undergo these selective and generally cosmetic surgical extraction and minimally invasive implantation procedures.

MATERIALS AND METHODS

Fifty cases of individuals were estimated for FTR. These include also those who have previously undergone FUT (8 cases) and FUE (10 cases). The remaining 32 cases are first time procedures for the statistical estimation of FTR. Two sites of ($3 \times 3cm^2$) from central part of occipital scalp donor area and two sites of ($3 \times 3cm^2$) from lateral part of scalp donor area from either side were subjected to the procedures for estimation of FTR. One hundred-punch excision of follicular units was made on each selected site and grafts were extracted with forceps. Manual punch size of 1mm was used. Complete and partial transection follicular units were determined by microscopic evaluation and average transection rates were calculated. FOX tcst was employed and each case was graded accordingly for the success potential of the procedure.

FOX Test

Bernstein and Rassman classified FOX test into five grades.

Grade 1

This classification is assigned when intact follicular units literally pop out of the scalp, or when there is only occasional transection of individual hair follicles within the unit.

Grade 2

Extraction from the patients may be relatively easy in the first session, but in subsequent procedures (when the donor area is slightly scarred) it becomes more problematic for extraction and the yields start to decline.

Grade 3

The emergent angle is in a difficult orientation for the extraction to be performed.

Grade 4-5

When it is almost impossible to predict the emergent angle, the yield is too low for the FUE procedure to be successful.

Transfection Rate:

Estimation of complete and partial transfection rates is performed. Definition of transection of a hair follicle is that its root portion or even the entire hair comprising follicle is excised intact along its body. The extracted follicle could be damaged, which may jeopardize its viability for subsequent engraftment on the occipital scalp.

Complete Follicular Transfection Rate (FTR)

If during the extraction of 2-hair follicles graft, they are held together despite excision and either of the hair follicle gets separated and remain intact, then it is called complete follicular transection.

Partial Transfection Rate (PTR)

If during the extraction of 2-hair follicles graft, only one hair follicle is excised out and the other one remains in the site of extraction, then following engraftment, it is labelled as a Partially Transfected Follicle.

RESULTS AND OBSERVATIONS

We have calculated the follicular transfection rates as affected by the hair and skin characteristics on each patient's scalp and correlated these values with the graft success following the completion of the procedure.

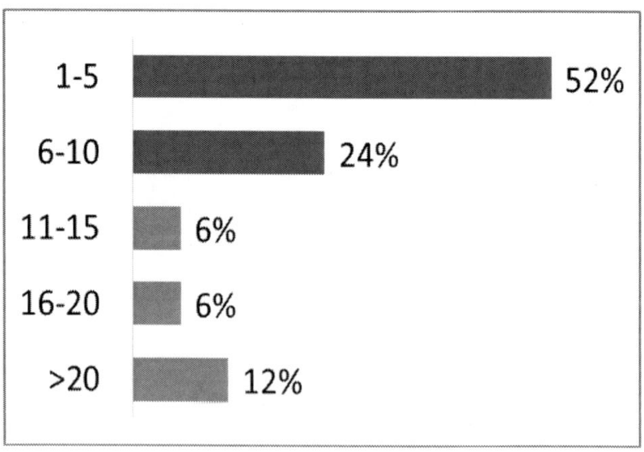

Figure 1. Partial Transfection Rates. A comparative bar graph histogram of the first and third columns (PTR and % patients) derived from Table 1.

Table 1. Partial Transfection Rates. The PTR value ranges are shown in the first column, middle column shows the number of patients that fall into those sequential PTR ranges, and the percentage of patients are denoted in the third column

Partial transection	No. of patients	%
0-5	2	4.0
6-10	18	36.0
11-15	13	26.0
16-20	2	4.0
21-25	4	8.0
26-30	2	4.0
31-35	4	8.0
36-40	2	4.0
>40	3	6.0
Total	50	100.0

Table 2. Complete Transfection Rates. The PTR value ranges are shown in the first column, middle column shows the number of patients that fall into those sequential PTR ranges, and the percentage of patients are denoted in the third column

Complete Transection	No. of patients	%
1-5	26	52.0
6-10	12	24.0
11-15	3	6.0
16-20	3	6.0
>20	6	12.0
Total	50	100.0

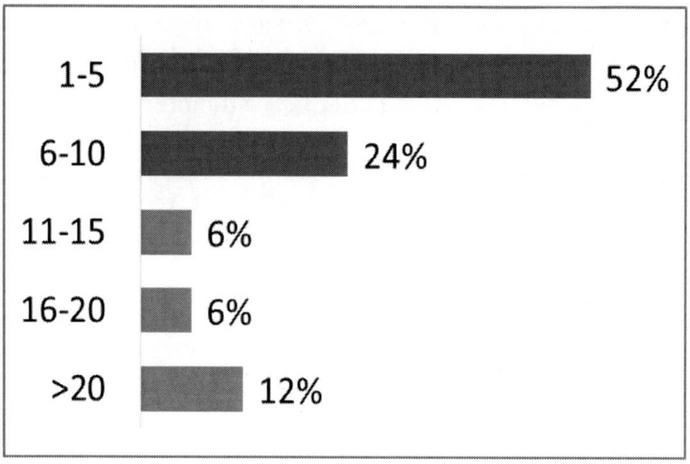

Figure 2. Complete Transfection Rates. Comparative bar graph histogram of the first and third columns (PTR and % patients) derived from Table 1.

Discussion

The skin characteristics that affect the transfections include: hydration level, degree of fatty tissue content, and different characteristics of the donor scalp area. Low FTR in occipital area as compared to temporal area, skin thickness (softness-hardness, elasticity or rigidity of the subcutaneous tissues), fibrosis of donor area post FUE or FUT, and bleeding in the donor area.

Apart from the effects of adipose tissue presence at the follicle sites on the success of transfections, the graft quality pre-transfection may be better maintained, but disrupted post-transfection. This is due to the reported presence of mesenchymal stem cells in the adipose tissue that may maintain the natural hair color.

Technical variations affect the transfection rates and graft quality. Instrumentation used for transfections include manual, power-assisted, semi-robotic and robotic devices.

Table 3. Transfection Rate. The determined transfection rates are classified into the five grades of the FOX test

Fox grade	No. of patients	%
Grade 1	18	36.0
Grade 2	14	28.0
Grade 3	12	24.0
Grade 4	3	6.0
Grade 5	3	6.0
Total	50	100.0

Table 4. Composite FOX grade table of PTR values derived from above Tables 1-3. The assigned FOX test grades and the percentage values in the PTR ranges on the extreme left column are denoted. Additional data include standard deviations calculated from the mean values and their comparison between the five grades produces a significant p value, P < 0.001

Partial transection	Fox grade					Total
	Grade 1	Grade 2	Grade 3	Grade 4	Grade 5	
0-5	2(11.1%)	0(0%)	0(0%)	0(0%)	0(0%)	2(4%)
6-10	15(83.3%)	3(21.4%)	0(0%)	0(0%)	0(0%)	18(36%)
11-15	1(5.6%)	11(78.6%)	1(8.3%)	0(0%)	0(0%)	13(26%)
16-20	0(0%)	0(0%)	1(8.3%)	1(33.3%)	0(0%)	2(4%)
21-25	0(0%)	0(0%)	4(33.3%)	0(0%)	0(0%)	4(8%)
26-30	0(0%)	0(0%)	2(16.7%)	0(0%)	0(0%)	2(4%)
31-35	0(0%)	0(0%)	2(16.7%)	1(33.3%)	1(33.3%)	4(8%)
36-40	0(0%)	0(0%)	1(8.3%)	1(33.3%)	0(0%)	2(4%)
>40	0(0%)	0(0%)	1(8.3%)	0(0%)	2(66.7%)	3(6%)
Total	18(100%)	14(100%)	12(100%)	3(100%)	3(100%)	50(100%)
Mean ±SD	7.27±2.74	11.78±2.04	27.33±8.93	30.33±10.78	41.33±7.02	16.78±12.01

$F=45.830$; $P<0.001$**

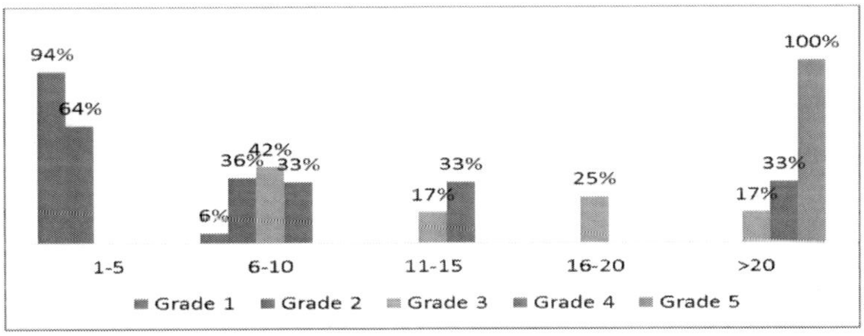

Figure 3. Normalized percentage values of the patients within each of the five different grades of PTR. The proportionate percentage patient values of each of the five grades also are denoted by the presence or absence of these 50 individuals undergoing the procedure. It is also interesting that grades 1 and 2 comprise a high percentage of patients with the most inferior grades 4 and 5 consist of significantly least number/percentage of patients.

The punch material is sharp (Titanium tipped) and/or blunt that will affect the size of punches, tumescence, magnification, and experience of the surgeon. Magnification with magnifying coupes was utilized. Idiopathic variables also affect the type of instrumentation employed for the procedures.

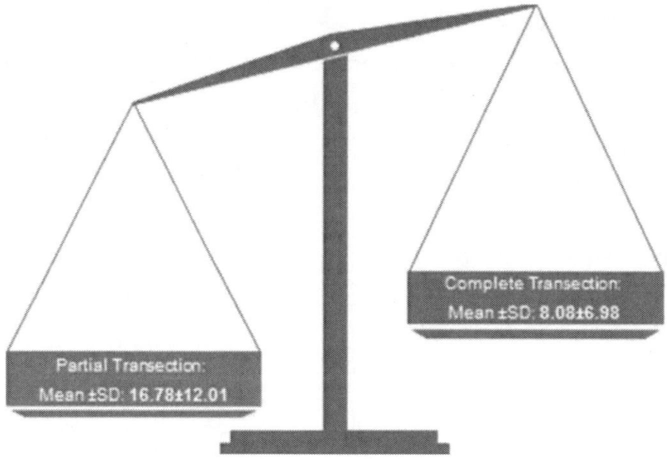

Figure 4. Statistical analysis of PTR and CTR values. The balance tilts towards the PTR values which suggest that the procedure success rate is about 50% lower, expectedly caused or affected by the adverse hair and skin characteristics.

Table 5. Clinical evaluation and determination of donor skin quality or physiological characteristics of the 50 patients who have undergone hair follicular transfections

Donor skin	No. of patients	%
Normal	28	56
Adipose excess	5	10
Post FUT & FUE scarring	17	34
Total	50	100

Conclusion

Low FTR is very essential for good harvest of grafts in FUE and better outcome. Newer methods and techniques should be employed for minimising transections, e.g., ARTAS, Piloscope etc. More studies are required to validate this study.

References

[1] Goh, C. L. A retrospective study on the characteristics of androgenetic alopecia among Asian races in the National Skin Centre, a tertiary dermatological referral centre in Singapore. *Ann. Acad. Med. Singapore*, 2002; 31(6): 751 - 755.

[2] Goh, C., Finkel, M., Christos, P. J., Sinha, A. A. Profile of 513 patients with alopecia areata: associations of disease subtypes with atopy, autoimmune disease

and positive family history. *J. Eur. Acad. Dermatol. Venereol.*, 2006; 20(9):1055 - 1060.

[3] Tan, E., Tay, Y .K., Goh, C. L., Giam, Y. C. The pattern and profile of alopecia areata in Singapore--a study of 219 Asians. *Intl. J. Dermatol.,* 2002; 41(11): 748 - 753.

[4] Vinay, K., Dogra, S. Stem cells in vitiligo: Current position and prospects. *Pigment Intl.*, 2014; 1(1): 8 - 12.

[5] Van Geel, N., Goh, B. K., Wallaeys, E., Keyser, S. D. and Lambert, J. A Review of Non-cultured Epidermal Cellular Grafting in Vitiligo. *J. Cutan. Aesthet. Surg.,* 2011; 4(1): 17 - 22. doi: 10.4103/0974-2077.79181

[6] Danilenko, D. M., Ring, B. D., Yanagihara, D., Benson, W., Wiemann, B., Starnes, C. O. and Pierce, G. F. Keratinocyte growth factor is an important endogenous mediator of hair follicle growth, development, and differentiation. Normalization of the nu/nu follicular differentiation defect and amelioration of chemotherapy-induced alopecia. *Am. J. Pathol.*, 1995; 147(1): 145 - 154.

[7] Booth, C., Potten, C. S. Keratinocyte growth factor increases hair follicle survival following cytotoxic insult. *J. Invest. Dermatol.,* 2000;114(4): 667 - 673.

[8] Moretti, S., Spallanzani, A., Amato, L., Hautmann, G., Gallerani, I. and Fabbri, P. Vitiligo and Epidermal Microenvironment: Possible Involvement of Keratino-cyte-Derived Cytokines. *Arch. Dermatol.,* 2002; 138(2): 273 - 274. doi: 10.1001/ archderm.138.2.26 6.

[9] Moretti, S., Spallanzani, A., Amato, L., Hautmann, G., Gallerani, I. and Fabbri, P. Vitiligo and Epidermal Microenvironment: Possible Involvement of Keratino-cyte-Derived Cytokines. *Arch. Dermatol.,* 2002; 138(2): 273 - 274. doi:10.1001/ archderm.138.2.266.

[10] Oh, J. H., Mohebi, P., Farkas, D. L., Tajbakhsh, J. Towards expansion of human hair follicle stem cells in vitro. *Cell Prolif.,* 2011; 44(3): 244 - 253. doi: 10.1111/ j.1365-2184. 2011.00754.x.

[11] Khatu, S. S., More, Y. E., Gokhale, N. R., Chavan, D. C. and Bendsure, N. Platelet-Rich Plasma in Androgenic Alopecia: Myth or an Effective Tool. *J. Cutan. Aesthet.* Surg., 2014; 7(2): 107 - 110. doi: 10.4103/0974-2077. 138352.

[12] Kim, D. H., Je, Y. J., Kim, C. D., Lee, Y. H., Seo, Y. J., Lee, J. H. and Lee, Y. Can Platelet-rich Plasma Be Used for Skin Rejuvenation? Evaluation of Effects of Platelet-rich Plasma on Human Dermal Fibroblast. *Ann. Dermatol.,* 2011; 23(4): 424 - 431. Published online 2011 Nov 3. doi: 10.5021/ad.2011.23.4.424.

[13] Dua, A. and Dua, K. Follicular Unit Extraction Hair Transplant, J. *Cutan. Aesthet. Surg.*, 2010 May-Aug.; 3(2):76 - 81.

[14] Bernstein, R. M., Rassman, W. R., Anderson, K. W. Follicular unit extraction mega sessions: Evolution of a technique. *Hair Transplant. Forum Intl.,* 2004; 14: 97 - 99.

[15] Bernstein, R. M., Rassman, W. R. New instrumentation for three step Follicular Unit Extraction. *Hair Transplant. Forum Intl.,* 2006; 16: 229 - 237.

[16] Ekrem, C., Aksoy, M., Koc, E., Aksoy, B. Evaluation of three instruments used in FUE. *Hair Transplant. Forum Intl.,* 2009: 14 - 15.

[17] Harris, J. A. The SAFE System: New instrumentation and methodology to improve follicular unit extraction (FUE) *Hair Transplant. Forum Int.*, 2004; 14: c163 - 164.

[18] Thadani, J., Kshatriya, P., Marathe, A., Vyas, R., Vyas, B. and Deb, K. Reversal of hair greying following autologous adipose mesenchymal stem cell transplantations: a coincidental finding. *Stem Cell Biol. Res.,* 2015; 2: 1 - 6. doi: 10.7243/2054-717X-2-3.

[19] Serdar, G., Nessin, B., Gurcan, A. Follicular Unit Extraction in hair transplantation: Personal experience. *Ann. Plast. Surg.,* 2008; 60: 127 - 133.

In: Stem Cells Origins of Cytokine Interactions
Editor: Prasad S. Koka
ISBN: 978-1-53615-302-6

Chapter 3

EFFECT OF GROWTH HORMONE–RELEASING HORMONE AND IGF-1 ON HUMAN PLACENTA-DERIVED MESENCHYMAL STEM CELLS' PROLIFERATION AND DIFFERENTIATION

***Avinash Kharat*[1], *Usha Padmanabhan*[2,*] *and S. V. Gadre*[1]**
[1]Department of Bio-analytical Sciences,
G.N. Khalsa College, Matunga East, Mumbai, India
[2]Department of Cell Biology, Haffkine Institute for Training,
Research and Testing, Acharya Donde Marg, Parel, Mumbai, India

ABSTRACT

Increasing evidence highlights the multipotent potential of mesenchymal stromal cells (MSCs), produced in peripheral tissues, bone marrow etc. to promote tissue regeneration and hence cell based therapies. In this regard, the human placenta has attracted attention for translational research due to its immunomodulatory properties, large stem cell number in tissue, and ease of availability for clinical experimentation.

The placenta plays the role of an endocrine organ in providing the foetus nutrition and also exerts some control over maternal hormones during gestation through expression of several hormones including the growth hormone releasing hormone (GHRH), placental growth hormone (GH) and insulin like growth factor (IGF-1). While on one hand, the role of GHRH in the placenta is unclear and there are only speculations of a paracrine role for GHRH in the placenta; on the other hand, GHRH and its analogues have been used to pre-condition bone marrow MSCs for cell based therapies in treatment of angiogenesis and GHRH antagonists are being considered for cancer therapy, due to its role in regulating somatic growth.

[*] Corresponding Author Email: u.padmanabhan@haffkineinstitute.org.

Given this scenario, we studied the effect of GHRH and IGF-1 on the proliferation and differentiation of human placenta derived MSCs. GHRH and IGF-1 by themselves and in combination stimulated faster proliferation of MSCs compared to controls. The combination of GHRH and IGF-1 also enhanced the rate of osetocytic and chondrocytic differentiation. GHRH was particularly effective in inducing chondrogenesis in p-MSCs. Our study opens the possibility for hitherto unidentified roles for GHRH in the placenta.

Keywords: placental mesenchymal stem cells, GHRH, IGF-1, differentiation, signaling

INTRODUCTION

Mesenchymal stem/stromal cells (MSC) have become an important component in stem cell-based neo-therapies for tissue regeneration and transplantation. MSCs are present in a variety of adult tissues such as adipose tissue, bone, lung, peripheral blood and the bone marrow [1, 2]. MSC have been generally used in experimental and clinical research because of their multipotency [3, 4], in bone marrow transplantation [5], tissue engineering [6] and cell therapy for example, in angiogenesis [7, 8].

The presence of MSCs in umbilical cord blood and placenta has been demonstrated [9-11].

The placenta has emerged as an attractive proposition as a source of MSCs due to its easy access without invasive procedures (contrary to bone marrow harvest) and large number of MSCs present in both the amnion and chorion [12-16].

The placenta is at the interface of fetal and maternal compartments where it facilitates nutrient and waste exchange to enable in utero existence. The trophoblasts in the chorionic villi accomplish this via invading and remodeling the maternal uterine vasculature and despite being fetal in origin, placental cells do not trigger a significant maternal immune response [17].

Growth hormone-releasing hormone (GHRH) is a factor released by the hypothalamus, which is the point of initiation of the hypothalamic-pitutary-gonadal/target organ (HPG) axis of systemic hormonal functions. GHRH induces secretion of the pitutary growth hormone (GH) into the blood, which affects a large number of tissue undergoing, growth, regeneration or metabolic homeostasis such as bone and muscle. It binds to growth hormone receptors on the liver and causes secretion of IGF-1. IGF-1 can itself regulate growth, regeneration or metabolic homeostasis in many organs. However, IGF-1 also acts as a feedback regulator and negatively regulates GHRH expression lowering GHRH levels [18-21].

GHRH expression has been detected in other non-neural extra-hypothalamic tissues such as heart, lung, ovaries, testis, in lymphocytes and several cancers such as prostate cancer [22-26]. Expression of GHRH in rat placenta identical to the hypothalamic GHRH has been reported [27]. GHRH has been shown to stimulate the growth of pituitary somatotrophs, the cells that secrete GH and cancer cells of the chondrion [28, 289] and although its function in pregnancy is unclear, the presence of the GHRH receptor on the placenta [30] suggests a paracrine function for GHRH. In an extension to this, GHRH receptor has been identified on MSCs isolated from mouse bone marrow and GHRH treatment of these cells increases their anigiogenic potential [31].

Apart from GHRH and its receptor, of interest to this study, the *placenta* also expresses considerable amounts of insulin and IGF-1 receptors at distinct locations on both *placental* surfaces. IGFs (IGF-1 and IGF-2) are low in the placenta in early gestation and increase as pregnancy progresses. IGFs bind to two receptor tyrosine kinases, the IGF-1 receptor (IGF-1R) and the insulin receptor (IR), and their hybrid receptors [32].

While the effects of IGF-1 are emerging [33], the role of GHRH on the proliferation and/or differentiation of placental MSCs has not yet been reported. We therefore studied the effect of GHRH and IGF-1 on the growth and differentiation of placental MSCs.

Material and Methods

Ethical Considerations

Full term placentas were collected from a local hospital after C-section deliveries. The study was conducted clearance from the Institutional Ethics committee and informed written consent of the patient.

Collection of Placentas

Human placenta from segment caesarean section deliveries of full term pregnancies were collected aseptically. Collected placental tissue were stored aseptically and transported at ambient temperature to the laboratory in collection medium containing antibiotics (penicillin 200 U/ml and streptomycin 200 μg/ml,) (Sigma Aldrich). All placental samples collected were negative for HIV (human immunodeficiency virus), toxoplasmosis, cytomegalovirus and rubella infection. Processing of placental tissue was made in the first four hours following collection to avoid any loss in cell viability.

Isolation of Cells from Placenta

Initially, the placenta tissue was washed with phosphate buffered saline, pH 7.2 (PBS). Blood clots and vessels were removed mechanically and the tissue was minced into small pieces and washed again with PBS. The minced tissues were then digested with 0.25% Trypsin-EDTA (Gibco-Invitrogen) for 45 min at 37°C for removal of the trophoblasts. The sample was then filtered through a 250 μm metal sieve. The remnants were collected and digested with collagenase- I (Gibco-invitrogen) 0.1% (1 hr at 37°C). Digested tissue was passed through a 250 μm metal sieve and 100 μm cell nylon membranes, to eliminate undigested tissue fragments. Cells were collected, following centrifugation of the filtrate at 300 g for 5 min, and were washed with 10ml 1X PBS (Sigma Aldrich). Cell suspension were loaded on Ficoll at a ratio of 1:2 and centrifuged at 300 g for 15 min. Mononuclear cells were collected and washed with 1X PBS and finally centrifuged at 300g for 5 min. The cell pellet was suspended in a complete culture medium consisting of DMEM (low glucose 1 g/l content, L-DMEM) supplemented with 20% FBS plus 100 U/ml penicillin, 100 μg/ml streptomycin, 1mM L-glutamine and seeded into two 75 cm^2 flasks.

Cultures were incubated at 37°C in a humidified 5% CO_2 incubators. Non adherent cells were removed (after 24 hr). Medium was replaced with fresh medium every three days, and adherent MSC were trypsinized upon reaching 80% confluence with Trypsin-EDTA (0.25% trypsin + 0.02% EDTA, Gibco-Invitrogen) after 14 days of culture.

Proliferation

Cells were cultured in culture medium (as described above) or in culture medium supplemented with 10 ng/ml IGF-1, 10 ng/ml GHRH and combination of both 10 ng/ml for 24, 48, 72 hrs, respectively.

Cell proliferation at these time points was assessed using the MTT [3-(4,5-dimethylthiazol-2-yl)-1,5-diphenyl tetrazolium bromide] (MTT, Sigma, USA) assay. In brief, cell were incubated with MTT (5 mg/ml in PBS) and incubated for 4 hrs at 37 ^{0}C. Post 4 hrs, the MTT solution was removed and cells were washed with PBS. Post washing, 0.5 ml of extraction solution (dimethylsulphoxide, DMSO) was added. The cells were thoroughly lysed washed by pipetting extensively to allow total colour release. The absorbance of the solution was read with a microplate reader at 540 nm. Each experiment was conducted three times in triplicate.

Differentiation

Cultures p-MSCs were trypsinized. Cells were then seeded into 6-well plates and were incubated for 24 hr for the cells to adhere. The medium was then replaced with the appropriate differentiation medium supplemented with either 10ng/ml IGF-1 or 10ng/ml GHRH or a combination of both. All differentiation studies were conducted at passage four of the p-MSC cells.

Chondrogenic Differentiation

Cells were seeded onto a medium consisting of L-DMEM supplemented with 10% FBS, 0.1 μM dexamethasone (GIBCO, Invitrogen), 10% Insulin, Transferrin, & Selenium solution (ITS, GIBCO-Invitrogen), 1 μM ascorbate-2-phosphate (Sigma Aldrich) to induce chondrocytic differentiation. Chondrocytic induction was confirmed by Safranin O staining (Alfa Aesar) using standard methods [33].

Osteogenic Differentiation

Osteocytic differentiation was induced in p-MSCs by culturing cells in an osteocytogenic differentiation medium containing L-DMEM, 10% FBS, 0.1 μM dexamethasone (Sigma-Aldrich), 10 mM β-glycerol phosphate (Sigma-Aldrich) in control conditions and with either, Ascorbic acid for 21 days and induction was confirmed by Alizarin Red S staining (Alfa Aesar) Alizarin red stain positive cells (differentiated MSC with calcium deposits) under microscope were stained bright orange-red, while negative cells (undifferentiated MSC without calcium deposits) were stained as faintly reddish cells [34].

RESULTS

Isolation of p-MSCs from Full Term Human Placentas

We isolated human placenta derived MSCs (p-MSCs) as described by various authors (Figure 1A -D) and in the Materials and Methods.

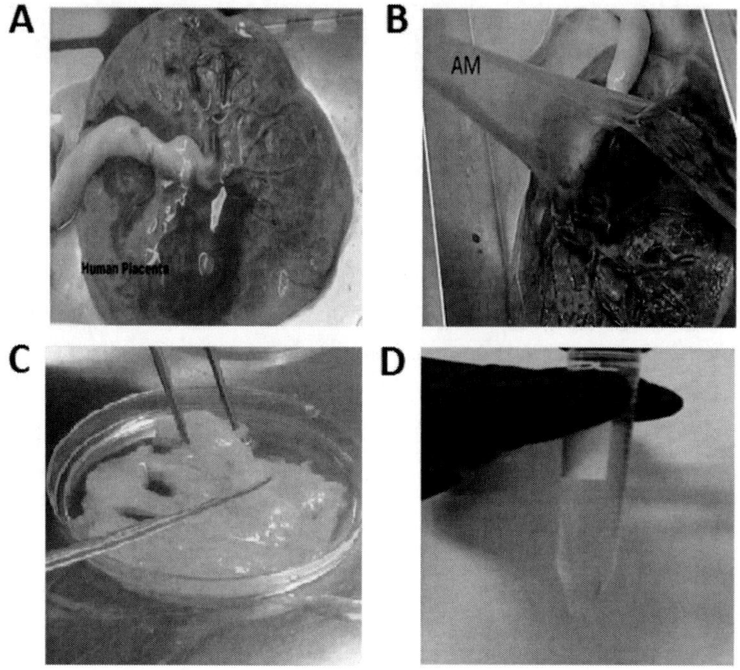

Figure 1. Cells took on a sigmoidal morphology after isolation and adherence to plastic (Figure 2A). After 7 days of culture, flasks were confluent (Figure 2B).

GHRH and IGF-1 Increase Proliferation in p-MSCs

To assess the effect of GHRH and IGF-1, placenta derived cells (third Passage) were cultured in the presence of IGF-1, GHRH and combination of both 10 ng/ml and compare with control after 24, 48, 72 hrs respectively. At all-time points, IGF-1 and GHRH increased cell proliferation by 2-fold, whereas the combination of IGF-1 and GHRH increased cell proliferation 4-fold (Figure 3).

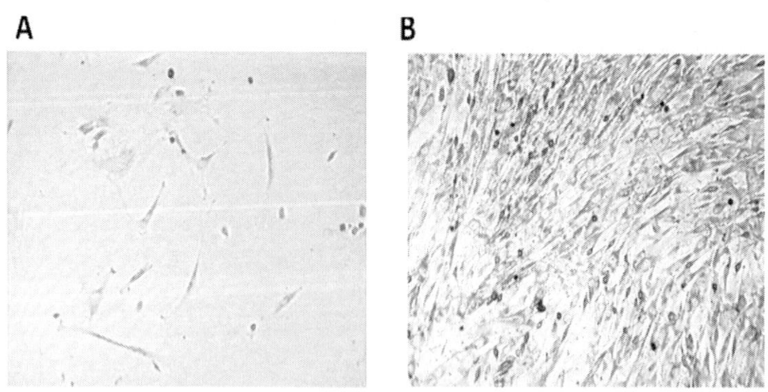

Figure 2. Isolated placental mesenchymal p-MSCs. A: Freshly isolated cells B: Cells after 7 days of culture in regular culture medium (L-DMEM + 20% FBS supp with 100 U/ml penicillin, 100 μg/ml streptomycin and 1mM L-glutamine). Microscope pictures were taken at 40X magnification.

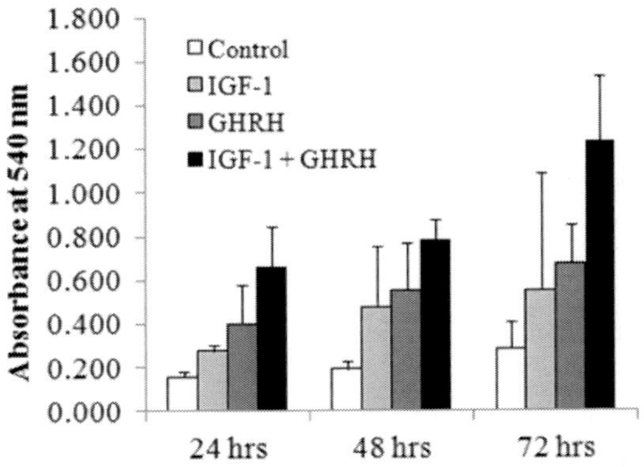

Figure 3. GHRH and IGF-1 increase proliferation of p-MSCs. Placental cells (Passage 3) were cultured in regular culture medium (DMEM with 20% FBS L-glutamine and penicillin, streptomycin) or in culture medium supplemented with IGF-1 (10 ng/ml), GHRH (10 ng/ml) or combination of both (each at 10 ng/ml). MTT assays were performed at the time intervals indicated. Proliferation experiments were conducted 3 times in triplicate. The data represent average of all 9 readings and the error bars represent the standard deviation.

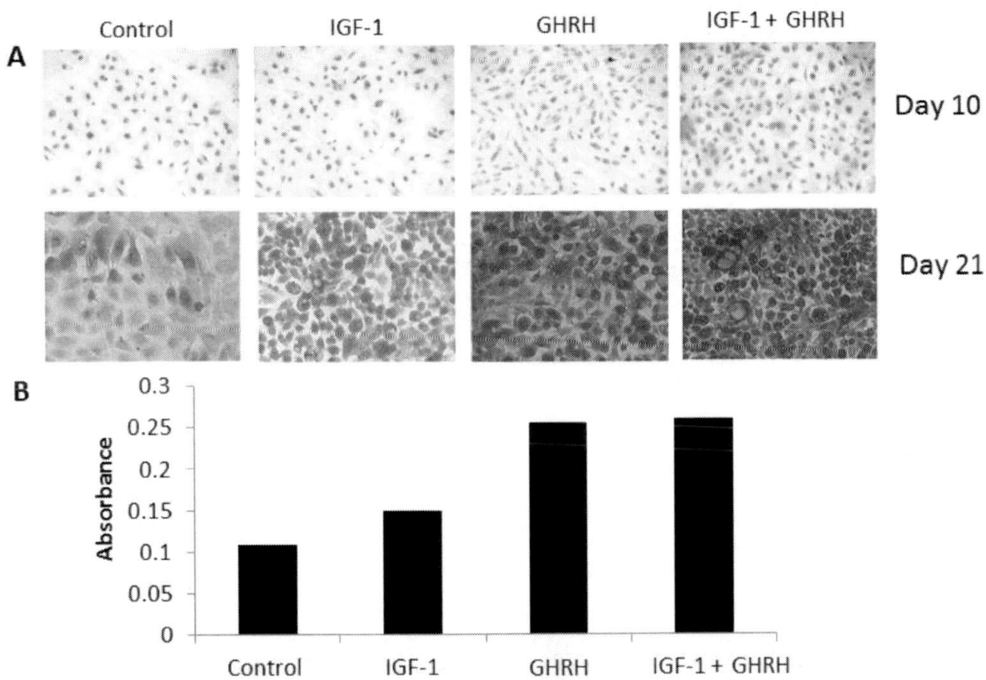

Figure 4. IGF-1 and GHRH increase chondrocytic differentiation of placental MSCs. Chondrocytic differentiation was induced by culturing cells in DMEM supplemented with 10% FBS, 0.1 μM dexamethasone (GIBCO, Invitrogen), 10% ITS (GIBCO-Invitrogen), 1 μM ascorbate-2-phosphate or in the same medium with with IGF-1 (10 ng/ml), GHRH (10 ng/ml) or combination of both (each at 10 ng/ml). A. Top panel and Bottom panel show chondrocytic differentiation in (1) Control (2) IGF-1 (3) GHRH and (4) IGF-1 + GHRH cells after 10 days and 15 days respectively. Cells were stained with Safrain O B. Shows quantitation of differentiation by measuring safranin O staining on day 10. Microscope pictures of top panel was at 10X magnification and those in the bottom panel were taken at 40X magnification.

GHRH and IGF-1 Increase Chondrocytic Differentiation in p-MSCs

The ability of MSC to differentiate into mesodermal cells such as chondrocytes was assessed. Chondrocytic differentiation was quantiated via safranin o staining (Figure 4A & B). Chondrocytic differentiation was seen to be developing faster in GHRH and IGF-1 + GHRH treated cells. IGF-1 also induced faster chondrocytic differentiation of p-MSCs than controls but lesser than that of GHRH and IGF-1 + GHRH.

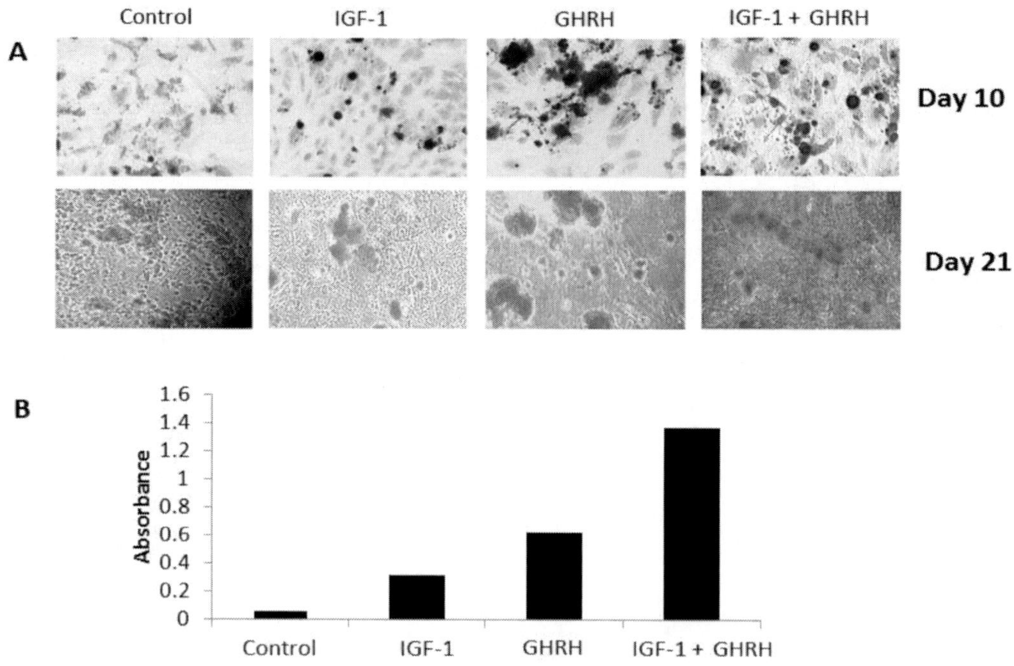

Figure 5. IGF-1 and GHRH increase osteocytic differentiation of placental MSCs. Osteocytic differentiation was induced by culturing cells in DMEM supplemented with 10% FBS, 0.1 µM dexamethasone (GIBCO, Invitrogen), 10% ITS (GIBCO-Invitrogen), 1 µM ascorbate-2-phosphate or in the same medium with with IGF-1 (10 ng/ml), GHRH (10 ng/ml) or combination of both (each at 10 ng/ml). A. Top panel and Bottom panel show chondrocytic differentiation in (1) Control (2) IGF-1 (3) GHRH and (4) IGF-1 + GHRH cells after 10 days and 21 days respectively. Cells were stained with Alizarin red B. Shows quantitation of differentiation by measuring Alizarin red staining on day 10. Microscope pictures of top panel was at 40X magnification and those in the bottom panel were taken at 10X magnification.

GHRH and IGF-1 Increase Osteocytic Differentiation in p-MSCs

Osteocytic differentiation was induced by culturing the cells in standard osteocytic differentiation medium. Alizarin red S staining was used to measure the calcium deposits induced by culturing p-MSCs in osteocytic differentiation medium Osteocytic differentiation followed the order IGF-1 + GHRH > GHRH > IGF-1 > Controls (Figure 5A & B).

DISCUSSION

In almost all protocols of isolation of mesenchymal cells from the placenta, the placenta is digested with trypsin-EDTA or other enzymes at 37°C, and placental mesenchymal stem cells (p-MSC) are harvested through physical methods (Figure 1A – D). We have followed standard procedures for culturing the isolated cells initial expansion takes 4-14 days, depending on the quality of the tissue, amount of starting material and efficiency and time of incubation during digestion [35-37]. Similar to that reported by these authors, our initial expansion took 7 days (Figure 2).

IGF-1 and GHRH Induce Proliferation in p-MSCs

In vitro IGFs regulate proliferation and survival of cyto-trophoblasts isolated from first trimester placental explants via the MAPK pathway and the PI3 K pathway, respectively [31]. In another study, IGF-1 at 10 ng/ml did not induce an increase in proliferation of MSCs derived from full term placentas [32]. On the other hand, GHRH induces proliferation in pituitary somatotrophs [28] and increase in GH production mainly by the PKA/cAMP-dependent pathway and to a lesser extent by the phospholipase C pathway (IP_3/DAG pathway) [38]. As shown in Figure 3, IGF-1 and GHRH together induced an additive effect on proliferation compared to IGF-1 or GHRH alone, suggesting that IGF-1 and GHRH act in synergy with each other. Such synergy between the MAPK and the c-AMP signalling pathways has also been observed in placental JEG-3 cells in response to hCG to induce the leptin gene expression [39].

GHRH and IGF-1 Induce Chondrocytic and Osteogenic Differentiation in p-MSCs

Chondrocytic (or *in vitro* cartilage) differentiation in MSCs depends on the coordinated activities of two fundamental parameters: cell density and growth factors [40-42]. Among growth factors, the TGF-β super family of proteins and their members, such as BMPs are established regulatory factors in chondrogenesis [41, 43, 44]. In this study, we did not use TGF-β or BMP so that we could independently assess the effect of GHRH on chondrogenesis. GHRH had a profound effect on chondrogenesis whereas in comparison IGF-1 had a small effect on chondrogenesis (Figure 4). Activation of PKA/cAMP has been suggested by many authors to be vital for efficient chondrogenesis [45]. Since GHRH as been reported to work through PKA/cAMP activation, whereas the downstream effectors of IGF-1 namely MAPK only have subtle effects on chondrogenesis, our results corroborate the work of other authors [46].

In vivo studies in mice have confirmed the role of IGF on longitudinal bone growth [47]. Whether insulin-like growth factor 1 (IGF1) inhibits or promotes the osteogenic differentiation in vitro remains controversial. IGF-1 at 100 ng/ ml has been shown to promote osteogenic differentiation in rat bone marrow derived MSCs [48] and in stem cells from apical papilla [49]. Similarly, osteogenesis can be inhibited or stimulated by cAMP/PKA signalling depending on the time duration and strength of signals [50]. In line with these reports, our results also showed that in comparison to chondrogenesis, both IGF-1 and GHRH had more modest effects on osteogenesis, reflecting the subtlety of the downstream effectors on osteogenesis. The additive effect of IGF-1 and GHRH seen during MSC proliferation was also seen in osteocytic differentiation (Figure 5).

To our knowledge while the role of GHRH in chondrocytic and osteocytic differnation has been speculated, ours is the first report of GHRH inducing chondrocytic and osteogenic differentiation in placental mesenchymal cells [51].

ACKNOWLEDGMENTS

This study was in part funded by the Pilot project grants from Department of Biotechnology, Govt. of India to Dr. Usha Padmanabhan (Grant no. 6242-P68/RGCB/PMD/DBT/UPMN/2015).

CONFLICT OF INTEREST

The authors have no conflicting interest to declare.

REFERENCES

[1] Minguell JJ, Conget P, Erices A. Biology and clinical utilization of mesenchymal progenitor cells. *Braz J Med Biol Res*. 2000. 33: 881-887.

[2] Uccelli A, Moretta L & Pistoia V Mesenchymal stem cells in health and disease. *Nat Revs Immunol*. 2008. 8, 726-736.

[3] Dominici M, Le Blanc K, Mueller I et al. Minimal criteria for defining multipotent mesenchymal stromal cells. The International Society for Cellular Therapy position statement. *Cytotherapy*. 2006. 8(4). 315–317.

[4] Kan I, Melamed E, Offen D. Integral therapeutic potential of bone marrow mesenchymal stem cells. *Curr Drug Targets.* 2005. 6:31-41.

[5] Thomas ED. Bone marrow transplantation: prospects for leukemia and other conditions. *Proc Inst Med Chic.* 1975. 30(8):256-8.

[6] Dazzi F, Ramasamy R, Glennie S, Jones SP, Roberts I. The role of mesenchymal stem cells in haemopoiesis. *Blood Rev.* 2006. 20: 161-17.

[7] Efimenko A, Starostina E, Kalinina N, Stolzing A. Angiogenic properties of aged adipose derived mesenchymal stem cells after hypoxic conditioning. *J Transl Med.* 2011. 9:10.

[8] Shyu KG, Wang BW, Hung HF, Chang CC, Shih DT. Mesenchymal stem cells are superior to angiogenic growth factor genes for improving myocardial performance in the mouse model of acute myocardial infarction. *J Biomed Sci.* 2006. 13:47-58.

[9] Erices A, Conget P, Minguell JJ. Mesenchymal progenitor cells in human umbilical cord blood. *Br J Haematol.* 2000. 109: 235-242.

[10] Ilancheran S, Moodley Y, Manuelpillai U. Human fetal membranes: a source of stem cells for tissue regeneration and repair? *Placenta.* 2009. 30: 2-10.

[11] Wolbank S, van Griensven M, Grillari-Voglauer R, Peterbauer-Scherb A. Alternative sources of adult stem cells: human amniotic membrane. *Adv Biochem Eng Biotechnol.* 2010. 123: 1-27.

[12] Fukuchi Y, Nakajima H, Sugiyama D, Hirose I, Kitamura T, Tsuji K. Human placenta-derived cells have mesenchymal stem/progenitor cell potential. *Stem Cells.* 2004. 22: 649-658.

[13] Bačenková D, Rosocha J, Tóthová T, Rosocha L, Šarisský M. Isolation and basic characterization of human term amnion and chorion mesenchymal stromal cells. *Cytotherapy.* 2011. 13.

[14] Rus Ciucă D, Soriţău O, Suşman S, Pop VI, Mihu CM. Isolation and characterization of chorionic mesenchyal stem cells from the placenta. *Rom J Morphol Embryol.* 2011. 52: 803-808.

[15] Pelekanos RA, Sardesai VS, Futrega K, Lott WB, Kuhn M, Doran MR. Isolation and Expansion of Mesenchymal Stem/Stromal Cells Derived from Human Placenta Tissue. *J. Vis. Exp.* 2016. 112: e54204, doi:10.3791/54204.

[16] Sundell IB, Koka PS. Placental membrane as a source of mesenchymal stem cells. *J Stem Cells.* 2011. 5(2): 83-88.

[17] Maltepe E, Fisher SJ. Placenta: The Forgotten Organ. *Ann. Rev. Cell. Develop. Biol.* 2015. 31: 523-552.

[18] *Fundamental Neuroscience* 4th edition. Edited by Squire, Berg, Bloom, du Lac, Ghosh, Spitzer. Academic Press, 2012.

[19] Bloom S du-Lac, Ghosh A, Squire LR, Spitzer NC, McConnell SK, Roberts JL, Zigmond MJ, Academic Press. Brazeau P, Ling N. Growth hormone releasing

factor, somatocrinn, releases pituitary growth hormone in vitro. *Proc. Natl. Acad. Sci.* (USA). 1982. 79(24): 7909-13.

[20] Plotsky PM, Vale W. Patterns of growth hormone-releasing factor and somatostatin secretion into the hypophysial-portal circulation of the rat. *Science.* 1985. 230 (4724): 461–3.

[21] Ceda GP, Davis RG, Rosenfeld RG, Hoffman AR. The growth hormone (GH)-releasing hormone (GHRH)-GH-somatomedin axis: evidence for rapid inhibition of GHRH-elicited GH release by insulin-like growth factors I and II. *Endocrinology.* 1987. 120(4): 1658-62.

[22] Mayo KE, Godfrey PA, Suhr ST, Kulik DJ, Rahal JO. Growth hormone-releasing hormone: synthesis and signaling. *Recent Prog Horm Res.* 1995. 50: 35–73.

[23] Matsubara S, Sato M, Mizobuchi M, Niimi M, Takahara J. Differential gene expression of growth hormone (GH)-releasing hormone (GRH) and GRH receptor in various rat tissues. *Endocrinology.* 1995. 136(9): 4147–4150, 1995.

[24] Bagnato A, Moretti C, Ohnishi J, Frajese G, Catt KJ Expression of the growth hormone-releasing hormone gene and its peptide product in the rat ovary. *Endocrinology.* 1992. 130:1097–1102.

[25] Stephanou A, Knight RA, Lightman SL Production of a growth hormone-releasing hormone-like peptide and its mRNA by human lymphocytes. *Neuroendocrinology.* 1991. 53:628–633.

[26] Plonowski A, Schally AV, Busto R, Krupa M, Varga JL, Halmos G. Expression of growth hormone-releasing hormone (GHRH) and splice variants of GHRH receptors in human experimental prostate cancers. *Peptides.* 2002. 23(6):1127-33.

[27] González-Crespo S, Boronat A. Expression of the rat growth hormone-releasing hormone gene in placenta is directed by an alternative promoter. *Proc. Natl. Acad. Sci. (USA).* 1991. 88(19):8749-8753.

[28] Billestrup N, Swanson LW, Vale W. Growth hormone-releasing factor stimulates proliferation of somatotrophs in vitro. *Proc. Natl. Acad. Sci. (USA).* 1986. 83(18): 6854-6857.

[29] Liu A-X, Zhang D, Zhu Y-M., Gao H-J, Jiang J-Y, Hu X-L, Lv P-P,. Leung P.C.K Huang H.-F. Impact of Axis of GHRH and GHRH Receptor on Cell Viability and Apoptosis of the Placental Choriocarcinoma Cell Line. *Curr Mol. Med.* 2016. 16(3): 299-311

[30] Ma Q, Xia X, Tao Q, Lu K, Shen J, Xu Q, Hu X, Tang Y, Block NL, Webster KA, Schally AV, Wang J, Yu H. Profound Actions of an Agonist of Growth Hormone–Releasing Hormone on Angiogenic Therapy by Mesenchymal Stem Cells. *Arteriosclerosis, Thrombosis, and Vascular Biology.* 2016. 36: 663-672.

[31] Forbes K, Westwood M, Baker PN, Aplin JD. Insulin-like growth factor I and II regulate the life cycle of trophoblast in the developing human placenta. *Am J Physiol Cell Physiol.* 2008. 294: C1313–1322.

[32] Youssef A, Han VK. Low Oxygen Tension Modulates the Insulin-Like Growth Factor-1 or -2 Signaling via Both Insulin-Like Growth Factor-1 Receptor and Insulin Receptor to Maintain Stem Cell Identity in Placental Mesenchymal Stem Cells. *Endocrinology.* 2001. 157(3): 1163-74.

[33] Kern S, Eichler H, Stoeve J, Klüter H, Bieback K. Comparative analysis of mesenchymal stem cells from bone marrow, umbilical cord blood, or adipose tissue. *Stem Cells.* 2006. 24(5): 1294-301.

[34] Kim BS, Kim JS, Chung YS, Sin YW, Ryu KH, Lee J, et al. Growth and osteogenic differentiation of alveolar human bone marrow-derived mesenchymal stem cells on chitosan/hydroxyapatite composite fabric. *J Biomed Mater Res A.* 2013. 101: 1550-1558.

[35] Bačenková D, Rosocha J, Tóthová T, Rosocha L, Šarisský M. Isolation and basic characterization of human term amnion and chorion mesenchymal stromal cells. *Cytotherapy.* 2011. 13.

[36] Rus Ciucă D, Soriţău O, Suşman S, Pop VI, Mihu CM. Isolation and characterization of chorionic mesenchyal stem cells from the placenta. *Rom J Morphol Embryol.* 2011. 52: 803 808.

[37] Pelekanos RA, Sardesai VS, Futrega K, Lott WB, Kuhn M, Doran MR. Isolation and Expansion of Mesenchymal Stem/Stromal Cells Derived from Human Placenta Tissue. *J. Vis. Exp.* 2016. 112, e54204, doi:10.3791/54204.

[38] Fridlyand LE, Tamarina NA, Schally AV, Philipson LH. Growth Hormone-Releasing Hormone in Diabetes. *Frontiers in Endocrinology.* 2016. 7:129.

[39] Maymó JL, Pérez Pérez A, Maskin B, Dueñas JL, Calvo JC, Sánchez Margalet V, et al. The Alternative Epac/cAMP Pathway and the MAPK Pathway Mediate hCG Induction of Leptin in Placental Cells. *PLoS ONE.* 2012. 7(10): e46216. doi:10.1371/journal.pone.0046 216).

[40] Eslaminejad MB, Nikmahzar A, Piriea A. The structure of Human Mesenchymal Stem Cells differentiated into cartilage in micro mass culture system. *Yakhteh.* 2006. 8(3): 162–171.

[41] Indrawattana N, Chen G, Tadokoro M, Shann LH, Ohgushi H, Tateishi T, et al. Growth factor combination for chondrogenic induction from human mesenchymal stem cell. *Biochem Biophys Res Commun.* 2004. 320(3): 914–919.

[42] Johnstone B. Mesenchymal stem cells and chondrogenesis. *Eur Cell Mater.* 2002. 4 (1): 27–27.

[43] Zhou S, Eid K, Glowacki J. Cooperation between TGF-beta and Wnt pathways during chondrocyte and adipocyte differentiation of human marrow stromal cells. *J Bone Miner Res.* 2004. 19(3): 463–470.

[44] Letamendia A, Labbé E, Attisano L. Transcriptional regulation by Smads: crosstalk between the TGF-beta and Wnt pathways. *J Bone Joint Surg Am.* 2001. 83-A 1(1): S31–S39.

[45] Lee Y-S, Chuong C-M, Activation of protein kinase A is a pivotal step involved in both BMP-2- and cyclic AMP-induced chondrogenesis. *J. Cell. Physiol.* 1997. 170(2): 153–165.

[46] Li J, Zhao Z, Liu J, Huang N, Long D, Wang J, Li X, Liu Y, MEK/ERK and p38 MAPK regulate chondrogenesis of rat bone marrow mesenchymal stem cells through delicate interaction with TGF-beta1/Smads pathway. *Cell Prolif.* 2010. 43(4): 333-43.

[47] Giustina A, Mazziotti G, Canalis E. Growth Hormone, Insulin-Like Growth Factors, and the Skeleton. *Endocrine Reviews*. 2008. 29(5): 535-559. doi:10.1210/er.2007-0036.

[48] Xue P, Wu X, Zhou L, Ma H, Wang Y, Liu Y, Ma J, Li Y. IGF1 promotes osteogenic differentiation of mesenchymal stem cells derived from rat bone marrow by increasing TAZ expression. *Biochem Biophys Res Commun.* 2013. 5, 433(2): 226-31.

[49] Sainan Wang, Jinquan Mu, Zhipeng Fan, Yan Yu, Ming Yan, Gang Lei, Chunbo Tang, Zilu Wang, Yangyu Zheng, Jinhua Yu, Guangdong Zhang, Insulin-like growth factor 1 can promote the osteogenic differentiation and osteogenesis of stem cells from apical papilla. *Stem Cell Res.* 2012. 8(3): 346-356.

[50] Siddappa R, Doorn J, Liu J, Langerwerf E, Arends R, van Blitterswijk C, de Boer, J. Timing, rather than the concentration of cyclic AMP, correlates to osteogenic differentiation of human mesenchymal stem cells. *J Tissue Eng Regen Med*. 2010. 4(5): 356-65.

[51] Xia X, Tao Q, Ma Q, Chen H, Wang J, Yu H. Growth Hormone-Releasing Hormone and Its Analogues: Significance for MSCs-Mediated Angiogenesis. *Stem Cells Int.* 2016, Article ID 8737589.

In: Stem Cells Origins of Cytokine Interactions
Editor: Prasad S. Koka
ISBN: 978-1-53615-302-6

Chapter 4

STAINING METHODS TO DISTINGUISH BETWEEN FUNGI AND ERYTHROCYTES IN ANCIENT TISSUE HISTOLOGY

***Inger Birgitta Sundell*[1,*] *and Teddi J. Setzer*[1,2]**
[1]Department of Anthropology, Wayne State University, Detroit, MI, US
[2]Department of Sociology, Anthropology, Social Work and Criminal Justice, Oakland University, Rochester, MI, US

ABSTRACT

Skeletal remains from ancient populations are usually the only material left for archaeologists to examine. Multidisciplinary research has revealed structures consistent with the same shape and size as modern erythrocytes in ancient bone marrow and skeletal tissue. These findings have been questioned because postmortem fungi contamination has a similar morphology to erythrocytes. Due to the risk of fungi contamination, identification of erythrocytes by specific proteins using immunohistochemical staining has been considered a requirement for accuracy. However, immunohistochemical staining only demonstrates that the target protein is present on the cell surface and if the protein is degraded, the researcher may misclassify an erythrocyte as a contaminant and loose information.

We provide an alternative approach using dyes that can verify erythrocytes, hemosiderin, and fungi but does not require specific cell-surface proteins. This method requires less time and is not as costly as immunohistochemical studies. Staining with dyes can provide important information of the quality of the sample before advanced procedures are employed. Paraffin-embedded demineralized bone sections from archaeological excavations of two Bronze Age burials and commercially available control *Aspergillus* slides were stained using haematoxylin and eosin (H&E), PAS staining with a Schiff

* Corresponding Author Email: du2557@wayne.edu.

reagent to detect fungi (carbohydrates), and an iron-specific stain that detects hemosiderin in degraded erythrocytes. Our research demonstrates that it is possible to distinguish between erythrocytes and fungi using a combination of staining techniques.

Keywords: archaeology, bone marrow, erythrocytes, histology, skeletal remains

INTRODUCTION

Structures consistent with erythrocytes found during the histological analysis of ancient tissue have been published in the scientific literature [1-7]. In the Iceman, for example, researchers found erythrocytes of the same cell size and shape approximated those of healthy recent erythrocytes [8]. Reports of erythrocytes in skeletal tissue have been critically assessed [9] because postmortem contaminants, such as fungi, have a similar structure.

The purpose of this experiment was to prove that it is possible to distinguish between erythrocytes and fungi using specific staining techniques. It has been argued [9] that staining is required to identify erythrocytes with confidence. This method relies on the presence of cell surface protein, like the erythrocyte specific Glycophorin A. The absence of this protein does not necessarily prove that erythrocytes are not present, only that the protein itself is not present. Our research provides an alternative, less time consuming and cheaper method of staining tissue using dyes that can result in differentiating erythrocytes from fungi in samples without requiring the preservation of one protein that is necessary for immunohistochemical studies.

Most cells are colorless and transparent; therefore, histological sections are stained to visualize the cells. Staining is usually performed using a dye. The dye stains some of the cells with a bright color; a counterstain that stains the rest of the cell a different color is often used in combination with the dye. The stains used can either be non-specific, staining most of the cells in the same way, or specific, selectively staining particular chemical groupings or molecules within cells or tissues. The most commonly used staining system is the combination of two dyes haemotoxylin and eosin (also known as called H&E). Eosin is an acidic dye: it is negatively charged and stains basic (or acidophilic) structures red or pink. Haematoxylin can be considered a basic dye and is used to stain acidic (or basophilic) structures a purplish blue. The PAS stain involves Schiff reagent is a bleached basic fuschin that results in a deep red (magenta color) in the section. PAS stains carbohydrates and carbohydrate rich macromolecules like glycogen, the intracellular storage form of carbohydrate in cells, and is widely used for the detection of fungi. The iron stain kit specific for the staining of ferric iron (Fe^{3+}) blue show degraded red blood cells with hemosiderin. Positive staining of hemosiderin is well established in sections of liver, lung, spleen, kidney, lymph node and bone marrow, and we here demonstrate that positive staining also occur in ancient tissue which confirm that the tissue is partially degraded bone marrow.

MATERIALS AND METHODS

Two human bones from the excavations at Serra 'e sa Caudeba of two Bronze Age tombs on the Campidano Plain of Sardinia, Italy during the early 1980s. Archaeologists, who have macroscopically examined these skeletal remains, described them as highly fragmented and of poor to average research quality [10]. Bone from this archaeological site has previously been investigated [11] and described as highly fragmented and of poor to average research quality [10]. Modern soil samples from this region are composed of 41.7% sand, 22.1% silt, 36.2% clay, with a pH = 7 and soil organic content of 1.6% with 2.7% soil organic matter, including high levels of the endemic species of the fungi genera *Fusarium* and *Tricoderma* [12]. *Trichoderma* is a hostile myco-parasite that can control both established and newly entered pathogens. *Trichoderma* metabolites result in the synthesis of compounds which increase resistance against plant pathogens, including fungi and bacteria [13, 14]. Its effects on buried human tissue are not known. The area receives an average of 741 mm of rainfall per year, with temperatures ranging from 7.1°C to 23.8°C [12].

"Tissue processing" describes the steps required to take human tissue from fixation to the state where it is completely infiltrated with a suitable histological wax and can be embedded ready for section cutting on the microtome. Tissue processing can be performed manually, or in an automated tissue processing machine. In both cases, the specimens are transferred from container to container to be processed, or the fluid-transfer types. Ancient tissue is frail so manual processing may be required. Paraffin processing is the common route to preparing slides for histological analysis. Paraffin processing provides excellent resolution of cellular structures, if tissues are correctly fixated. Processing allows paraffin to infiltrate down to the cellular level, which gives the tissue enough even rigidity to be thinly sectioned and placed on a slide for further analysis. The tissues are then placed in xylene, which acts as a linking agent between the alcohols and wax. Finally the tissues are impregnated with paraffin wax at a temperature just above the melting point of the wax being used. How much time and heat needed, at each stage, is dependent on the type of tissues being processed. Microscopic analysis of cells and tissues requires the preparation of very thin (usually 5um), high quality sections mounted on glass slides and appropriately stained to demonstrate normal and abnormal structures. Impregnating soft tissue with formalin is a standard procedure; however impregnating skeletal tissue with paraffin requires significant manipulations of the bone tissue.

Table 1. Work used to prepare skeletal remains for staining in this study

Step	Method	Note	Time
Demineralization	Cut samples to appropriate size and soak in EDTA or HCL	Rinse in PBS after fixation	12 hrs
Fixation	Cut samples to appropriate size and soak in fixative NBF, PFA	Fixative volume should be 5-10x of tissue volume	12 hrs
Dehydration	70% ethanol	2 changes,	1h each
	80% ethanol	2 changes,	1h each
	95% ethanol	2 changes,	1h each
	100% ethanol	2 changes,	1h each
	Xylene or substitute	3 changes	1 hr each
Embed tissues	Place in embedding cassette, place cassette in mold and pour on melted paraffin		
Cut and mount	Trim paraffin blocks to an optimal cutting surface		
	Cut 3-10 μm slices (5 μm is commonly used)		
	Place paraffin ribbon or slice in 40-45°C water bath	It will expand and wrinkles will vanish	
	Fish out floating paraffin section on to the glass slides	Use "plus"-slides	
	Dry sections O/N at 37°C		
Rehydrate	Xylene	3+ changes	3-10 min
	100% ethanol	2 changes	3 min each
	95% ethanol	2 changes	3min each
Rinse in water			
Stain slides	Follow manufacturer's instructions		

Bone samples were fixed in 10% neutral buffered formalin (NBF). After 12 hours, they were washed in phosphate buffer (PBS, pH 7.4), then demineralized with EDTA or hydrochloric acid washed subsequently in PBS pH7.4, and transferred to 70% ethanol. As described in Table 1 standard histological techniques involving fixation, paraffin embedding, sectioning and staining were applied.

In the present study we used ThermoScientific Gram Stain Kit Film, Richard Allan Scientific, Kalamazoo, MI, USA and a Gram Stain Control Slide, AlphaTec, Vancouver, WA, USA. Periodic acid-optimized Schiff's stain (PAS) kit procedure, American Master Tech, Lodi, CA, USA and a *Aspergillus*/Fungus Tissue Control Slide (TCS0005-IFU), ScyTek, Logan, UT, USA.

RESULTS

Isolation of cellular components and DNA can reveal information of past populations. Erythrocytes are the most frequently occurring blood cell. Good preservation of erythrocytes in bone marrow indicates that other cells, such as DNA containing hematopoietic stem cells will also be preserved. In the present study cells with a typical shape of erythrocytes and within a typical size range of 5–8 um, were detected in material isolated from the medullary cavity and the demineralized bone section. Using H&E stain the erythrocytes stained intensely red. Basic cellular elements such as hemoglobin molecules have an affinity for the acidic dye, eosin.

Gram staining of demineralized bone may detect bacteria or possibly a yeast or mold infection. Fungus either in the form of yeasts or molds may be seen on a Gram stain. Yeast will appear as single round cells, while molds may appear as branches so called hyphae. Preceding the paraffin-embedding of the hard-tissue specimens for microscopical investigations, decalcification was performed, there is a risk of false-negative scores in decalcified hard-tissue sections [15]. In this study we did not detect mold or yeast in the Gram stained samples.

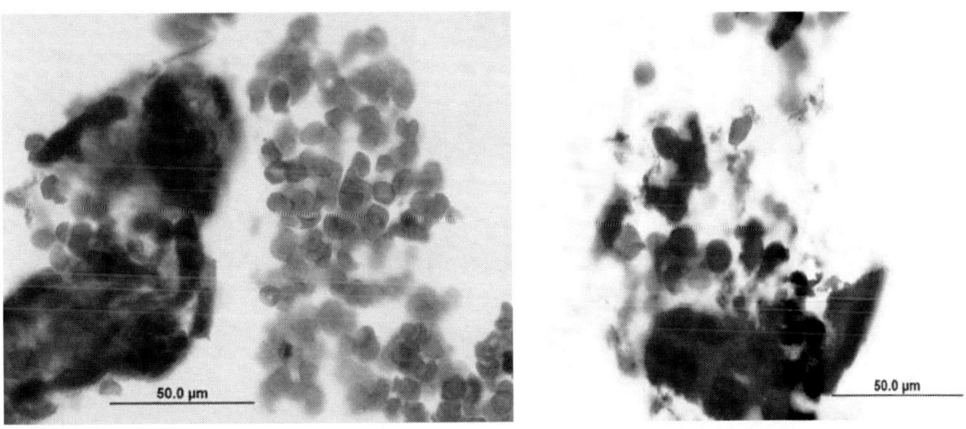

Figure 1. H&E stained demineralized bone containing unidentified structures, and erythrocytes. A distinctive feature of erythrocyte aging is their conversion into echinocytes. They are characterized by the loss of biconcave disc morphology and emergence of spiculations. Images are magnified 1000x.

The presence of fungus in the tissue sections provides evidence of invasive infection. In the case of ancient tissue this invasion may have occurred either before or after death. Fungi absorb nutrients from living or dead organic matter that they grow on and contain DNA which can contaminate human DNA. Because of their size and morphologic diversity, many fungi can be seen in tissue sections by conventional light microscopic examination of H&E stained sections. Fungi can be identified by their size and specific morphology (Figure 2).

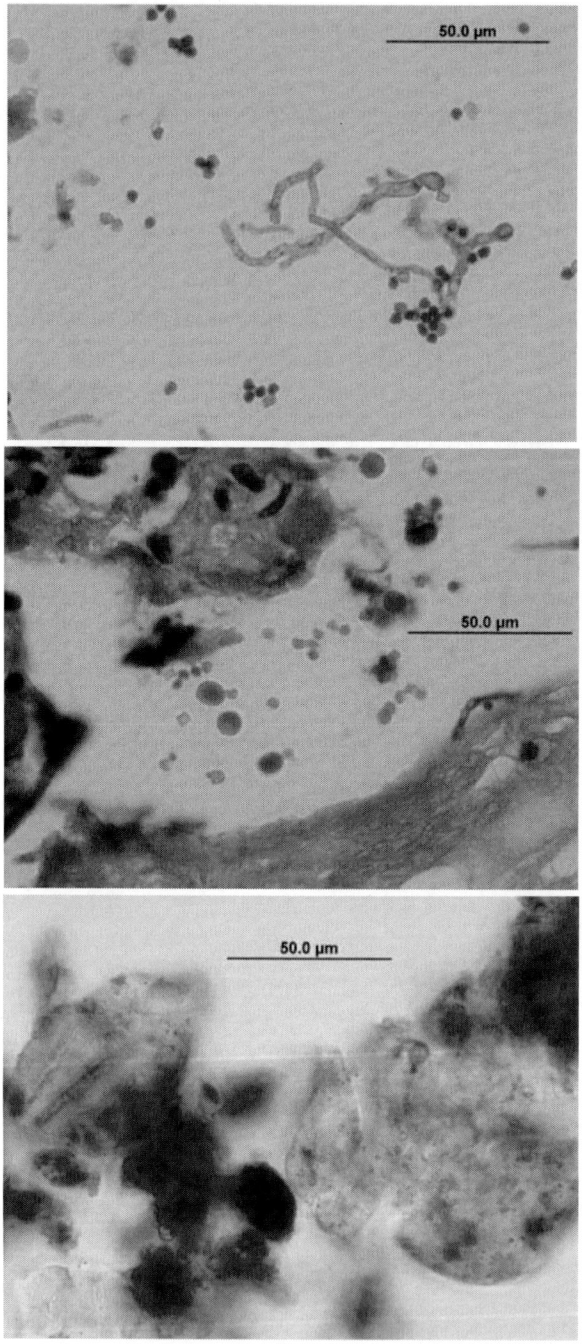

Figure 2. Upper: PAS stained *Aspergillus* control slide (ScyTek, Logan, UT, USA), the fungus are bright red. An erythrocyte appear gray (top left corner of photo).

Middle: H&E stained *Aspergillus* control slide. The control slide contains both erythrocytes and spherical fungi, allowing a size comparison. Erythrocytes appears redder and larger that the fungi.

Lower: PAS stained bone marrow from human bones from the excavations at Serra 'e sa Caudeba of two Bronze Age tombs on the Campidano Plain of Sardinia, Italy.

One drawback to using just the H&E stain for fungal diagnosis is the difficulty to distinguish poorly stained fungi from tissue components, even at higher magnifications. Morphologic features may not be evident and sometimes even misleading when investigating ancient skeletal remains. In modern medicine PAS can stain degenerated fungi that may not be visible on H&E stain, in ancient material PAS can be helpful in distinguishing fungi from erythrocytes (Figure 2). Some fungi like the *Aspergillus* species from fresh tissue are autofluorescent when examined under ultraviolet light source [16, 17], in the ancient bones there may be other fluorescent materials so this method was not employed to the samples.

There is always a concern that ancient samples are contaminated by bacteria and fungus during the taphonomic processes which occur in an organism after its death. While the Gram stains do not detect all infectious material in a sample a PAS stain was performed (Figure 2). Most fungi are fairly large and their walls are rich in polysaccharides (1,2-glycol groups), which can be demonstrated with Schiff's reagent of the PAS stain. PAS staining is among the most frequently employed techniques in routine diagnostics by histology laboratories. It is simple to perform and reliable, and will demonstrate the polysaccharide-laden wall of most fungal organisms. The specimen used in the present study had traces of fungus localized as spots. Using the H&E stained *Aspergillus* control slide we show that our staining system can detect a difference between the fungus and the red blood cells in the slide (Figure 2).

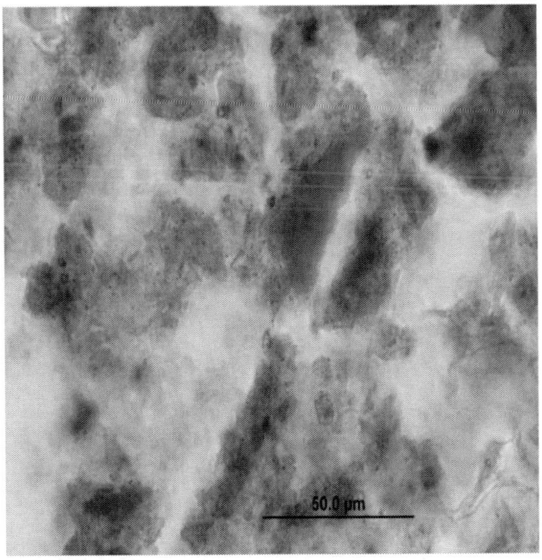

Figure 3. Ferric iron detection (blue dots) in the material from the bone cavity from the skeletal remains used in the present study. Although the bone marrow appears degraded without clearly defined cells the staining pattern achieved is typical for new bone marrow, distinct blue dots can be observed. The ferric iron detection supports the observations that this material is degraded ancient bone marrow.

PAS for fungal staining, however, will also stain polysaccharides wherever they appear in tissues. This method will stain many intracellular components in various types of normal cells; this staining appears weaker than the bright pink fungal staining because of the difference in concentration of polysaccharides. Using an iron stain on the bone marrow specimen, ferric iron (found in oxidized hemoglobin) is detected as expected in bone marrow. Hemosiderin may be present in areas of old hemorrhage or be deposited in tissues with high iron contents, hemosiderin is also known as storage iron granules. Hemosiderin stem from erythrocytes that became congested in an area where they die. Dying erythrocytes release hemoglobin as they rupture. In living tissue hemoglobin will be engulfed by local white blood cells, events in necrotic tissue are unknown. There is a trend in loss of stainable hemosiderin in decalcified samples causing a reduction in the intensity of hemosiderin staining [18] however cutting sections on a microtome require the samples to be decalsified. Despite the possible reduction of hemosiderin due to the decalsification procedure applied, positive staining can be observed (Figure 3).

Discussion

Histological analysis is a popular tool for the study of ancient skeletal remains. Ancient bone obtained at excavations has typically been exposed to elements; this exposure can cause diagenic changes in bone composition and structure. Paleohistologists must find methods that will allow for abnormalities that occurred in living tissue to be distinguished from those that occurred postmortem. Investigations of ancient bone have been performed for over 150 years and factors that cause changes to the bone have been identified. Modern science and paleohistologists can contribute new and improved methods for use in ancient skeletal remains.

Collagen and red blood cells have also been observed in bones by early paleohistologists [19] and modern paleohistologists [1, 2, 20]. The age of the bones is a less important factor in the preservation of organic components than the conditions where they have been found. Soil condition needs to be recorded by archaeologists during excavation since it is known that soil conditions may have an impact on the histological structure of the bones [21]. For example, water seems to be more unfavorable than dry soil [22] but the effect of the environment is not yet fully documented. Changes due to the environmental factors has been suggested to be caused mainly by chemical replacement of mineral ions [23, 24], while the morphological structure remains unaltered [25]. Organisms that thrive in damp conditions, such as fungi, may enter the tissue and even alter bone morphology. DNA extraction should not be performed on bones with fungus infection, fungus infected bones should be considered contaminated. This will complicate the histological identification of structures in the bone. Osteoclast-like destruction of bone by mold invasion was recognized already in 1864 (Schaffer

1895). Fungi of the genus *Mucor* is an osteolytic agent (26), fungi of other genera, including *Cladosporium*, *Candida*, and *Demtiacea*, may also alter the structure of dead tissue including ancient skeletal remains.

The PCR (polymerase chain reaction) method for DNA amplification was developed by Kary Mullis et al. in 1984 and was rapidly adapted to detect a variety of infectious agents. Despite this success, PCR has not been widely adopted to detect fungal pathogens in modern diagnostic medicine. Archaeologists may seek assistance from scientists with skills in molecular biology to develop or to adapt techniques to be used on ancient samples. However, the diagnosis of specific human fungal infections continues to be a challenge even to skilled molecular biologists. Scientists from different fields often have different requirements on controls included for comparison to the tested samples. The small amount of sample obtained from ancient material can make such requirements hard to satisfy.

In conclusion, by applying a series of dye staining techniques to histological samples of ancient skeletal tissue we demonstrate a method to distinguish between erythrocytes and fungi.

ACKNOWLEDGMENTS

Authors thank the Soprintendenza per i Beni Archeologici per le Province di Cagliari e Oristano, the Museo de Villanovaforru, E. Atzeni, A. Usai, M. Perra, R. Forresu, O. Fonzo, L. Tait, D. Sullivan, P. Hortolà, L. Lai, R. Tykot, and T. W. Killion.

REFERENCES

[1] Schweitzer MH, Wittmeyer JL, Horner JR, Toporski JK. 2005. Soft-tissue vessels and cellular preservation in Tyrannosaurus rex. *Science* 25:1952–1955.

[2] Schweitzer MH, Wittmeyer JL, Horner JR. 2007. Soft tissue and cellular preservation in vertebrate skeletal elements from the Cretaceous to the present. *Proc R Soc B* 274:183–197.

[3] Stout SD, Teitelbaum SL. 1976. Histological analysis of undecalcified thin sections of archaeological bone. Am *J Phys Anthropol* 44:263-269.

[4] Stout. SD. 1978. Histological Structure and Its Preservation in Ancient Bone. *Current Anthropology* 19(3)3: 601-604.

[5] Maat GJR. 1991. Ultrastructure of normal and pathological fossilized red blood cells compared with pseudopathological biological structures. *Int J of Osteoarchaeol* 1:209-214.

[6] Charlier P, Georges P, Bouchet F, Huynn-Charlier I, Carlier R, Mazel V, Richardin P, Brun L, Blondiaux J, Lorin de la Grandmaison G. 2008. The microscopic (optical and SEM) examination of putrefaction fluid deposits (PFD). Potential interest in forensic anthropology. *Virchows Archiv* 453:377-386.

[7] Prats-Muñoz G, Malgosa A, Armentano N, Galtés I, Esteban J, Bombi JA, Tortosa M, Fernández E, Jordana X, Isidro A, Fullola JM, Petit MÀ, Guerrero VM, Calvo M, Fernández PL. 2012. A paleoneurohistological study of 3,000-year-old mummified brain tissue from the Mediterranean Bronze Age. *Pathobiology* 79: 239-46.

[8] Janko M, Stark RW, Zink A. Preservation of 5300 year old red blood cells in the Iceman. *J R Soc Interface* 2012;9:2581–90.

[9] Cappella A, Stefanelli S, Caccianiga M, Rizzi A, Bertoglio B, Sforza C, Cattaneo C. 2015. Blood or spores? A cautionary note on interpreting cellular debris on human skeletal remains. *Int J Legal Med* 129(4):919-26.

[10] Atzeni E, Usai A, Bellintani P, Fonzo O, Lai L, Tykot R, Setzer TJ, Congiu R, Simbula S. 2012. Le tombe Nuragiche di Sa Sedda 'e Sa Caudela (Collinas – CA). Scavi 1982–84. *Quaderni* 23:25–50.

[11] Setzer TJ, Sundell IB, Dibbley SK, Les C. 2013. Technical note: a histological technique for detecting the cryptic preservation of erythrocytes and soft tissue in ancient human skeletonized remains. *Am J Phys Anthropol* 152(4):566-8.

[12] Balamas V, Migheli Q, Scherm B, Garau P. 2010. Multilocus phylogenetics show high levels of endemic fusaria inhabiting Sardinian soils (Tyrrhenian Islands). *Mycologia* 102:803–812.

[13] Benítez T, Rincón AM, Limón MC, Codón AC. 2004. Biocontrol mechanisms of Trichoderma strains. *Int Microbiol* 7:249-260.

[14] Jayalakshmi SK, Raju S, Usha R, Benagi VI, Sreeramulu K. 2009. Trichoderma harzianumL1as a potential source for lytic enzymes and elicitor of defense responses in chickpea (*Cicer arietinum* L.) against wilt disease caused by Fusarium oxysporum f. sp. ciceri. *Aust J Crop Sci* 3(1): 44-52.

[15] Wijnbergen M, Van Mullem PJ. 1987 Effect of histological decalcifying agents on number and stainability of gram-positive bacteria. *J Dent Res* 66(5):1029-31

[16] Graham AR. Fungal autofluorescence with ultraviolet illumination. *Am J Clin Pathol* 1983; 79:231-34.

[17] Mann JL. 1983. Autofluorescence of fungi: an aid to detection in tissue sections. *Am J Clin Pathol* 79:587-90.

[18] Byard RW, Bellis M. 2010. The effect of decalcifying solutions on hemosiderin staining. *J Forensic Sci* 55(5):1356-8.

[19] Graf W. 1949. Preserved histological structures in Egyptian mummy tissues and ancient Swedish skeletons. *Acta Anatomica* 8:236–250.

[20] McNamara ME, Orr PJ, Kearns SL, Alcala L, Anadon P, Penalver E. 2006. High-fidelity organic preservation of bone marrow in ca. 10 Ma amphibians. *Geology* 34:641– 644.

[21] Shackleford A. 1966. The ultrastructure of Mississippian and Archaic Indian bones from various soil and drainage conditions. *American Journal* of *Physical Anthropology* 25:291-98.

[22] Solomon CD, Hasse N. 1967. Histological and histochemical observations of decalcified sections of ancient bones from excavations in Israel. *Israel J Med Sci* 3: 747-754.

[23] Race, GJ, Fry, EI, Mathews, JL, Wagner, MJ, Martin, JH, Lynn, JA. 1968. Ancient Nubian human bone: A chemical and ultrastructural characterization including collagen. *American Journal of Physical Anthropology* 28:157-62.

[24] Race, GJ, Wendorf, F, Humphrey, SB, Fry EI. 1972. Paleopathology of ancient Nubian human bone studied by chemical and electron microscopic methods. *Journal of Human Evolution* 1:263-79.

[25] Posner AS. 1969. Crystal chemistry of bone mineral. *Physiological Review* 49: 760-92.

[26] Marchiafava V, Bonucci E, Ascenzi A. 1974. Fungal osteoclasia: a model of dead bone resorption. *Calcified Tissue Res* 14: 195-210.

In: Stem Cells Origins of Cytokine Interactions
Editor: Prasad S. Koka
ISBN: 978-1-53615-302-6

Chapter 5

POTENTIAL ORIGINS OF CANCER STEM CELLS IN THE DISEASE EVOLUTION AND ETIOLOGY

Prasad S Koka[1,*], Inger Birgitta Sundell[2], Usha Padmanabhan[1], Boris Minev[3,4], Teddi J. Setzer[2] and Abhay Chowdhary[5]

[1]National Professor & Visiting Scientist of Virology and Immunology, Haffkine Institute, Mumbai, India
[2]Department of Anthropology, Wayne State University, Detroit, Michigan, US
[3]StemImmune, Inc., San Diego, California, US
[4]Department of Radiation Medicine and Applied Sciences, Rebecca & John Moores Comprehensive Cancer Center, University of California, San Diego, California, US
[5]Department of Microbiology, Sir J J Hospital, Mumbai, Maharashtra, India

ABSTRACT

In this report we will discuss the early formation of the cancer stem cells (CSCs) both in terms of human genetic evolution and also in the role of stem cell differentiation from totipotency via pluri- and multi-potency to terminal differentiation. In addition, we will address the issues of therapies that target specifically CSCs, even within the conglomeration of cancer cells as a whole and the onset and development of this disease. We will address the possibility that interactions between normal hematopoietic stem cells (HCSCs) and cancer cells leads to the formation of cancer stem cells to monitor emergence of malignant transformation that give rise to or from cancer stem cells. If and whether indeed the HCSCs lose their stem cell characteristics over time during the course of the disease progression, would be a matter of great interest in cancer etiology. But this does not seem to happen in leukemias. The reason as to why HCSCs in preference to other types of stem cells would play a role in the formation of cancer stem cells is that the

[*] Corresponding Author Email: kokaprasad005@gmail.com.

HCSCs are the components of blood formation, even if these cells are not solely involved only in the formation of primarily leukemias *in vivo,* due to their imperative circulation in the body. As we have previously reported, it is plausible for the cell biologically engineered generation and characterization of hematopoietic cancer stem cells, *in vivo*. We had observed evidence for non-leukemic malignancies such as tumorigenesis in these humanized mice. These would be expected to lead to potential anti-cancer therapies due to the generally perceived implication that the CSCs are therapy-resistant as opposed to their non- CSC counterparts. Presumably these malignancies exist in latency and cause a relapse and metastasis despite the use of currently available anticancer therapies which may yet kill CSCs in a selective manner, or a replenishment of CSCs will occur post-treatment. A drug dependent susceptibility possibly partially and transiently may also be likely. The anticancer treatment resistance could potentially be phenotype dependent of the CSC *in toto*. This will not only help in understanding the (stem cell) origin of cancer, but could also be used in translational and clinical treatments that may ultimately benefit many patients with cancer.

BACKGROUND

The term "cancer" describes a group of diseases that are characterized by uncontrolled cellular growth, cellular invasion into adjacent tissues, and the potential to metastasize, even more so if not treated at a sufficiently early stage of the disease etiology. In cancer, cellular aberrations arise from accumulated genetic modifications, either via changes in the underlying genetic sequence or from epigenetic alterations (e.g., modifications that do not affect the genetic sequence itself) [1, 2, 3]. Recent studies indicate that cancers are continually replenished by a small population of cancer stem cells (CSCs) that are capable of self-renewal. It is presumed that these CSCs are resistant to therapies and are responsible for the occurrence of a relapse.

History of Cancer

A fragmentary copy of a papyrus dated ca. 2500 and several others dated ca. 1600 BC describe cancer and are probably the oldest references to this disease. Hippocrates (ca. 460 BC – ca. 370 BC) made drawings of outwardly visible tumors on the skin, nose, and breasts and described the appearance of the cut surface of a solid malignant tumor, with "the veins stretched on all sides as the crab has its feet", or greek *karkinos* from where it has been named as "carcinoma" in English language. The genetic basis of cancer was recognized in 1902 by the German zoologist Theodor Boveri [4]. The first causes of cancer that are reported are nasal cancer in snuff users in 1761, scrotal cancer from chimney sweeps in 1775, and Hodgkin's disease in 1832. After investigating hundreds of Egyptian mummies and finding only one case of prostate cancer with very few literature references to cancer, David & Zimmerman have stated that cancer was extremely rare in antiquity and that cancer was a man-made disease, down to pollution and changes to diet and lifestyle [5].

Evolution and Cancer

Similar to low rates of cancer incidence in older human populations, a number of works have reported that incidence of cancer in non-human primates is very low. This fact is especially evident for epithelial neoplasms such as breast, prostate or lung carcinomas, which are responsible for more than 20% of human deaths but whose incidence in great apes such as chimpanzees (who are most closely related to humans) is less than 2% [6-10]. The difference in cancer incidence between humans and chimpanzees can mainly derive from three facts: (i) exposure to different environmental factors, including diet and habits, (ii) differences in life expectancy, and (iii) genetic differences that might result in humans being more susceptible to cancer development. The last may be likely for being prone to all round innovation for temptations for luxuries in survival. For example, a comparison of the DNA isolated from Chimpanzee and human brains has revealed that DNA from human brains had lower levels of methylation compared to that of chimps and the genes that were lower in methylation were drivers of cell metabolism [11]. However whether such methylation differences between the primates and humans are enlarged to the detriment of humans over passage of time is not studied.

Self-Renewal and Stem Cells

Cancer cells must have a growth advantage that enables them to out-proliferate the rest of the normal cells within a tissue. Usually a cellular hierarchy, in which a small population of stem cells is responsible for the production of the mature cells of the organ, exists in most (if not all) tissues and they maintain the functional homeostasis within a given tissue by replacing damaged cells during injury to tissue or replacing lost/ discarded cells from apoptosis or necrosis, with newer cells.

For this to happen, stem cells undergo changes in cell cycle status and differentiate into other cells by changing their genetic landscape and sacrificing their own pluri- or multi- potency, to cater to the needs of the tissue. Stem cells carry the characteristic property of self-renewal. Self-renewal of stem cells is different from normal proliferation of somatic cells in that self-renewal is usually asymmetric division, i.e., regulators of self-renewal are localized asymmetrically during mitosis so that they are inherited by only one of the two daughter cells and the other daughter cell becomes a more differentiated cell than the parent cell [12]. Thus self-renewal (with implied retention of phenotypes) of embryonic or the further differentiated pluripotent stem cells consists of two events: (a) formation of a new stem cell with retention of phenotype that can undergo self-renewal and (b) formation of a progenitor cell that can undergo proliferation and differentiation but not further self-renewal.

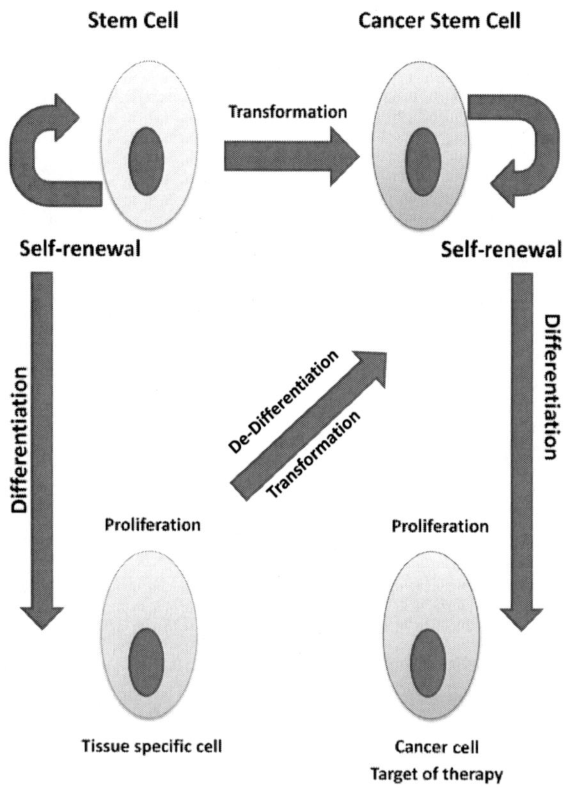

Figure 1. Similarities and differences between stem cells and cancer stem cells.

Self-renewal involves several classes of genes: proto-oncogenes (promoting self-renewal), e.g., *c-mpl*, and tumor suppressor genes (e.g., *p53*) which limit self-renewal and genes that maintain genome integrity. For example, expression of the proto-oncogene, *Bmi-1*, is consistently required for the self-renewal of diverse adult stem cells, as well as for the proliferation of cancer cells in the same tissues. *Bmi-1* expression promotes stem cell self-renewal partly by repressing the expression of *Ink4a* and *Arf*, which are tumor suppressor genes that are commonly deleted in cancer [13, 14].

Molecular Mechanisms of Cancer

In direct contrast to self-renewal of stem cells, somatic cell proliferation constitutes of symmetric division through mitosis with daughter cells retaining all characteristics of the parent cell. Cancer cells must be capable of continuous proliferation and self-renewal to retain the many mutations required for carcinogenesis and to sustain the growth of a tumor, since differentiated cells constrained by the Hayflick Limit and cannot divide indefinitely. The Hayflick limit (or Hayflick phenomenon) is the number of times (40-60) a normal human cell population will divide until cell division stops. This is because the telomeres associated with each cell's DNA will get slightly shorter with each new cell division until they shorten to a critical length [15, 16]. In other words, a normal cell lacks a sustained division as opposed to the cancer cell.

Cancer Stem Cells

The concept of a cancer cell (carrying a mutation) with the ability of self-renewal gave rise to the idea of a cancer stem cell, which were first identified by John Dick in acute myeloid leukemia in the late 1990s. Most tumors therefore contain a minority population of CSCs that maintain the tumor [17]. For example, abnormal regulation of self-renewal leads to leukemia (blood cancer) or other types of cancer [18]. Similar to that of stem cells, CSCs can give rise to both more CSCs as well as cancer cells (CCs), Figure 1. CCs have either no capacity or a markedly diminished capacity, to form new tumors (discussed above). It is the prevailing paradigm that chemotherapy targets the CCs and the tumor will rapidly recur if the therapy eliminates cancer cells but spares a significant population of the CSCs [19-26].

In a recent study it concludes that the number of stem cell divisions determine the lifetime incidence of various cancers. With every division, there's a risk of a cancer-causing mutation in a daughter cell, and hence the tissues that host the greatest number of stem cell divisions are those most vulnerable to cancer [27]. Random errors occurring during DNA replication in normal stem cells are a major contributing factor in cancer development. While inherited gene mutations and random DNA changes made during stem cell divisions cannot be controlled, environmental exposures and healthy lifestyles can minimize erratic cell divisions and activation of oncogenes. Any age-associated increase in cancer risk implies that individual vulnerability to cancer increases with age, and aging-related processes in a living being may be responsible for this increased incidence [28-32]. Therefore, to treat cancer effectively, CSCs generated within the tumor must be eliminated. It is hypothesized that *in vivo* that there exists few cancer stem cells that cause *de novo* proliferation even in the presence of therapies that target and kill most of the existing cancer cells.

Stem Cells Upstream or Downstream to CSCs

Due to the presumed resistance or "latency" of the CSCs, we expect that the multipotent mesenchymal stem cells (MSCs) are the upstream progenitors of the CSCs

[33-35]. The downstream progenitors of the CSCs are expected to be restricted to the yet multipotent but not at least initially such as at Stage I to be the pluripotent hematopoietic stem cells (HSCs). We propose that the multipotent HSCs then undergo terminal differentiation into the cancer cells of various types. These could include breast, ovarian, prostate, colon, melanoma, neuronal, and other cancer cells. Thus whether MSCs which are increasingly found to be beneficial are a double edged sword is in question and warrant a thorough investigation.

Genes Involved in CSC Formation

Mutations in the genes that cause self-renewal like STAT5A, and those promote multi-lineage differentiation of CD34+ progenitor stem cells such as c-Mpl [36], may be inducers of CSCs. Mutations in JAK2 and c-Mpl induce myelo-proliferative disorders [37]. A metastasis model of human prostate cancer in NOD/SCID-hu mice has been established [38]. Constitutive activation of STAT5 is associated with several leukemias, chromosomal aberrations, and the transformation of hematopoietic cells to cytokine independent growth [39]. The SCID-hu Thy/Liv chimeric small animal model system [40] appears to be ideally suited for induction of human cancer stem cells and their early detection. This model system consists of functional human hematopoietic organ that maintains hematopoiesis and allows the reconstitution of human CD34+ progenitor hematopoietic stem cells (HSCs) in a fresh stromal microenvironment, or in the host circulatory system [40, 41]. A NOD/SCID-hu Thy/Liv system has been used to generate humanized immune responses *in vivo* utilizing the host vascular and cytokine physiology [8]. Importantly, this model system allows long-term systemic human T-cell reconstitution *in vivo*. The HLA-restricted immune responses utilize the human T-cells (CD8 and CD4), B cells, dendritic cells, monocytes and macrophages. Thus, this humanized mouse model provides both human antigen-presenting proliferation regulatory cells (APCs such as CD4, HLA, DC, macrophages) and cytotoxic effector cells (CD8 or CTL), for the induction of anti-tumor immune responses [41].

As an example, cancer of the cervix is one of the second most common cancers in women in the developing world. Evidence suggests that there is a link between high-risk human papillomaviruses (HPVs) and cervical cancer. The association of certain types primarily with normal tissues and benign lesions as opposed to cancer associated types has led to the concept of low and high risk HPV's respectively. Studies have shown that persistence of high risk HPVs may determine progression to more severe stages of cervical disease, while the majority of HPV infections are transient and do not seem to be important in cervical carcinogenesis. Cancer of the cervix is initiated and sustained by high risk papillomaviruses HPV 16 and 18 and the continued expression of E6 and E7 oncogenes. Additional events or mutations such as deregulated signaling pathways are required to complement the functions of these oncogenes. Cancer initiating cells (CIC's) are proposed to be responsible for the generation of metastasis and resistance to therapy. Accumulating evidence indicates CIC's are found among different human cancers and cell lines derived from them. Therapeutic intervention for cervical cancer is targeted therapy with respect to cancer stem cells and also for effective treatment of cervical cancer.

The Evolutionary Origins of CSCs

Given the low incidence of cancers in ancient humans and in non-human primates, it would be worthwhile to see whether the ancient DNA (aDNA) samples contain the present day oncogenes. Then it would suggest that our generation of oncogenes date back to ancient periods of time. Then the next step would be to know the similarities and differences in the nucleotide sequences and three dimensional structures of the aDNA "derived oncogenes", or their peptide gene products, and recent generation oncogenes such as *ras (H, K, N), myc, neu (HER2), hox, etc.* Further steps would be to compare the sequences of the current generation "activated" oncogenes to those of the ancient, aDNAs. Any findings of similarities or those identical in genetic structure in the process of comparison of these sequences, implies whether cancer existed in ancient times, or is a result of mutations due to the consequences of the factors resulting from the technological growth, or industrialization of recent times. Even in the absence of any significant homologies between the aDNA and modern day DNA (designated as "mDNA") samples, it is entirely possible that less than major homologies between these aDNA and mDNA samples do exist and are the result of mutational events occurring over the generations in time. In the case of such mutational events, then their functional events as determined using cloned cDNAs of aDNA vs mDNA should be considered for the generation of tumors through introduction of recombinants of these DNAs. Such recombinants may also be useful for novel tumor antigen discovery and in the development of tumor biomarker targeting. It would also be interesting to investigate the biomarker genes in the aDNAs vs cDNAs of the modern day oncogenes [42].

The mutational or evolutionary factors could include: environmental such as air pollution, deforestation, food processing, genetically modified foods, usage of fertilizers, industrial waste causing mutations in seafood, river and ocean water pollution causing harm to life (humans, animals, water born species), drinking water pollution in less developed countries, oil – coal – natural gas pollution, etc.

Conclusion

We have engrafted the human CD34 positive hematopoietic cancer stem cells (HCSCs) and the CD34 negative cancer cells into the SCID-hu Thy/Liv [19]. This is to determine if there is a greater degree of metastasis in mice injected with the HCSCs compared to those injected with the original cancer cells lacking the CD34 hematopoietic stem cell phenotype. We will also determine whether the HCSCs are more resistant to killing by vaccine-induced *survivin*-specific CTLs. It is important to establish whether the CSCs are more resistant to killing by vaccine-induced cytotoxic T lymphocytes (CTLs) than their further differentiated counterparts (CCs'), as suggested by maturation or loss of the stem cell characteristics. We also hypothesized that peptides derived from the sequence of the universal tumor antigen *survivin* [43] are naturally expressed on the surface of the CSCs in association with MHC Class I molecules, where they can serve as targets for CTL. We propose to utilize *survivin*-derived Class I-restricted peptides encapsulated in biodegradable nanoparticles to induce CTLs that are specifically directed against the cancer stem cells. We confirmed recently that our nanoparticles significantly enhance the delivery of peptides to the cytoplasm of the antigen-presenting cells, demonstrating the potential of this approach for delivery of bio-therapeutics and vaccines. By utilizing our new nanoparticle-based vaccines we were able to induce very potent and specific anti-tumor responses. The identification of CSCs and the design of nanoparticle-based vaccines for specific targeting of these HCSCs will undoubtedly contribute to the development of new therapeutic strategies aimed at eradicating the tumorigenic stem cell subpopulation within cancer. The ability to prospectively identify, isolate and study cancer stem cells will significantly alter the way we think about, prevent, and treat cancer. The use of the NOD/SCID-hu model system for studying the CSCs and the design of nanoparticle vaccines targeting cancer stem cells is a novel approach that have not been used previously. Therefore, the findings of this study may be translated to clinical applications, which can benefit many patients with cancer to help reduce the human suffering and the impact of this disease.

References

[1] Michael, F. Clarke; John, E. Dick; Peter, B. Dirks; Connie, J. Eaves; Catriona, HM. Jamieson; D. Leanne, Jones; Jane, Visvader; Irving, L. Weissman; Geoffrey, M. Wahl. Cancer Stem Cells—Perspectives on Current Status and Future Directions: AACR Workshop on *Cancer Stem Cells*. Cancer Research DOI: 10.1158/0008-5472.CAN-06-3126 Published 1 October 2006.

[2] Andrew, P. Feinberg; Rolf, Ohlsson; Steven, Henikoff. The epigenetic progenitor origin of human cancer. *Nature Reviews Genetics*, 7, 21-33 (January 2006) doi: 10.1038/nrg1748.

[3] Peter, A. Jones; Stephen, B. Baylin. *The Epigenomics of Cancer Cell.*, 2007, Feb 23, 128(4), 683–692. doi: 10.1016/j.cell.2007.01.029.

[4] Boveri, T. Concerning the Origin of Malignant Tumours by Theodor Boveri. Translated and annotated by Henry Harris. *J Cell Science*, 2008, 121, 1-84.

[5] David, RA; Zimmerman, MR. Science and society: Cancer: an old disease, a new disease or something in between? *Nature Reviews Cancer*, 10, 728-733.

[6] Varki, A. A chimpanzee genome project is a biomedical imperative. *Genome Res.*, 2000, 10, 1065-1070.

[7] Waters, DJ; Sakr, WA; Hayden, DW; Lang, CM; McKinney, L; Murphy, GP; Radinsky, R; Ramoner, R; Richardson, RC; Tindall, DJ. Workgroup 4: spontaneous prostate carcinoma in dogs and nonhuman primates. *Prostate,* 1998, 36, 64-67.

[8] Beniashvili, DS. An overview of the world literature on spontaneous tumors in nonhuman primates. *J Med Primatol.* 1989, 18, 423-437.

[9] McClure, HM: Tumors in nonhuman primates: observations during a six-year period in the Yerkes primate center colony. *Am J Phys Anthropol.*, 1973, 38, 425-429.

[10] Seibold, HR; Wolf, RH. Neoplasms and proliferative lesions in 1065 nonhuman primate necropsies. *Lab Anim Sci.*, 1973, 23, 533-539.

[11] Zeng, J; Konopka, G; Hunt, BG; Preuss, TM; Geschwind, D; Soojin, VY. Divergent Whole-Genome Methylation Maps of Human and Chimpanzee Brains Reveal Epigenetic Basis of Human Regulatory Evolution. *The American Journal of Human Genetics*, Volume 91, Issue 3, 455 – 465.

[12] Betschinger, J; Knoblich, JA. Dare to be different: asymmetric cell division in Drosophila, C. elegans and vertebrates *Curr. Biol.*, 14 (2004), pp. R674–R685.

[13] He, S; Nakada, D; Morrison, SJ. Mechanisms of stem cell self-renewal. *Annu Rev Cell Dev Biol.*, 2009, 25, 377-406. doi: 10.1146/annurev.cellbio.042308.113248.

[14] Pardal, R; Molofsky, AV; He, S; Morrison, SJ. Stem cell self-renewal and cancer cell proliferation are regulated by common networks that balance the activation of proto-oncogenes and tumor suppressors. *Cold Spring Harb Symp Quant Biol.*, 2005, 70, 177-85.

[15] Hayflick, L; Moorhead, PS. (1961). "The serial cultivation of human diploid cell strains." *Exp Cell Res.*, 25 (3), 585–621.

[16] Hayflick, L. (1965). "The limited *in vitro* lifetime of human diploid cell strains." *Exp. Cell Res.*, 37 (3), 614–636.

[17] Jamieson, CHM; Weissman, IL; Passegue, E. Chronic versus acute myelogenous leukemia: A question of self-renewal. *Cancer Cell*, 6, 531-533, 2004.

[18] Zou, GM. Cancer stem cells in leukemia, recent advances. *J Cell Physiol*, 213, 440-447, 2007.

[19] Zhang, M; Dias, P; Minev, B; Koka, PS. Induction, isolation and characterization human fetal hematopoietic cancer stem cells *in vivo*. *J Stem Cells*, 2010, 5(1), 1-7.

[20] Bhargav, H; Huilgol, V; Metre, K; Sundell, IB; Tripathi, S; Nagaratna, R; Jadhav, M; Raghuram, N; Nagarathna, R; Nagendra, HR; Koka, PS. Evidence for extended age dependent maternal immunity to virus in infected children: mother to child transmission of HIV infection and potential interventions including role of sulfatides of the human fetal adnexa and complementary or alternative medicines. *J Stem Cells*, 2012, 7(3), 127-153.

[21] Duggal, R; Minev, B; Geissinger, U; Wang, H; Chen, NG; Koka, PS; Szalay, AA. Biotherapeutic approaches to target cancer stem cells. *J Stem Cells*, 2013, 8(3/4), 135-150.

[22] Ramdass, B; Chowdhary, A; Koka, PS. Hematological malignancies: disease pathophysiology of leukemic stem cells. *J Stem Cells*, 2013, 8(3/4), 151-188.

[23] Ramdass, B; Duggal, R; Minev, B; Chowdhary, A; Koka, PS. Functional role of solid tumor stem cells in disease etiology and susceptibility to therapeutic interventions. *J Stem Cells*, 2013, 8(3/4), 189-232.

[24] Ramdass, B; Chowdhary, A; Koka, PS. Cancer initiating cells as target for prevention of recurring disease etiology: role of these malignant putative progenitor cells in relapse or metastasis of human cervical carcinoma. *J Stem Cells*, 2013; 8(3/4), 233-251.

[25] Ram, A; Nagarathna, R; Rao, RM; Bhargav, H; Koka, PS; Tripathi, S; Raghuram, NV; Kodaganur, GS; Nagendra, HR. Development and validation of a need based integrated yoga program for cancer patients: a retrospective study. *J Stem Cells*, 2012, 7(4), 269-282.

[26] Metri, K; Bhargav, H; Chowdhury, P; Koka, PS. Ayurveda for chemo-radio therapy related side effects in cancer patients. *J Stem Cells*, 2013, 8(2), 115-130.

[27] Cristian, Tomasetti; Bert, Vogelstein. Variation in cancer risk among tissues can be explained by the number of stem cell divisions. *Science*, 2015, 347(6217), 78-81.

[28] Anisimov, VN. The Effect of Anti-aging Intervention on Cancer Risk. 1987; pp 122-130. In The Book: *Extending the Lifespan: Biotechnical*, *Gerontological*, *and Social Problems*. Edited by Klaus Sames, Sebastian Sethe, Alexandra Stolzing.

[29] Vladimir, N. Anisimov. Insulin/IGF-1 signaling pathway driving aging and cancer as a target for pharmacological intervention. *Experimental Gerontology*, 38, (2003), 1041–1049.

[30] Sherr, CJ; DePinho, RA. Cellular senescence: mitotic clock or culture shock? *Cell.*, 2000 Aug 18, 102(4), 407-410.

[31] Harry, Rubin. Selected Cell and Selective Microenvironment in Neoplastic Development. *Cancer Research*, 2001, 61(3), 799-807.

[32] Krtolica, A; Campisi, J. Cancer and aging: a model for the cancer promoting effects of the aging stroma. *Int J Biochem Cell Biol.*, 2002 Nov, 34(11), 1401-1414.

[33] Sundell, IB; Koka, PS. Placental membrane as a source of mesenchymal stem cells. *J Stem Cells*, 2010, 5(2), 83-88.

[34] Das, M; Sundell, IB; Koka, PS. Potency of adult mesenchymal stem cells in the cell based therapy. *J Stem Cells*, 2013, 8(1), 1-16.

[35] Ramdass, B; Koka, PS. Ligament and tendon repair through regeneration using mesenchymal stem cells. *Curr Stem Cell Res Ther*, 2014, 10(1), 84-88, 2014 Oct 2 [Epub ahead of print].

[36] Hitchcock, IS; Chen, MM; King, JR; Kaushansky, K. YRRL motifs in the cytoplasmic domain of the thrombopoietin receptor regulate receptor internalization and degradation. *Blood*, 112, 2222-2231, 2008.

[37] Gibson, SE; Schade, AE; Szpurka, H; Bak, B; Maciejewski, JP; His, ED. Phospho-STAT5 expression pattern with the MPL W515L mutation is similar to that seen in chronic myeloproliferative disorders with JAK2 V617F. *Human Pathol*, 39; 1111-1114, 2008.

[38] Yonou, H; Yokose, T; Kamijo, T; et al. Establishment of a novel species- and tissue-specific metastasis model of human prostate cancer in humanized non-obese diabetic/severe combined immunodeficient mice engrafted with human adult lung and bone. *Cancer Research*, 61, 2177-2182, 2001.

[39] Wierenga, ATJ; Vellenga, E; Schuringa, JJ; Maximal STAT5-induced proliferation and self-renewal at intermediate STAT5 activity levels. *Molec Cell Biol*, 28, 6668-6680, 2008.

[40] Sundell, IB; Koka, PS. Chimeric SCID-hu Model as a Human Hematopoietic Stem Cell Host that Recapitulates the Effects of HIV-1 on Bone Marrow Progenitors in Infected Patients. *J Stem Cells*, 1, 283-300, 2006.

[41] Melkus, MW; Estes, JD; Padgett-Thomas, A; Gatlin, J; Denton, PW; Othieno, FA; Wege, AK; Haase, AT; Garcia, JV. Humanized mice mount specific adaptive and innate immune responses to EBV and TSST-1. *Nature Med*, 12, 1316-1322, 2006.

[42] Koka, PS. Biomarker Discovery and Biotherapeutics Applications of Photo-synthetic Light-Harvesting and Bioluminescence Light-Emitting Chromophore-Protein Complexes in Stem Cell Biology and Regenerative Medicine. *J Stem Cells*, 2014, 9(3), 127-133.

[43] Andersen, MH; Thor, SP. Survivin--a universal tumor antigen. *Histol Histopathol*, 17, 669-675, 2002.

In: Stem Cells Origins of Cytokine Interactions
Editor: Prasad S. Koka
ISBN: 978-1-53615-302-6

Chapter 6

IMPACT OF SHORT-TERM RESIDENTIAL YOGA-BASED LIFESTYLE INTERVENTION ON COGNITIVE FUNCTION IN HYPERTENSIVE PATIENTS

Kashinath G. Metri**, Balaram Pradhan and H. R. Nagendra
Department of Yoga and Life Sciences, Swami Vivekananda Yoga Anusandhana Samsthana, Eknath Bhavan, Gavipuram Circle, Kempegowda Nagara, Bangaluru, India

ABSTRACT

Background: Hypertension (HTN) is a highly prevalent disease and a leading cause of morbidity and mortality worldwide. It leads to several long term complications including cognitive impairment. Yoga is a form of alternative and complementary therapy known to be effective in improving cognitive function. There is however, a lack evidence of the role of yoga on cognitive function in hypertensive patients.

Aim: To study the impact of short term residential yoga-based lifestyle intervention on cognitive functions in hypertensive patients.

Methodology: Sixty hypertensive subjects, within age range 30 to 60 years (Mean ± SD; 46.92 ± 9.2 yrs) were enrolled in this study. Thirty subjects (IAYT) followed 6 days of residential yoga program, consisting of 5 hrs of different yoga sessions including physical practices, meditation, devotional songs, a low salt & calorie diet and counseling sessions at SVYASA University campus in Bengaluru. Thirty subjects (Non-IAYT) followed their daily routines. Subjects with diabetes, psychiatric problem, stroke, head injury, neuro-deficits, history of recent surgery, and previous exposure to yoga in last one year were excluded from the study. All the subjects were administered Digit Letter Substitution Test and Digit Forward and Backward Test at the baseline and after 6 days.

Results: For the IAYT group, there was a significant improvement in Digit Forward and Backward Test (p>0.001) scores and in DLST test (p>0.001) after 6 days of intervention compared to baseline. Whereas, non-IAYT group showed, no significant improvement in any

* Corresponding Author Email: kgmhetre@gmail.com

of the variables assessed. Therefore, compared to the Non-IAYT group, the IAYT group showed significantly better improve in DF ($p < 0.001$), DB ($p < 0.001$) and DLST scores ($p < 0.001$).

Conclusion: Short term residential yoga practice helps in enhancing cognitive function in hypertensive patients.

INTRODUCTION

Hypertension (HTN) is one among the most common public health problems across the world. It is a leading cause of morbidity and mortality in both developing and developed countries. It accounts for about 13% of all global deaths/year [1].

HTN affects approximately 25% of the world population with an estimated increase of 60% cases by 2025 [2]. Furthermore, HTN is an independent and modifiable risk factor of cardiac disease, chronic kidney disease, Alzheimer's disease and stroke [3, 4]. The brain is a major target organ affected by hypertension.

Long term hypertension leads to damage of cerebral blood vessels and contributes to the narrowing of lumens of arteries and arterioles that nourish deep tissues in the brain. This process accelerates the development of atherosclerotic plaques in the cerebral blood vessels contributing to cognitive decline [5, 6, 7].

Many cross-sectional and longitudinal studies have confirmed the strong association of hypertension with cognitive decline [8]. Studies have also reported that HTN increases the risk of onset of dementia in elderly [9].

Midlife hypertension is known to be responsible for poor cognition and increased risk of development of dementia in later age of life [10].

Alzheimer's disease and vascular cognitive impairment are the two most common causes of cognitive [11].

Optimal blood pressure control is known to reduce the risk of Alzheimer's disease and generalized neurodegenerative changes responsible for cognitive impairment [10].

Yoga is an ancient system of mind-body medicine dedicated to bring balance to physical, mental, and emotional health of individual. Yoga prescribed by ancient Indian sages includes positive mental attitude, compassionate behavior, a simple and healthy diet along with physical postures (asanas), breathing practices (pranayama), and meditation [12].

Several scientific investigations shown that practice of yoga postures, breathing techniques, meditation and relaxation techniques have positive influence on cardiac parameters and cognitive function [14, 15]. Evidence suggests that practice of yoga helps in reducing systolic and diastolic blood pressure, cardiac risk factors and improves general well-being in hypertension individuals [16].

It has been demonstrated that any injury to the brain tissue attracts the neural stem/progenitor cells towards injured area and activates the process of neurogenesis [17].

Oxidative stress is strongly associated with several neurodegenerative conditions such as Alzheimer's disease, dementia, Parkinson's disease. Oxidative stress is considered to be one of the most important causes of neurodegeneration and it has been demonstrated that increased oxidative stress is associated with decreased neural cell proliferation [18]. Several scientific investigations have demonstrated that practice of yoga the oxidative stress and degeneration process [19]. Stem cell implantation is also an emerging therapy in the medical field. This has been used in a variety of clinical conditions including neurodegenerative conditions such as Alzheimer's disease, Parkinson's disease, and cardiovascular diseases. In an animal experimental study it was demonstrated that stem cell implantation in infarcted areas of the brain improved cognitive abilities after such an implantation treatment [20].

METHODOLOGY

Subjects

Sixty hypertensive individuals within age range 30 to 60 years (with mean ± SD; 46.9 2 ± 9.2 yrs) were divided into 2 groups; i.e., IAYT (30, 5 females) and non-IAYT group (30, 7 females). No significant difference was observed between the groups in terms of gender distribution and age (Table 1).

Table 1. Shows the baseline comparison between IAYT and Non-IAYT group

Variables	IAYT	Non-IAYT	P value
No of subjects	N = 20	N = 20	NS
	F = 5, M = 15	F = 7, M = 13	
Age (Mean) years	46.62 ± 9.9	47.08 ± 9.69	0.13
HR (beats/minute)	78.21 ± 2.45	76.64 ± 2.18	0.45
SYS BP (mmHg)	135.23 ± 5.27	130.72 ± 3.51	0.08
Dia BP (mmHg)	77.38 ± 3.177	74.14 ± 1.917	0.09
DLST Score	39.52 ± 11.70	38.27 ± 11.02	0.479
DF Score	4.30 ± 1.23	4.67 ± 1.27	0.113
DB score	3.73 ± 1.01	3.93 ± 1.20	0.433
BMI (kg/M^2)	25.45 ± 5.86	26.27 ± 4.15	0.13

F-Female subjects; M-Male subjects; NS-Non significant difference in between groups: HR-Heart rate; SYS BP-Systolic blood pressure; Dia BP- Diastolic blood pressure; DLST-Digit Letter Substitution Test: Used to assess Selective attention, concentration, psychomotor speed and scanning memory; DF-Digit forward- A neurocognitive test to measure short term memory; DB-Digit Backward- A neurocognitive test to measure working memory; BMI- Body Mass Index.

Inclusion and Exclusion Criteria

We included the subjects; hypertensive subjects within age range 30-60 years; of any gender willing to participate in the study.

Subjects were excluded if they had psychiatric problem; history of epilepsy; diabetes mellitus; resistance hypertension; head injury; neuro-deficits; history of recent surgery; previous exposure to yoga in last one year and secondary hypertension.

Ethical Considerations

Study was approved by Institutional Ethical Committee (IEC) of SVYASA Yoga University, Bangalore, India.

Written informed consent was obtained from all the subjects before the commencement of the study.

ASSESSMENTS

Cognitive Measures

Memory

Short term and working memory were assessed before and after intervention using digit forward and digit backward span of Wechsler's memory scale

The forward/backward digit span (Wechsler Adult Intelligence Scale-Revised (WAIS-R) (Wechsler, 1981) 'REF' was used to measure immediate memory span. The dependent measures are total forward digit span and total backward digit span. In the forward digit span, subjects can correctly repeat back a span of up to16 digits. In the backward version, subjects recite a span of upto 14 digits backwards. Higher scores are indicative of higher memory recall [19].

Attention, Concentration and Processing Speed Were Assessed Using Digit Letter Substitution Test (DLST)

DLST

It is one of the most widely used paper pencil test consists of 8 rows × 12 columns of randomly arrayed digits. The key at the top of each sheet, pairs each of the 9 digits with 9 selected letters. Subjects have to write the corresponding letters in the empty box below each digit. Choice of strategy for substituting letters is up to each subject: horizontally, vertically, or selecting one digit at a time. Subjects have to substitute as many letters for digits as possible in the test time of 90 sec. Test supervisors timed each test on a standard stopwatch [21].

Blood Pressure

Heart rate, systolic blood pressure and diastolic blood pressure were assessed using FinaPres, NIBP system.

Intervention

IAYT Intervention

Subjects in IAYT group underwent a specialized 6 day residential yoga-based lifestyle intervention program at SVYASA university campus in Bangalore. Their daily programs included 3 hrs of specific yoga practices designed for hypertension, 2 sessions of volunteer regulation of breath (*Pranayama*), one meditation session and two sessions of relaxations and devotional songs, a low salt and low calorie diet along with counseling sessions. In contrast, the control Non-IAYT group followed the typical (non-specialized) daily routines. All the subjects were on anti-hypertensive medication [22].

Data Extraction and Analysis

Data extraction of digit forward, backward test and Digit Letter Substitution test was done followed the instruction given in the manual of Digit Forward Backward Test.

Table 2. Shows daily activates of subjects in IAYT group during intervention

Daily activities	Time (Hrs)	Daily activities	Time (Hrs)
Cyclic meditation	15:00	Ablution	5:00
HTN Special techniques	16:00	Om Meditation	5:30
Tuning to nature	17:00	HTN Special technique	6:00
Devotional session	18:00	Spiritual discourse	7:15
MSRT	18:45	Breast fast	8:00
Dinner	19:30	Yogic counseling	9:30
Happy assembly	20:30	Pranyama (Breathing techniques)	10:30
Group discussion/Self-study	21:15	Milk or herbal tea	11:30
Lights off	22:00	HTN Special technique	12:05
		Lunch & rest	13:00

HTN Special techniques: Includes special yoga practices designed for hypertension; Om Meditation: A Meditative technique; MSRT-Mind Sound Resonance Technique: It is a kind of yoga-based relaxation techniques includes chanting of mantra in supine posture; Happy Assembly- Playing indoor games, laughter, singing etc.

RESULTS

Group Comparisons

IAYT Group

Pre-post comparison showed a significant improvement in systolic ($p = 0.004$) and diastolic ($p = 0.006$) BP, digit forward ($p < 0.001$) and digit backward ($p < 0.001$) test, DLST total attempt ($p = 0.005$) and DLST net attempt ($p = 0.006$) after 6 days of the program as compared to the baseline.

Non-IAYT Group

No significant improvement found in non-IAYT group but there was a significant decrease in DLST net score ($p = 0.03$).

Table 3. Within group and between groups comparison of variable at baseline and after 6 days

Groups- Variable	Pre [Mean ± SD]	Post [Mean ± SD]	Mean diff.	% change	df	Within group comparison	Between group comparison
Yoga HR	78.21 ± 2.45	77.11 ± 2.20	2.51	-1.41	29	0.66	0.579
Control HR	76.64 ± 2.18	75.51 ± 1.799	1.131	-1.47	29	0.72	
Yoga SYS BP	135.23 ± 5.27	122.73 ± 3.92	-12.5	-9.24	29	0.004	.038*
Control SYS BP	130.72 ± 3.51	138.62 ± 5.89	-7.878	6.04	29	0.12	
Yoga Dia BP	77.38 ± 3.17	71.45 ± 2.537	-5.92*	-7.66	29	0.006	0.95
Control Dia BP	74.14 ± 1.917	76.16 ± 2.13	2.017	2.72	29	0.36	
Yoga DF	4.30 ± 1.23	6.60 ± 1.35	2.3	53.49	29	<0.001	<0.001**
Control DF	4.67 ± 1.27	4.83±1.18	0.16	3.43	29	0.902	
Yoga DB	3.73 ± 1.01	5.60 ± 1.35	1.87	50.13	29	<0.001	<0.001**
Control DB	3.93 ± 1.20	4.03 ± 1.10	0.1	2.54	29	0.889	
Yoga DLST TS	39.52 ± 11.70	42.52 ± 12.53	2.93	7.59	29	0.005	<0.001**
Control-DLST TS	38.27 ± 11.02	35.88 ± 11.65	-2.39	-6.25	29	0.052	
Yoga DLST WS	0.14 ± 0.35	0.12 ± 0.4	-0.02	-14.29	29	0.12	0.052
Control DLST WS	0.30 ± 0.47	0.60 ± 0.67	0.1	33.3	29	0.072	
Yoga DLST NS	39.98 ± 11.8	42.38 ± 12.5	3	7.62	29	0.006	<0.001**
Control DLST NS	37.27 ± 10.94	35.28 ± 11.78	-2.69	-7.08	29	0.03	

* Statistical significance at 0.05 level; ** Statistical significance at 0.001 level; HR-Heart rate; SYS BP- Systolic blood pressure; Dia BP- Diastolic blood pressure; DLST-Digit Letter Substitution Test: Used to assess Selective attention, concentration, psychomotor speed and scanning memory; DLST TS- DLST Total score- domain of DLST test measuring total number of attempts made by subjects; DLST WS; DLST Wrong Score- total wrong attempts; DLST NS- DLST Net sore-Total number of correct attempt; DF-Digit forward- A neuro-cognitive test to measure short term memory; DB-Digit Backward- A neuro-cognitive test to measure working memory.

Between Group Comparison

Between group comparison showed a significant improvement in systolic BP ($p = 0.038$), digit forward ($p < 0.001$) and digit backward ($p < 0.001$) test, DLST total attempt ($p < 0.001$) and DLST net attempt ($p < 0.001$) in the IAYT group as compared to non-IAYT group.

DISCUSSION

The present study was aimed to assess the impact 6 days residential yoga-based lifestyle intervention on cognitive function in hypertensive subjects.

We observed a significant improvement from the baseline in DBF test score and DLST score after one week in intervention group as compared to control group.

The present results thus suggest that short term residential yoga program is beneficial in improving cognitive function in hypertensive individuals.

Several previous studies have reported the significant improvement in cognitive function following yoga practice in different subject populations. Similarly in our study, we also found significant improvement in cognitive function following 6 days residential yoga intervention [23].

Most of the previous studies performed to assess the impact of yoga on cognitive function were in school children, healthy adults, elderly people, and few were in different medical conditions such as diabetes, multiple sclerosis etc. To the best of our knowledge this is the first study to investigate the efficacy of yoga intervention on cognitive function in hypertensive population.

A study by Kyizom et al. (2010) assessed the impact of 45 days Hatha yoga intervention on working memory and attention. Significant improvement in working memory and attention was reported after 45 days compared to the baseline. This present study also support the results observed by Kyizom. Both these studies differ in term of frequency and duration of yoga intervention, assessment tools used and subjects enrolled in the study. Compared to our study, the previous study was of longer duration (45 days), frequency of yoga sessions were less (1 session/day) and subjects were diabetic patients; whereas in our study, subjects were hypertensive patients and duration of intervention was shorter (6 days) and frequency of session was more (6 session/day). Even though subjects enrolled in both the studies differ, diabetes mellitus and hypertension are associated with cognitive decline [24].

Another study by Velikonja et al. (2010) assessed impact of 10 weeks Hatha yoga intervention on attention in subjects with multiple sclerosis and found a significant improvement in attention (using D2 test) in post assessment compared to the baseline.

Similarly in our study, we also found significant improvement in attention assessed (using DLST), but the subjects in our study were hypertensive patients whilst the previous study's subjects were patients of multiple sclerosis. Intervention period in the previous study was 10 weeks, one session/day, whereas in our study intervention shorter (6 days) and frequency of yoga sessions were more/day as compared to previous study [25].

A study by Bowden (2012) [26], reported a significant improvement in working memory assessed using 2 back tests, following 5 weeks of Iyengar intervention twice a week in healthy adults. The present study is in support of Bowden's findings, in the previous study (Bowden), duration of yoga intervention was longer (5 weeks) but the frequency of yoga session was less (2 days/week) and the subjects were healthy adults, whereas in our study, duration of yoga intervention was short (6 days) but the frequency of yoga sessions were more (4 sessions/day).

MECHANISM

The exact mechanism of yoga in improving cognition is not known. The following possible reasons might have contributed in improving cognition in present study:

1. The intervention included various yoga postures and loosening practices. Increased physical activity is known to improve cognition [27] by enhancing neural plasticity of brain areas which are involved in cognition [28, 29].
2. Stress is known to have negative effects on cognition [30] and yoga intervention is known to alleviate the stress and its effects [31]. Reduced stress is associated with improved cognition [32].
3. Various yoga postures needs constant attention and body awareness, practice of postures improves the ability of concentration and attention [35].
4. Different components of yoga such as meditation known to enhance neuro-plasticity of brain area which are associated with attention, concentration, sensory processing and memory [34].

Other reasons for improvement of memory in hypertensive subjects might be earnest sensitivity and acceptance to such interventions coupled with the intervention environment. A serene atmosphere free from daily possible psychological triggers, which are known to affect cognition, is a confounding variable. Short duration can also motivate commitment to the lifestyle modifications.

Strengths of Study

To the best of our knowledge, this is the first study to assess the impact of yoga on cognitive function in hypertensive patients. The intervention which included physical activity, breathing practices and diet modification adopted was a feasible and comprehensive lifestyle for hypertension. The study includes the domains of medicine, Yoga, and neuropsychiatry field.

Limitation of the Study

Small sample size and non-RCT study; IAYT intervention given may be not practical to implement along with daily routines. Further, environments of both groups were different. More measuring tools could be incorporated as only paper pencil tests were used to assess primary outcome.

Future Suggestions

Future studies should be carried out using robust study design with larger sample size. Control group should be provided similar environment as that of intervention group. Introduction of latest and different computer-based cognitive tools could be used.

Acknowledgments

We are grateful to all the hypertensive patients who participated in this study.

References

[1] Robitaille C, Dai S, Waters C, Bancej C, Quach S, Ellison J, et al. Diagnosed hypertension in Canada: incidence, prevalence and associated mortality. *CMAJ.* 2012;184(1):E49-56.

[2] Ferreira SR, Moura EC, Malta DC, Sarno F. Frequency of arterial hypertension and associated factors: Brazil, 2006. *Rev Saúde Pública.* 2009;43 Suppl 2:98-106.

[3] Messerli FH, Williams B, Ritz E. Essential hypertension. *Lancet.* 2007;370:591–603.

[4] Kelley BJ, Petersen RC. Alzheimer's disease and mild cognitive impairment. *Neurol Clin.* 2007;25:577– 609.

[5] Baumbach GL, Heistad DD. Cerebral circulation in chronic arterial hypertension. *Hypertension.* 1988;12:89–95.

[6] Guo Z, Viitanen M, Fratiglioni L, Winblad B. Low blood pressure and dementia in elderly people: the Kungsholmen project. *BMJ.* 1996;312(7034):805–808.

[7] Skoog I. A review on blood pressure and ischaemic white matter lesions. *Dement Geriatr Cogn Disord.* 1998; 9(Suppl 1): 13–19.

[8] Li, X., Liang, Y., Chen, Y., Zhang, J., Wei, D., Chen, K & Zhang, Z. (2015). Disrupted Fronto-parietal Network Mediates White Matter Structure Dysfunction Associated with Cognitive Decline in Hypertension Patients. *The Journal of Neuroscience*. 35(27), 10015-10024.

[9] Fujishima M, Tsuchihashi T. Hypertension and dementia. *Clinical and Experimental Hypertension.* 1999 Jan 1;21(5-6):927-935.

[10] Qiu C, Winblad B, Fratiglioni. The age-dependent relation of blood pressure to cognitive function and dementia. *Lancet Neurol. 2005 4(8):487-499.*

[11] O'Brien JT, Erkinjuntii E, Reisberg B, Roman G, Sawada T, Pantoni L et al. Vascular cognitive impairment. *Lancet Neurol.* 2003; 2: 89–93.

[12] Taimini IK: *The Science of Yoga.* Madras, India: The Theosophical Publishing House, 1961.

[13] Sharma VK, Das S, Mondal S, Goswami U, Gandhi A. Effect of Sahaj Yoga on neuro-cognitive functions in patients suffering from major depression. *Indian journal of physiology and pharmacology*. 2006 Oct 4;50(4):375.

[14] Velikonja O, Čurić K, Ožura A, Jazbec SŠ. Influence of sports climbing and yoga on spasticity, cognitive function, mood and fatigue in patients with multiple sclerosis. *Clinical neurology and neurosurgery*. 2010 Sep 30;112(7):597-601.

[15] Ornish D, Brown SE, Billings JH, Scherwitz LW, Armstrong WT, Ports TA, McLanahan SM, Kirkeeide RL, Gould KL, Brand RJ. Can lifestyle changes reverse coronary heart disease? The Lifestyle Heart Trial. *The Lancet.* 1990 Jul 21; 336(8708):129-133.

[16] Cohen DL. Yoga and hypertension. *Journal of yoga & physical therapy*. 2013 Nov 13;3(4).

[17] Picard-Riera N, Nait-Oumesmar B, Baron-Van Evercooren A. Endogenous adult neural stem cells: Limits and potential to repair the injured central nervous system. *J Neurosci Res*. 2004; 76:223–231.

[18] Hamilton A, Holscher C. The effect of ageing on neurogenesis and oxidative stress in the APP swe/PS1 deltaE9 mouse model of Alzheimer's disease. *Brain research*. 2012 Apr 17;1449:83-93.

[19] Rao RM, Telles S, Nagendra HR, Nagarathna R, Gopinath K, Srinath S, Chandrashekara C. 2008. Effects of yoga on natural killer cell counts in early

breast cancer patients undergoing conventional treatment. *Med Sci Monit*; 14(2):LE3-4.

[20] *Stem Cell Clinics Online: The Direct-to-Consumer Portrayal of Stem Cell Medicine* Darren Lau, Ubaka Ogbogu, Benjamin Taylor, Tania Stafinski, Devidas Menon,1 and Timothy Caulfield.

[21] Conklin HM, Curtis CE, Katsanis J, Iacono WG. Verbal working memory impairment in schizophrenia patients and their first-degree relatives: evidence from the digit span task. *American Journal of Psychiatry*. 2000 Feb 1.

[22] Pradhan B. Effect of kapalabhati on performance of six-letter cancellation and digit letter substitution task in adults. *International journal of yoga*. 2013 Jul;6(2):128.

[23] Tekur P, Singphow C, Nagendra HR, Raghuram N. Effect of short-term intensive yoga program on pain, functional disability and spinal flexibility in chronic low back pain: a randomized control study. *The journal of alternative and complementary medicine*. 2008 Jul 1;14(6):637-44.

[24] Gothe NP, McAuley E. Yoga and cognition: a meta-analysis of chronic and acute effects. *Psychosomatic medicine*. 2015 Sep 1;77(7):784-97.

[25] Kyizom T, Singh S, Singh KP, Tandon OP, Kumar R. *Effect of pranayama & yoga-asana on cognitive brain functions in type 2 diabetes-P3 event related evoked potential (ERP).*

[26] Velikonja, K, Čurić, A Ožura, SŠ Jazbec. Influence of sports climbing and yoga on spasticity, cognitive function, mood and fatigue in patients with multiple sclerosis. *Clin Neurol Neurosurg*. 2010;112:597–601.

[27] Bowden D, Gaudry C, An SC, Gruzelier J. A comparative randomized controlled trial of the effects of brain wave vibration training, Iyengar yoga, and mindfulness on mood, well-being, and salivary cortisol. *Evidence-based complementary and alternative medicine*. 2011 Dec 15; 2012.

[28] SmithPJ, Blumenthal JA, Hoffman BM, et al. Aerobic exercise and neurocognitive performance: a meta-analytic review of randomized controlled trials. *Psychosom Med*. 2010;72:239–252.

[29] Colcombe SJ, Erickson KI, Scalf PE, et al. Aerobic exercise training increases brain volume in aging humans. *J Gerontol*. 2006;61:1166–1170.

[30] Knaepen K, Goekint M, Heyman EM, Meeusen R. Neuroplasticity exercise-induced response of peripheral brain-derived neurotrophic factor: a systematic review of experimental studies in human subjects. *Sports Med*. 2010; 40:765–801.

[31] Lupien SJ, McEwen BS, Gunnar MR, Heim C. Effects of stress throughout the lifespan on the brain, behavior and cognition. *Nature Reviews Neuroscience*. 2009 Jun 1; 10(6):434-45.

[32] Balaji Deekshitulu PV. Stress and Yoga. *Journal of Yoga & Physical Therapy*. 2012 Apr 16.

[33] Crocker PR, Alderman RB, Smith FM. Cognitive-Affective Stress Management Training with high performance youth volleyball players: Effects on affect, cognition, and performance. *Journal of Sport & Exercise Psychology*. 1988 Dec 1;10(4).

[34] Lazar SW, Kerr CE, Wasserman RH, et al. Meditation experience is associated with increased cortical thickness. *Neuroreport*. 2005;16:1893–1897.

In: Stem Cells Origins of Cytokine Interactions
Editor: Prasad S. Koka
ISBN: 978-1-53615-302-6

Chapter 7

Hypoxia Decreases Osteogenic Differentiation Potential but Enhances Vasculogenic Differentiation Potential of Human Adipose Stem Cells on Biphasic Calcium Phosphate Scaffolds in Fibrin Gels

Fransisca A. S. van Esterik[1,2], Jolanda M. A. Hogervorst[1], Pieter Koolwijk[3] and Jenneke Klein-Nulend[1,*], PhD

[1]Department of Oral Cell Biology, University of Amsterdam and Vrije Universiteit Amsterdam
[2]Department of Dental Materials Science, Academic Centre for Dentistry Amsterdam (ACTA), University of Amsterdam and Vrije Universiteit Amsterdam, MOVE Research Institute Amsterdam, Amsterdam, The Netherlands
[3]Department of Physiology, Institute of Cardiovascular Research, VU University Medical Center, Amsterdam, The Netherlands

Abstract

Background: In bone tissue engineering, biphasic calcium phosphate (BCP) with different hydroxyapatite (HA)/β-tricalcium phosphate (β-TCP) ratios can be combined with regeneration-competent stem cells, and bioscaffolds like fibrin that trigger neovascularization. We have shown recently that human adipose stem cells (hASCs) on BCP with a HA/β-TCP ratio of 20/80 (BCP20/80) incorporated in fibrin gel show

* Corresponding Author Email: j.kleinnulend@acta.nl.

enhanced osteogenic and/or vasculogenic differentiation potential compared to BCP with a HA/β-TCP ratio of 60/40 (BCP60/40) *in vitro*. A hypoxic microenvironment for the cells might occur in these hASC-seeded BCP-based composites after implantation *in vivo*. This limits oxygen supply which is essential for successful bone tissue engineering.

Objectives: Here we aimed to test the effect of hypoxia on osteogenic and/or vasculogenic differentiation of hASC-seeded BCP60/40 or BCP20/80 scaffolds in fibrin gels as well as fibrin gel degradation.

Methods: BCP60/40-based composites and BCP20/80-based composites were cultured under normoxia (20% O2) or hypoxia (1% O2) for 12 days, and total DNA content, alkaline phosphatase activity, nitric oxide production, vascular endothelial growth factor production, and fibrin gel degradation were assessed during culture.

Results: Hypoxia similarly decreased total DNA content at day 7 (1.3-fold) and 12 (1.2-fold) in both composites. It also decreased alkaline phosphatase activity at day 5 (5.0 and 4.7-fold), day 7 (4.5 and 6-fold), and 12 (7.1 and 6.8-fold) in BCP60/40-based composites and BCP20/80-based composites. Hypoxia similarly increased vascular endothelial growth factor production at day 5 (3.9-fold) and day 7 (2.6-fold) by both BCP-based composites. Hypoxia did not affect nitric oxide production and fibrin gel degradation in both BCP-based composites at all-time points.

Conclusion: Hypoxia decreased osteogenic differentiation but enhanced vasculogenic differentiation of cultured BCP-based composites. These *in vitro* results implicate that cell-seeded BCP-based composites implanted *in vivo*, might enhance vascular endothelial growth factor production resulting in sprouting and tube formation leading to increased survival of cells in these composites, and stimulation of bone formation.

Keywords: adipose stem cells, biphasic calcium phosphate, hypoxia, osteogenic differentiation, vasculogenic differentiation

INTRODUCTION

Bone tissue engineering is based on combinations of scaffolds, cells (including stem cells), and mechanical and/or chemical stimuli [1]. For clinical bone augmentation, such as maxillary sinus floor elevation, autologous bone is the golden standard [2]. An alternative to the golden standard is biphasic calcium phosphate (BCP), which is composed of hydroxyapatite (HA) and β-tricalcium phosphate (β-TCP), and its chemical composition resembles the inorganic part of the natural bone matrix [3]. In bone tissue engineering, BCP with different HA/β-TCP ratios can be combined with regeneration-competent stem cells, such as freshly isolated autologous stromal vascular fraction of adipose tissue and adipose tissue-derived mesenchymal stem cells (ASCs) [4-6]. These cell-based bone constructs have been successfully used for bone augmentation in e.g., maxillary sinus floor elevation [6].

After implantation of cell-seeded BCP constructs *in vivo*, the disrupted blood vessels at the implant site may rapidly lead to a hypoxic microenvironment until neovascularization occurs. *In vitro* simulation of this microenvironment might offer a realistic scenario of how cell-seeded BCP constructs may preserve their features and achieve their function *in vivo*. Hypoxia is an important feature of the cellular microenvironment which maintains stem cells in an undifferentiated state, but no consensus has been reached about the influence on DNA repair [7-9]. Understanding the role of a hypoxic microenvironment is essential for the development of successful tissue engineering approaches including bone regeneration, since the physiological oxygen concentration in bone can be as low as 1.3% [10]. The effects of hypoxia on human ASC (hASC) expansion and differentiation are contrasting and no consensus has been reached. Although the effect of hypoxia on ASC differentiation is strongly dependent on cultivation conditions, the differentiation of ASCs towards the osteogenic lineage is likely to be inhibited by hypoxia [11, 12]. On the other hand, the combination of hypoxia and osteogenic stimuli accelerates ASC differentiation and mineralized matrix production *in vitro* [13]. Implantation of BCP with a HA/β-TCP ratio of 20/80 (BCP20/80) in patients undergoing unilateral maxillary sinus floor elevation resulted in more ingrowth of newly formed bone tissue at the most cranial side compared to patients treated with BCP60/40 (unpublished data), but a correlation with blood vessel formation has not yet been established. Cell-seeded BCP60/40 implanted in patients undergoing bilateral maxillary sinus floor elevation also resulted in more active bone formation at the cranial side compared to the middle part of the grafted area, which might be correlated to increased blood vessel formation [6].

For clinical application of these cell-based bone constructs, a functional blood vessel network is necessary to allow nutrient diffusion and waste removal, which is crucial for cell recruitment, survival, proliferation, differentiation, and tissue remodeling. Osteogenesis and angiogenesis are tightly coupled processes, and vascular development needs to be induced before osteogenesis can take place. This complexity has been a major challenge for engineering viable and functional bone constructs [14, 15]. Cell-based bone constructs might be combined with bioscaffolds to enhance vascular development [16]. hASCs seeded on BCP20/80 incorporated in a fibrin bioscaffold show enhanced osteogenic and vasculogenic differentiation potential compared to BCP60/40 [17].

Understanding the effect of hypoxia on the osteogenic and/or vasculogenic differentiation of composites consisting of hASC-seeded BCP60/40 or BCP20/80 scaffolds incorporated in fibrin gel is essential for bone tissue engineering. Therefore, we aimed to test the effect of hypoxia on osteogenic and/or vasculogenic differentiation potential as well as degradation of BCP60/40-based composites and BCP20/80-composites. We hypothesized that hypoxia would decrease osteogenic differentiation, and enhance vasculogenic differentiation potential in both composites. hASCs were seeded on BCP60/40 or BCP20/80 scaffolds, with similar pore size and porosity, and then incorporated in fibrin gel, and cultured for 12 days. Total DNA content, osteogenic and vasculogenic differentiation potential, and fibrin degradation were assessed up to 12 days. *In vitro* tube formation was determined at day 19. This study reports the effect of hypoxia on the osteogenic and/or vasculogenic differentiation potential of BCP20/80 and BCP60/40-based composites, and demonstrates that hypoxia decreases osteogenic differentiation potential and enhances vasculogenic differentiation potential in both BCP-based composites.

Materials and Methods

Biphasic Calcium Phosphate Scaffolds

Two different calcium phosphate scaffolds were used: (1) Straumann® BoneCeramic 60/40 (Institut Straumann AG, Basel, Switzerland), a porous BCP scaffold composed of 60% HA and 40% β-TCP (BCP60/40), and (2) Straumann® BoneCeramic 20/80 (Institut Straumann AG), a porous BCP scaffold composed of 20% HA and 80% β-TCP (BCP20/80). To avoid differences in osteogenic and vasculogenic potential of hASCs seeded on BCP60/40 or BCP20/80 incorporated in fibrin gels caused by the BCP fabrication process, the two different BCPs were produced by the same company, and had similar pore size and porosity [17].

Donors

Subcutaneous adipose tissue was harvested from residues of abdominal wall resections of five healthy female donors (aged 33, 40, 47, 50, and 54), who underwent elective abdominal wall correction at the Tergooi Hospital Hilversum and a clinic in Bilthoven, The Netherlands. The Ethical Review Board of the VU Medical Center, Amsterdam, The Netherlands, approved the protocol. Informed consent was obtained from all patients.

Isolation and Culture hASCs

hASCs were isolated from the resection material from above mentioned donors as described with minor modifications [18]. In brief, adipose tissue was cut into small pieces, and enzymatically digested with 0.1% collagenase A (Roche Diagnostics GmbH, Mannheim, Germany) for 45 min at 37°C in phosphate-buffered saline (PBS) containing 1% bovine serum albumin (Roche Diagnostics GmbH) under continuous stirring. Ficoll® density-centrifugation step (Lymphoprep™; Axis-Shield, Oslo, Norway; 1000xg for 20 min, ρ = 1.077 g/ml Ficoll®, osmolarity 280 ± 15 mOsm) was performed to remove remaining erythrocytes from the stromal vascular fraction. The resulting stromal vascular fraction pellet containing hASCs was resuspended in Dulbecco's modified Eagle's medium (Life Technologies™ Europe BV, Bleiswijk, The Netherlands), counted, frozen and stored in liquid nitrogen until further use. Heterogeneity studies including cell characterization and multipotent differentiation potential of these cells have been reported previously by our group [18].

Cryopreserved stromal vascular fraction-containing cell suspensions were pooled and cultured in α-Minimum Essential medium (α-MEM; Life Technologies™, Waltham, MA, USA) with 5% platelet lysate (see below), 100 U/ml penicillin (Sigma-Aldrich Biochemie GmbH, Hamburg, Germany), 100 μg/ml streptomycin sulfate (Sigma-Aldrich Biochemie GmbH), and 10 IU/ml heparin (LEO Pharma A/S, Ballerup, Denmark) to prevent coagulation at 37°C in a humidified atmosphere with 5% CO_2. The medium was refreshed three times/week. After reaching confluency, cells were harvested with 0.25% trypsin (Life Technologies™) and 0.1% ethylenediaminetetraacetic acid (Merck, Darmstadt, Germany) in PBS at 37°C, replated, cultured until passage 2 (P2), and stored in liquid nitrogen until further use. Cryopreserved pooled hASCs-containing cell suspensions were thawed and seeded at $0.5x10^5$ cells per T-225 culture flask (Greiner Bio-One, Kremsmuenster, Austria) in α-MEM with 2% platelet lysate, antibiotics, and 10 IU/ml heparin. Cells were cultured until P3-P4, and used for composites preparation (see below).

Platelet-Lysate

Pooled platelet products from five donors were obtained from the Bloodbank Sanquin (Sanquin, Amsterdam, The Netherlands). Platelet lysate was obtained by lysing the platelets through temperature-shock by freezing at -80°C, thawing, and centrifugation at 600xg for 10 min to eliminate remaining platelet fragments. The supernatant was added at 2% (v/v) to the medium.

Fibrin Gel

Human fibrinogen plasminogen-depleted protein (Enzyme Research Laboratories, South Bend, IN, USA) was dissolved in Medium 199 (M199; Life Technologies™) with antibiotics at 37°C for 1 h. Solubilized fibrinogen was filtered through a 0.2 μm filter (Millipore, Amsterdam, The Netherlands), and the concentration measured with a Synergy HT® spectrophotometer (BioTek Instruments Inc., Winooski, VT, USA). To prepare fibrin gel, 2 mg/ml fibrinogen solution was polymerized with 1.0 IU/ml bovine α-thrombin (Enzyme Research Laboratories) in a buffer containing 50 mM sodium citrate, 0.2 M sodium chloride, and 0.1% polyethylene glycol-8000, for 1 h at room temperature followed by 1 h at 37ºC, and used for preparation of BCP60/40 and BCP20/80-based composites (see below), as well as for fibrin gel coating of polystyrene 48-well culture plates (Cellstar, Greiner Bio One International GmbH, Frickenhausen, Germany) used for culture of these composites.

Composites Preparation and Culture

Cultured hASCs were washed three times with PBS to remove platelet lysate, and either seeded on BCP60/40 or BCP20/80 scaffolds and then incorporated in fibrin gels to prepare BCP60/40 and BCP20/80-based composites. Twenty-five to 30 mg BCP60/40 or BCP20/80 was hydrated in PBS for 30 min. After PBS removal, $1x10^5$ ASCs in 100 μl α-MEM were allowed to attach for 30 min at room temperature. Cell-seeded BCP scaffolds were embedded in fibrin gels and placed on fibrin gel-coated plates.

BCP60/40-based composites and BCP20/80-based composites were cultured in α-MEM (Life Technologies™) with 2% platelet lysate, antibiotics, 50 μg/ml 2-phospho-L-ascorbic acid trisodium salt (Sigma-Aldrich Chemie GmbH, Steinheim, Germany), and 10 IU/ml heparin for 12 days at 37ºC, at normoxic culture conditions in a humidified atmosphere of 20% O_2 and 5% CO_2 or at hypoxic culture conditions (see below) in a humidified atmosphere of 1% O_2 and 5% CO_2. Medium was refreshed after 7 days.

Hypoxic Cell Culture

Hypoxic cell culture conditions were maintained inside a custom designed hypoxic workstation (TCPS, Rotselaar, Belgium), with a O_2 (via injection of N_2), and CO_2 controlled, humidified incubator (Sanyo, Ettenleur, The Netherlands), placed inside a T4 glovebox (Jacomex, Dagneux, France) equipped with an O2X1 oxygen transmitter (GE Infrastructure Sensing, Inc., Billerica, MA, USA). The oxygen concentration inside the incubator was continuously monitored with an internal zirconia sensor and periodically checked with O_2 test tubes (Drager Safety, Zoetermeer, The Netherlands). To prevent reoxygenation during hypoxic culture, all media were pre-incubated for 3 h before use.

Total DNA Content in Composites

Total DNA content was assessed in BCP60/40-based composites and BCP20/80-based composites at day 1, 5,7, and 12 by using alamar Blue® fluorescent assay (Invitrogen, Frederick, MD, USA), according to the manufacturer's instructions. We found a linear relationship between alamar Blue® fluorescence and DNA content (data not shown). Fluorescence was read in medium samples at 530 nm with a Synergy HT® spectrophotometer.

Alkaline Phosphatase Activity

Alkaline phosphatase (ALP) activity was measured to assess the osteoblastic phenotype of hASCs in BCP60/40-based composites and BCP20/80-based composites at day 1, 5, 7, and 12. Both composites were transferred to 24-well culture plates (Cellstar), washed with PBS, crushed in 300 μl Milli-Q water, and stored at -20ºC prior to further use. ALP activity was measured in the cell lysate using 4-nitrophenyl phosphate disodium salt (Merck) as a substrate at pH 10.3, according to the method described by Lowry [19]. The absorbance was read at 405 nm with a Synergy HT® spectrophotometer. ALP activity was expressed as μM per ng DNA.

Nitric Oxide

NO production by BCP60/40-based composites and BCP20/80-based composites was measured as nitrite (NO_2^-) accumulation in conditioned medium at day 1, 5, 7, and 12 by using Griess reagent containing 2% sulfanilamide, 0.2% naphtylethelene-diamine-dihydrochloride, and 5% H_3PO_4. Serial dilutions of 0.1 M $NaNO_2$ in α-MEM were used as a standard curve. The absorbance was measured at 540 nm with a Synergy HT® spectrophotometer.

Vascular Endothelial Growth Factor Production

Human vascular endothelial growth factor (VEGF) production by BCP60/40-based composites and BCP20/80-based composites was measured in conditioned medium at day 1, 5, 7, and 12 by an enzyme-linked immunosorbent assay (DuoSet® ELISA, R&D Systems, Inc., Minneapolis, MN, USA). The absorbance was measured at 490 nm with a Mithras LB 940 Multimode Microplate Reader (Berthold Technologies GmbH & Co. KG, Bad Wildbad, Germany). VEGF production was expressed in ng/ml.

Fibrin Degradation

Fibrin degradation products were quantified using an enzyme-linked immunosorbent assay as described [20]. Briefly, the antibody fibrin degradation products-14 (FDP-14;

TNO, Quality of Life, Leiden, The Netherlands) recognizing different epitopes of fibrin degradation products was used as catching antibody. Fibrin degradation product concentrations in conditioned medium of BCP60/40-based composites and BCP20/80-based composites were investigated at day 1, 5, 7, and 12, and Biopool standard (Trinity Biotech, Wicklow, Ireland) was used as a reference. Finally, monoclonal antibody D-dimer-13 (DD-13; TNO) labeled with horseradish peroxidase was used as tagging antibody. The coloring reaction was performed using 3,3’,5,5’-tetramethybenzidine (Sigma-Aldrich, St. Louis, MO, USA) and stopped with 1 M H_2SO_4. The optical density was read at 450 nm with Synergy HT® spectrophotometer.

Statistical Analysis

Data were obtained from quadruple cultures of two independent experiments for BCP60/40-based composites and BCP20/80-based composites. Data are presented as mean ± SEM. To compare DNA content, ALP activity, NO production, VEGF production, and fibrin degradation for BCP60/40-based composites and BCP20/80-based composites at normoxic and hypoxic culture conditions, two-tailed unpaired t-test was performed. Differences were considered significant if $p < 0.05$. Statistical analysis was performed using IBM® SPSS® Statistics version 21 software package (SPSS Inc., Chicago, IL, USA) and GraphPad Prism® 5.0 (GraphPad Software Inc., La Jolla, CA, USA).

RESULTS

Hypoxia Decreased Total DNA Content and ALP Activity in BCP60/40-Based Composites and BCP20/80-Based Composites

Total DNA content in BCP60/40-based composites and in BCP20/80-based composites was similar under normoxia and hypoxia at day 1 and 5, and decreased compared to normoxia at day 7 (BCP60/40: 1.3-fold; BCP20/80: 1.3-fold decrease) and day 12 (BCP60/40: 1.2-fold; BCP20/80: 1.2.-fold decrease; Figures 1a and 1b). Cell proliferation in BCP60/40-based composites increased by 1.2-fold under normoxia, and by 1.3-fold under hypoxia from day 1 to day 12. In BCP20/80-based composites, cell proliferation increased by 1.5-fold under normoxia, and by 1.3-fold under hypoxia from day 1 to day 12. ALP activity in BCP60/40-based composites and in BCP20/80-based composites under hypoxia was decreased compared to normoxia at day 5 (BCP60/40: 5.0-fold; BCP20/80: 4.7-fold decrease), day 7 (BCP60/40: 4.5-fold; BCP20/80: 6.0-fold decrease), and day 12 (BCP60/40: 7.1-fold; BCP20/80: 6.8-fold decrease; Figures 1c and 1d).

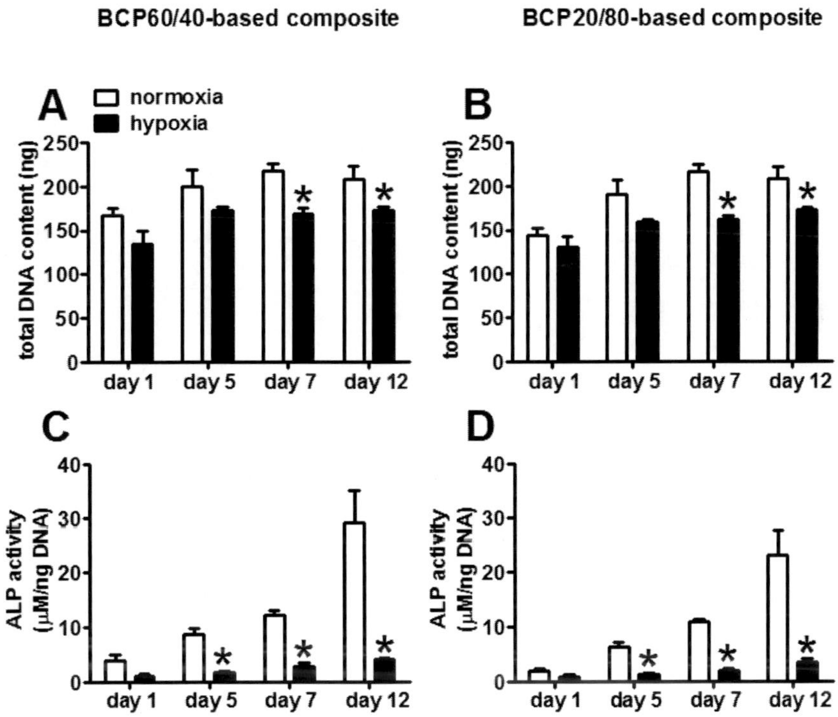

Figure 1. Total DNA content in (a) BCP60/40 based-composites and (b) BCP20/80-based composites under normoxia and hypoxia at day 1, 5, 7, and 12. Hypoxia decreased total DNA content in (a) BCP60/40-based composites and (b) BCP20/80-based composites compared to normoxia at day 7 and day 12. ALP activity in (c) BCP60/40-based composites and (d) BCP20/80-based composites under normoxia and hypoxia at day 1, 5, 7, and 12. Hypoxia decreased ALP activity in (c) BCP60/40-based composites and (d) BCP20/80-based composites. Values are mean ± SEM (n = 4-8). *Significantly different from normoxia, $p < 0.05$. BCP, biphasic calcium phosphate; ALP, alkaline phosphatase.

Hypoxia Did Not Affect NO Production by BCP60/40-Based Composites and BCP20/80-Based Composites

NO production by BCP60/40-based composites was similar under normoxia and hypoxia at day 1, 5, 7, and 12 (Figure 2a).

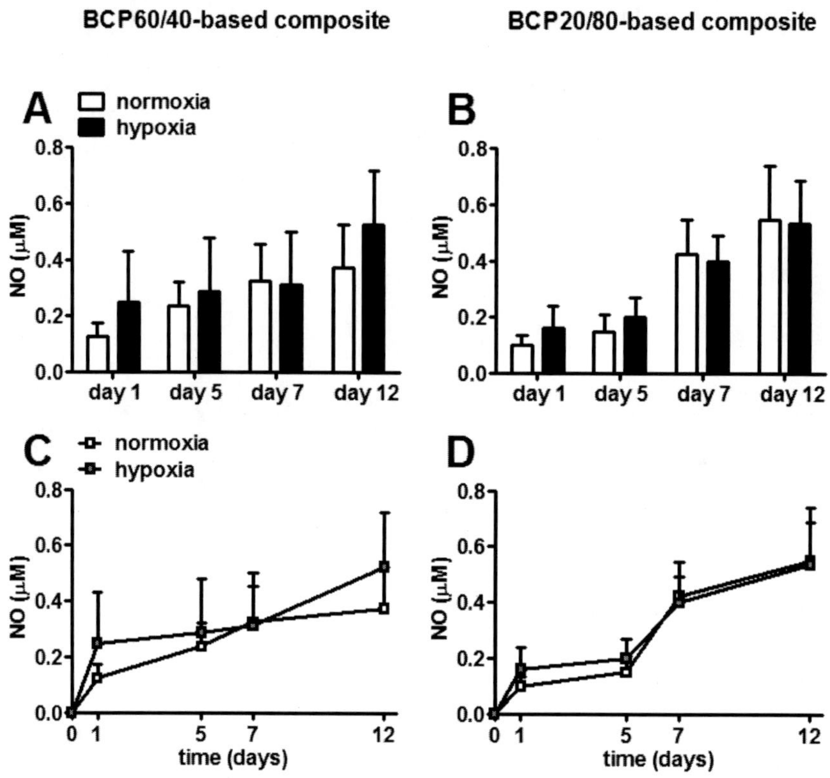

Figure 2. NO production by (a) BCP60/40-based composites and (b) BCP20/80-based composites under normoxia and hypoxia at day 1, 5, 7, and 12. Hypoxia did not significantly affect NO production by (a) BCP60/40-based composites and (b) BCP20/80-based composites. Values are mean ± SEM (n = 6-8). BCP, biphasic calcium phosphate; NO, nitric oxide; cum, cumulative.

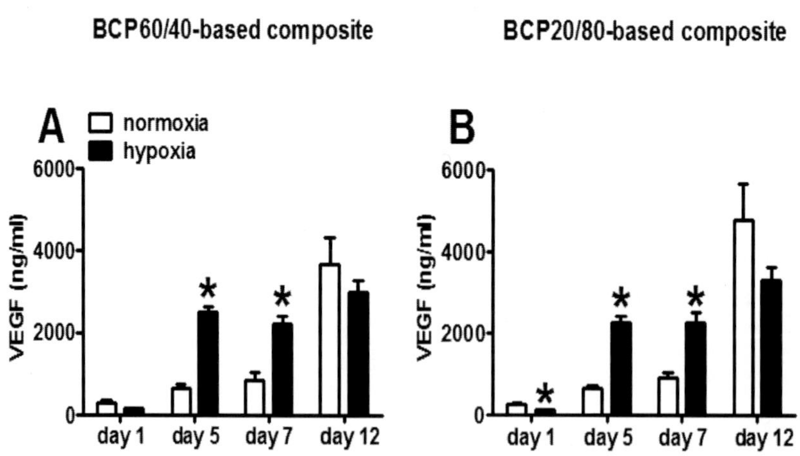

Figure 3. VEGF production by (a) BCP60/40-based composites and (b) BCP20/80-based composites under normoxia and hypoxia at day 1, 5, 7, and 12 days. Hypoxia enhanced VEGF production by (a) BCP60/40-based composites and (b) BCP20/80-based composites. Values are mean ± SEM (n = 8). *Significantly different from normoxia, $p < 0.05$. BCP, biphasic calcium phosphate; VEGF, vascular endothelial growth factor; cum, cumulative.

Under normoxia, NO production by BCP60/40-based composites increased by 3.0-fold from day 1 to day 12, while under hypoxia, NO production was increased by 2.1-fold (Figure 2a). There were also no differences in NO production by BCP20-based composites under normoxia and hypoxia at day 1, 5, 7, and 12 (Figure 2b). NO production by BCP20/80-based composites increased by 5.5-fold under normoxia from day 1 to day 12, while NO production was increased by 3.3-fold under hypoxia (Figure 2b).

Hypoxia Enhanced VEGF Production by BCP60/40-Based Composites and BCP20/80-Based Composites

VEGF production by BCP20/80-based composites under hypoxia decreased by 2.1-fold, but not by BCP60/40-based composites, compared to normoxia at day 1 (Figure 3). VEGF production by BCP60/40-based composites and BCP20/80-based composites under hypoxia was increased at day 5 (BCP60/40: 3.9-fold; BCP20/80: 3.8-fold), and day 7 (BCP60/40: 2.6-fold; BCP20/80: 2.5-fold), but reached similar levels at day (Figure 3).

Hypoxia Did Not Affect Fibrin Degradation by BCP60/40-Based Composites and BCP20/80-Based Composites

Similar levels of fibrin degradation products in the medium of BCP60/40-based composites under hypoxia was observed compared to normoxia at day 1, 5, 7, and 12 (Figure 4a). Concentration of fibrin degradation products in the medium of BCP60/40-based composites increased by ~ 4000-fold under normoxia and hypoxia from day 1 to day 12 (Figure 4a). There were also no differences in concentration of fibrin degradation products in the medium of BCP20/80-based composites under hypoxia compared to normoxia at day 1, 5, 7, and 12 (Figure 4b). The level of fibrin degradation products in the medium of BCP20/80-based composites increased by ~ 13,000-fold under normoxia, and by ~ 4000-fold under hypoxia, from day 1 to day 12 (Figure 4b).

DISCUSSION

We studied the effect of hypoxia on osteogenic and/or vasculogenic differentiation of hASC-seeded BCP60/40 or BCP20/80 scaffolds in fibrin gels as well as fibrin gel degradation. We found that (i) Hypoxia decreased total DNA content and ALP activity in BCP60/40-based composites and BCP20/80-based composites; (ii) Hypoxia did not affect NO production by BCP60/40-based composites and BCP20/80-based composites; (iii) Hypoxia increased VEGF production by BCP60/40-based composites and BCP20/80-based composites; (iv) Hypoxia did not affect fibrin degradation by BCP60/40-based composites and BCP20/80-based composites. Therefore, our results showed that hypoxia decreased osteogenic differentiation, and enhanced vasculogenic differentiation of BCP-based composites *in vitro*. This suggests that the hypoxic host microenvironment that cells might encounter after implantation *in vivo*, might enhance sprouting and tube formation by VEGF production, which increases survival of cells in BCP-based composites and stimulates bone formation.

Stem cells *in vivo* reside in a hypoxic microenvironment in which they must proliferate and differentiate. Hypoxia enhances hASC proliferation [12], and maintains stemness of hASCs [12, 13, 21]. We found similar proliferation under normoxia and hypoxia in BCP60/40-based composites and BCP20/80-based composites, which indicates cell survival and maintenance of stemness of these cells in these composites. Hypoxia increases glucose consumption and lactate production of hASCs [22]. This suggests that the hASCs in our BCP-composites under hypoxia compensate their metabolism more towards anaerobic glycolysis.

Hypoxia inducible factors (HIFs) are transcription factors, and their signaling pathway mediates the effects of hypoxia on cells. These transcription factors are essential for the maintenance of stemness of stem cells including hASCs and human embryonic stem cells [13, 23, 24]. The stemness maintenance of hASCs under hypoxia allows stem cell differentiation towards specific lineages [13]. Osteogenic differentiation potential of hASCs is reduced under hypoxia [25]. This is in accordance with our results showing that hypoxia decreased ALP activity of hASCs in BCP60/40-based composites and BCP20/80-based composites, suggesting decreased osteogenic differentiation of hASCs in both BCP-based composites, and maintenance of stemness of hASCs under hypoxia.

HIF-1 is a heterodimeric transcription factor composed of two subunits: HIF-1α and HIF-1ß. Low oxygen-induced HIF-1 increases VEGF expression [26]. VEGF is a growth factor that plays an important role in e.g., vascular development, angiogenesis, and wound healing. Increased VEGF expression induces proliferation, sprouting, migration, and tube formation of endothelial cells [27, 28]. The upregulated VEGF production in BCP60/40-based composites and BCP20/80-based composites under hypoxia at earlier time points suggests that sprouting and tube formation are increased under hypoxia. We have attempted to determine the effect of hypoxia on formation of tubular structures in both composites after co-culture with primary human endothelial cells as described [29]. This assay proved unsuitable for our composites, and future studies are needed to show possible differences in sprouting and tube formation in BCP-based composites under hypoxia compared to normoxia.

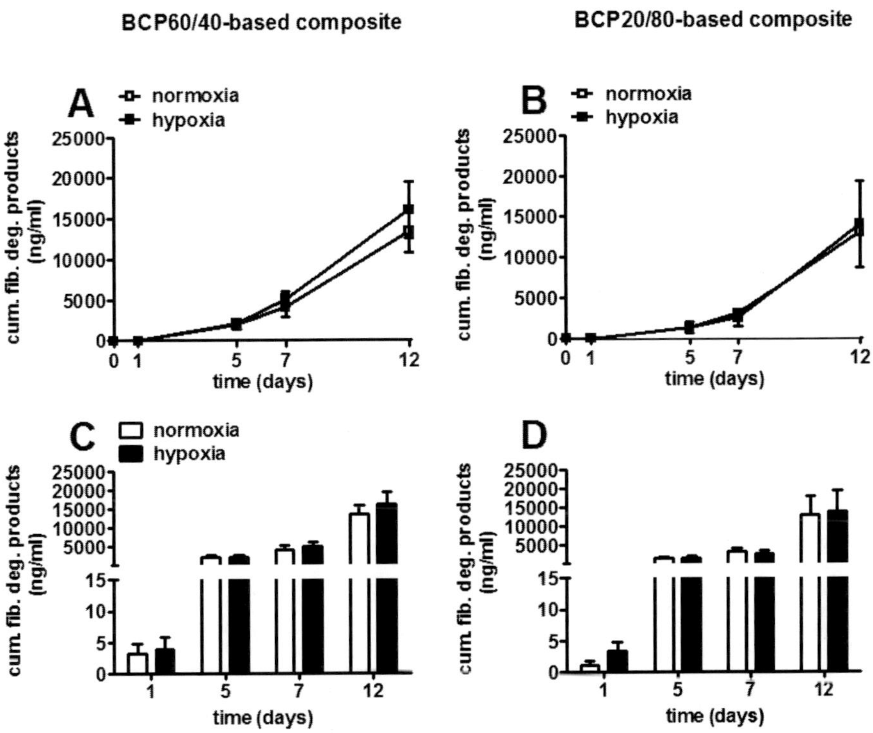

Figure 4. Fibrin degradation of (a) BCP60/40-based composites and (b) BCP20/80-based composites under normoxia and hypoxia at day 1, 5, 7, and 12 days. Hypoxia did not significantly affect fibrin degradation of (a) BCP60/40-based composites and (b) BCP20/80-based composites. Values are mean ± SEM (n = 7-8). BCP, biphasic calcium phosphate; cum, cumulative; fib, fibrin; deg, degradation.

VEGF is expressed by hASCs as well as by osteoblasts [30-33]. Hypoxia influences osteoblastic function, i.e., it up-regulates VEGF gene and protein expression, and increases ALP activity [32]. The induction of VEGF expression by osteoblasts in response to hypoxia is modulated by an oxygen-sensing mechanism [34]. Since osteogenesis and angiogenesis are tightly coupled processes, it might be expected that an upregulation of VEGF expression under hypoxia causes increased osteogenic differentiation of hASCs. However, we did not find such an increase in ALP activity as a result of hypoxia. This suggests that hypoxia influenced maintenance of cell stemness strongly, which prevents that a possible effect by VEGF on osteogenic stimulation can be expressed. We found similar VEGF production in BCP60/40-based composites and BCP20/80-based composites at the latest time point. This could be due to a hypoxic microenvironment for hASCs in BCP-composites under normoxia and hypoxia, which is in line with our previous study suggesting that VEGF expression increased in BCP20/80-based composites due to a hypoxic microenvironment for hASCs [17].

NO is a signaling molecule involved in many processes like inflammation, angiogenesis, and bone mass regulation. The molecule is synthesized by NO synthase (NOS) or by reduction of nitrite which can spontaneously occur under hypoxic and/or acidic conditions [35]. Three different NOS isoforms exist: endothelial NO synthase (eNOS), neuronal NO synthase (nNOS), both calcium-dependent, and inducible form of NO synthase (iNOS) which is calcium-independent. All three isoforms are expressed in bone tissue [36]. Low-oxygen-induced HIF-1 transcription factor increases iNOS and eNOS mRNA and protein expression [37-39]. Therefore, an upregulation of NO production by BCP60/40-based composites and BCP20/80-based composites under hypoxia was expected, but similar NO production by BCP-based composites under normoxia and hypoxia was found. Since hypoxia did not affect NO production in BCP-based composites, this suggests similar spontaneous reduction of nitrite as well as induction of HIF-1 regulating iNOS and eNOS expression in both BCP-based composites under both conditions which is appealing for bone tissue engineering approaches.

Cells sense and respond to mechanical properties of their surroundings by locally pulling and pushing on the matrix, and at the same time actively change the stiffness and tension of their surroundings in a mechanical feed-back loop [40, 41]. Differences in matrix stiffness affect cell-mediated fibrin remodelling [42], which might have impact on this mechanical feed-back loop. The similar fibrin remodeling rates we found in BCP60/40-based composites and in BCP20/80-based composites under normoxia and hypoxia indicates no effect of hypoxia on matrix stiffness suggesting no impact on the mechanical feed-back loop between cells and matrix. This finding supports that our composites seem promising candidates for bone tissue engineering approaches like bone augmentation *in vivo* [17].

In summary, hypoxia decreased DNA content and ALP activity in BCP60/40-based composites and BCP20/80-based composites, and increased VEGF production by both BCP-based composites. Therefore, we conclude that hypoxia decreased osteogenic differentiation potential, and enhanced vasculogenic differentiation potential of BCP60/40-based composites and BCP20/80-based composites *in vitro*. This suggests that the hypoxic host microenvironment that cells will encounter after implantation *in vivo* might enhance sprouting and tube formation by VEGF production which increases cell survival in BCP-based composites, and stimulates bone formation.

ACKOWLEDGMENTS

The work of Fransisca A. S. van Esterik was supported by a grant from the University of Amsterdam for the stimulation of a research priority area Oral Regenerative Medicine. The authors thank Henk-Jan Prins and Benno Naaijkens for adipose stem cell isolation, Michiel van Wijhe for co-culturing human endothelial cells with BCP-based composites, René Musters for confocal microscopy of tubular structure formation in BCP-based composites, and Lenie van den Broek for making the vascular endothelial growth factor ELISA available at their location.

REFERENCES

[1] Helder, MN; Knippenberg, M; Klein-Nulend, J; Wuisman, PIJM. Stem cells from adipose tissue allow challenging new concepts for regenerative medicine. *Tissue Engineering*, 2007, 13, 1799-1808.

[2] Zijderveld, SA; Zerbo, IR; van den Bergh, JP; Schulten, EAJM; ten Bruggenkate, CM. Maxillary sinus floor augmentation using a beta-tricalcium phosphate (Cerasorb) alone compared to autogenous bone grafts. *The International Journal of Oral & Maxillofacial Implants*, 2005, 20, 432-440.

[3] Frenken, JW; Bouwman, WF; Bravenboer, N; Zijderveld, SA; Schulten, EAJM; ten Bruggenkate, CM. The use of Straumann Bone Ceramic in a maxillary sinus floor elevation procedure: a clinical, radiological, histological and histomorphometric evaluation with a 6-month healing period. *Clinical Oral Implants Research,* 2010, 21, 201-208.

[4] Guha, AK; Singh, S; Kumaresan, R; Nayar, S; Sinha, A. Mesenchymal cell response to nanosized biphasic calcium phosphate composites. *Colloids and Surfaces B: Biointerfaces*, 2009, 73, 146-151.

[5] Vats, A; Bielby, RC; Tolley, NS; Nerem, R; Polak, JM. Stem cells. *Lancet,* 2005, 366, 592-602.

[6] Prins, HJ; Schulten, EAJM; ten Bruggenkate, CM; Klein-Nulend, J; Helder, MN. Bone regeneration using the freshly isolated autologous stromal vascular fraction of adipose tissue in combination with calcium phosphate ceramics. *Stem Cells Translational Medicine,* 2016, 2015, 0369.

[7] Rodríguez-Jiménez, FJ; Moreno-Manzano, V; Lucas-Dominguez, R; Sánchez-Puelles, JM. Hypoxia causes downregulation of mismatch repair system and genomic instability in stem cells. *Stem Cells*, 2008, 26, 2052-2062.

[8] Li, TS; Marbán, E. Physiological levels of reactive oxygen species are required to maintain genomic stability in stem cells. *Stem Cells*, 2010, 28, 1178-1185.

[9] Tsai, CC; Chen, YJ; Yew, TL; Chen, LL; Wang, JY; Chiu, CH; Hung, SC. Hypoxia inhibits senescence and maintains mesenchymal stem cell properties through down-regulation of E2A-p21 by HIF-TWIST. *Blood,* 2011, 117, 459-469.

[10] Spencer, JA; Ferraro, F; Roussakis, E; Klein, A; Wu, J; Runnels, JM; Zaher, W; Mortensen, LJ; Alt, C; Turcotte, R; Yusuf, R; Cote, D; Vinogradov, SA; Scadden, DT; Lin, CP. Direct measurement of local oxygen concentration in the bone marrow of live animals. *Nature,* 2014, 508, 269-273.

[11] Merceron, C; Vinatier, C; Potron, S; Masson, M; Amiaud, J; Guigand, L; Cherel, Y; Weiss, P; Guicheux, J. Differential effects of hypoxia on osteochondrogenic potential of human adipose-derived stem cells. *American Journal of Physiology - Cell Physiology*, 2010, 298, C355-C364.

[12] Weijers, EM; van den Broek, LK; Waaijman, T; van Hinsbergh, VW; Gibbs, S; Koolwijk, P. The influence of hypoxia and fibrinogen variants on the expansion and differentiation of adipose tissue-derived mesenchymal stem cells. *Tissue Engineering Part A,* 2011, 17, 2675-2685.

[13] Fotia, C; Massa, A; Boriani, F; Baldini, N; Granchi, D. Hypoxia enhances proliferation and stemness of human adipose-derived mesenchymal stem cells. *Cytotechnology*, 2015, 67, 1073-1084.

[14] Hutton, DL; Moore, EM; Gimble, JM; Grayson, WL. Platelet-derived growth factor and spatiotemporal cues induce development of vascularized bone tissue by adipose-derived stem cells. *Tissue Engineering Part A,* 2013, 19, 2076-2086.

[15] Correia, C; Grayson, WL; Eton, R; Gimble, JM; Sousa, RA; Reis, RL; Vunjak-Novakovic, G. Human adipose-derived cells can serve as a single-cell source for the *in vitro* cultivation of vascularized bone grafts. *Journal of Tissue Engineering and Regenerative Medicine*, 2014, 8, 629-639.

[16] Moore, MC; Pandolfi, V; McFetridge, PS. Novel human-derived extracellular matrix induces *in vitro* and *in vivo* vascularization and inhibits fibrosis. *Biomaterials*, 2015, 49, 37-46.

[17] Van Esterik, FAS; Zandieh-Doulabi, B; Kleverlaan, CJ; Klein-Nulend J. Enhanced osteogenic and vasculogenic differentiation potential of human adipose stem cells on biphasic calcium phosphate scaffolds in fibrin gels. *Stem Cells International*, 2016, 2016, 1934270.

[18] Oedayrajsingh Varma, MJ; Breuls, RGM; Schouten, TE; Jurgens, WJFM; Bontkes, HJ; Schuurhuis, GJ; van Ham, S; van Milligen, FJ. Phenotypical and functional characterization of freshly isolated adipose-tissue-derived stem cells. *Stem Cells and Development*, 2007, 16, 91-104.

[19] Lowry, OH. Micromethods for the assay of enzyme ll specific procedure. Alkaline phosphatase. *Methods in Enzymology*, 1955, 4, 371.

[20] Weijers, EM; van Wijhe, MH; Joosten, L; Horrevoets, AJG; de Maat, MP; van Hinsbergh, VWM; Koolwijk, P. Molecular weight fibrinogen variants alter gene expression and functional characteristics of human endothelial cells. *Journal of Thrombosis and Haemostasis*, 2010, 8, 2800-2809.

[21] Valorani, MG; Montelatici, E; Germani, A; Biddle, A; D'Alessandro, D; Strollo, R; Patrizi, MP; Lazzari, L; Nye, E; Otto, WR; Pozzilli, P; Alison, MR. Pre-culturing human adipose tissue mesenchymal stem cells under hypoxia increases their adipogenic and osteogenic differentiation potentials. *Cell Proliferation*, 2012, 45, 225-238.

[22] Hutton, DL; Grayson WL. Hypoxia inhibits de novo vascular assembly of adipose-derived stromal/stem cell population but promotes growth of preformed vessels. *Tissue Engineering Part A*, 2016, 22, 161-169.

[23] Covello, KL; Kehler, J; Yu, H; Gordan, JD; Arsham, AM; Hu, CJ; Labosky, PA; Simon, MC; Keith, B. HIF-2alpha regulates Oct-4: effects of hypoxia on stem cell function, embryonic development, and tumor growth. *Genes & Development*, 2006, 20, 557-570.

[24] Forristal, CE; Wright, KL; Hanley, NA; Oreffo, RO; Houghton, FD. Hypoxia inducible factors regulate pluripotency and proliferation in human embryonic stem cells cultured at reduced oxygen tensions. *Reproduction*, 2010, 139, 85-97.

[25] Choi, JR; Pingguan-Murphy, B; Abas, WABW; Azmi, MAN; Omar, SZ; Chua, KH; Safwani, WKZW. Impact of low oxygen tension on stemness, proliferation and differentiation potential of human adipose-derived stem cells. *Biochemical and Biophysical Research Communications,* 2014, 448, 218-224.

[26] Vincent, KA; Shyu, KG; Luo, YX; Magner, M; Tio, RA; Jiang, CW; Akita, GY; Gregory, RJ; Isner, JM; Goldberg, M. Angiogenesis is induced in a rabbit model of hind limb ischemia by naked DNA encoding an HIF-1 alpha/VP16 hybrid transcription factor. *Circulation*, 2000, 102, 2255-2261.

[27] Ferrara, N; Gerber, HP; LeCouter, J. The biology of VEGF and its receptors. *Nature Medicine*, 2003, 9, 669 676.

[28] Gerhardt, H; Golding, M; Fruttiger, M; Ruhrberg, C; Lundkvist, A; Abramsson, A; Jeltsch, M; Mitchell, C; Alitalo, K; Shima, D; Betsholtz, C. VEGF guides angiogenic sprouting utilizing endothelial tip cell filopodia. *The Journal of Cell Biology*, 2003, 161, 1163.

[29] Koolwijk, P; van Erck, MGM; de Vree, WJA; Vermeer, MA; Weich, HA; Hanemaaijer, R; van Hinsbergh, VWM. Cooperative effect of TNFα, bFGF, and VEGF on the formation of tubular structures of human microvascular endothelial cells in a fibrin matrix. Role of Urokinase activity. *The Journal of Cell Biology,* 1996, 132, 1177-1188.

[30] Chen, G; Shi, X; Sun, C; Li, M; Zhou, Q; Zhang, C; Huang, J; Qiu, Y; Wen, X; Zhang, Y; Zhang, Y; Yang, S; Lu, L; Zhang, J; Yuan, Q; Lu, J; Xu, G; Xue, Y; Jin, Z; Jiang, C; Ying, M; Liu, X. VEGF-mediated proliferation of human adipose tissue-derived stem cells. *PLoS One*, 2013, 8, e73673.

[31] Hsiao, ST; Asgari, A; Lokmic, Z; Sinclair, R; Dusting, GJ; Lim, SY; Dilley, RJ. Comparative analysis of paracrine factor expression in human adult mesenchymal stem cells derived from bone marrow, adipose, and dermal tissue. *Stem Cells and Development*, 2012, 21, 2189-2203.

[32] Steinbrech, DS; Mehrara, BJ; Saadeh, PB; Chin, G; Dudziak, ME; Gerrets, RP; Gittes, GK; Longaker, MT. Hypoxia regulates VEGF expression and cellular proliferation by osteoblasts *in vitro*. *Plastic and Reconstructive Surgery*, 1999, 10, 738-747.

[33] Street, J; Lenehan B. Vascular endothelial growth factor regulates osteoblast survival – evidence for an autocrine feedback mechanism. *Journal of Orthopaedic Surgery and Research,* 2009, 4, 19.

[34] Steinbrech, DS; Mehrara, BJ; Saadeh, PS; Greenwald, JA; Spector, JA; Gittes, GK; Longaker, MT. VEGF expression in an osteoblast-like cell line is regulated by a hypoxia response mechanism. *American Journal of Physiology - Cell Physiology*, 2000, 278, C853-C860.

[35] Klein-Nulend, J; van Oers, RFM; Bakker, AD; Bacabac, RG. Nitric oxide signaling in mechanical adaptation of bone. *Osteoporosis International*, 2014, 25, 1427-1437.

[36] Van't Hof, RJ; Ralston, SH. Nitric oxide and bone. *Immunology,* 2001, 103, 255-261.

[37] Lowenstein, CJ; Padalko, E. iNOS (NOS2) at a glance. *Journal of Cell Science*, 2004, 117, 2865-2867.

[38] Thompson, L; Dong, Y; Evans, L. Chronic hypoxia increases inducible NOS-derived nitric oxide in fetal guinea pig hearts. *Pediatric Research*, 2009, 65, 188-192.

[39] Hoehn, T; Stiller, B; McPhaden, AR; Wadsworth, RM. Nitric oxide synthases in infants and children with pulmonary hypertension and congenital heart disease. *Respiratory Research*, 2009, 10, 110.

[40] Jansen, KA; Bacabac, RG; Piechoka, IK; Koenderink, GH. Cell actively stiffen fibrin networks by generating contractile stress. *Biophysical Journal*, 2013, 105, 2240-2251.

[41] Notbohm, J; Lesman, A; Tirrell, DA; Ravichandran, G. Quantifying cell-induced matrix deformation in three dimensions based on imaging matrix fibers. *Integrative Biology,* 2015, 7, 1186-1195.

[42] De Jong, T; Weijers, EM; Bakker, AD; Koolwijk, P; Smit TH. Matrix remodeling and osteogenic differentiation of human adipose-derived stem cells increases with higher fibrin matrix stiffness. *Journal of Biomaterials and Tissue Engineering*, 2016, 6, 729-738.

In: Stem Cells Origins of Cytokine Interactions
Editor: Prasad S. Koka
ISBN: 978-1-53615-302-6

Chapter 8

THE CULTURE MEDIUM CONDITIONED BY ADIPOSE TISSUE IS ABLE TO REGULATE THE LNCRNA HOTAIR, HOXC11, HOXC12, OSX AND SATB2 GENE EXPRESSION IN MSCS DURING OSTEOBLASTS' DIFFERENTIATION

P. Piombino[1], F. Riccitiello[2], L. Califano[1], S. Rengo[2], G. Spagnuolo[2] and A. Procino[3,*], PhD
[1]Department of Maxillofacial Surgery
[2]Department of Neurosciences, Reproductive and Odontostomatologic Sciences, University of Naples Medical School "Federico II," Italy
[3]Sorriso s.c.a r.l. Limited Liability Consortium
Bio-Nanotechnology for Human Health, Naples, Italy

ABSTRACT

The osteogenesis is a complex process that involves an accurate control of bone development, growth as well as remodeling during postnatal life. In particular, osteoblasts arise from mesenchymal stem cells through a process regulated in several steps that leads to the osteogenic differentiation. Class I homeobox genes (HOX in human and Hox in mice), are 39 transcription factors, mainly involved in the regulation of embryonic development. It has been shown that HOXA10 was able to activate several osteoblast related genes like Osx. The HOXA10 gene was considered a key factor able to control osteogenesis. In the last decade, several non-coding RNAs (ncRNAs), have been identified like crucial factors during osteogenesis. HOTAIR is an lncRNA related to Osteosarcoma and able to induce the Epithelial Mesenchymal Transition. The aim of this paper was understood the role of

* Corresponding Author Email: alfredo.procino@unina.it.

HOTAIR and HOXC locus in the control of Osteoblasts development in MSCs treated with adipose tissue obtained with Lipogems method.

Keywords: osteoblasts, HOX genes, HOXC, HOTAIR, osteogenesis, memory program

INTRODUCTION

Bone tissue is a dynamic structure capable of supporting mechanical variations, due to different environmental conditions. Moreover, the old tissue can be replaced with new matrix using the high potential of bone remodeling. Specific cells, like Osteoclasts and Osteoblasts, lead the resorption process and deposit of new bone matrix. The osteogenesis is a complex process that involves an accurate control of bone development, growth as well as remodeling during postnatal life. In particular, osteoblasts arise from the osteogenic differentiation of mesenchymal stem cells through a process regulated in different steps. Although, the understanding of the osteogenesis transcriptional control is increased considerably, the molecular regulatory basis is still poorly understood (Zhu et al. 2013). In the near future, the increase of knowledge about the role of specific transcriptional factors in the control of Osteoblasts differentiation consequent to post-genomic era will be expected, in order to identify the molecular mechanisms useful in the tissue regeneration and tissue engineering methodologies to apply in clinical practice (Nair et al. 2014). 500 genes regulate bone development; particularly homeobox-containing genes like Osterix (Osx), special AT-rich sequence binding protein-2 (SATB2), HOX-family are the best candidates in the control of bone tissue organization (Weiss et al. 1995).

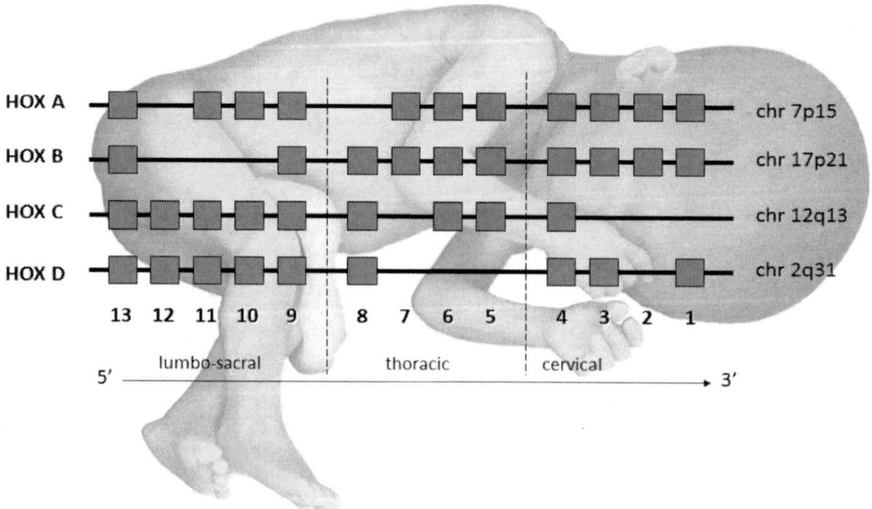

Figure 1. The HOX gene arranging during human embryonic development (see the text).

Class I homeobox genes (HOX in human and Hox in mice), are 39 transcription factors involved in the regulation of embryonic development program; The HOX genes are characterized by a sequence of 183 nucleotides encoding a hoemodomain of 61 amino acid that binds to DNA, as biological gripper, activating or repressing specific genes (Ferber., et al. 2000; Foucher., et al. 2002). Moreover, the HOX genes, conversely of others gene networks, are not spread along the genome, but are located on four chromosomes and are arranged in specific chromosomal clusters or loci (HOXA chr 7p15.3, HOXB chr 17q21.3, HOXC chr 12q13.3 and HOXD chr 2q31), each having from 9 to 11 genes. Based on the position within the locus and similarity of homeobox sequence, corresponding genes of the four clusters, can be aligned with each other in 13 paralogous groups (Figure 1) (Barnes., et al. 1995). The HOX network is active in adult human tissues and organs, controls the spatial-temporal generation of biological structure expected during embryonic development, regulates the cell memory program and control the cell phenotype. Epigenetic change in the memory program is critical in the onset and evolution of normal organs and tissue, but deregulation of genes related to cell memory program plays a key role in the change of cell phenotypes and in the determination of several pathologies (Bantignes and Cavalli 2006).

Three genes families control the cell memory program: Polycomb, able to block the interaction, DNA-chromatin leading to silencing of HOX genes. Trithorax genes are able to induce the mRNA transcription through an open configuration of DNA-chromatin interaction, and lead the activation of HOX genes. Finally, the HOX family is involved in the control of the cells phenotype, through a specific gene program mainly by means of an accurate regulation of the mRNAs transcription (Figure 2) (Ghering and Hiromi 1986, Procino A. 2014).

Many evidences have shown the involvement of Homeobox genes in bone formation: the upregulation of HOXA2 is crucial for repress osteogenesis. (Dobreva, et al. 2006). Recently, It has been studied the role of HOXA10 in the control of osteoblasts differentiation; the hyper-expression of HOXA10, was able to activate several osteoblast related genes like Osx; this transcription factor controls bone mineralization and osteoblasts differentiation. Therefore, HOXA10 gene was considered as a key factor for: i) the proper timing expression of specific osteogenesis markers; ii) correct mineral and matrix deposition during osteoblasts maturation (Gordon, et al. 2011)

It has already demonstrated that Osx plays a crucial role during bone formation. Osterix is able to manage the expression of specific genes related to osteogenesis. His zinc-finger domain is crucial during the transition from pre-osteoblasts to functional osteoblasts (Baek, et al. 2009; Nakashima et al. 2002).

Osteogenesis is a process controlled by several transcriptional factors, one of the most important titled special AT-rich sequence binding protein-2 (SATB2); SATB2 appears to exert different roles in craniofacial patterning, osteoblast differentiation and maturation, bone formation. SATB2 is able to activate or repress several osteoblasts related genes, controlling the osteoblast differentiation (Yamaguchi., et al. 1991; Wozney., et al. 1995; Zhao., et al. 2013).

Several non-coding RNAs (ncRNAs), have been identified like crucial factors during osteogenesis. In details, miR-29b, 22 miR-204/211 e miR-182, 24 are able to control bone development during Mesenchymal Stem Cells (MSCs) differentiation. Moreover, it has been demonstrated the ability of lncRNA MEG3, H19 and ANCR to regulate, in MSCs, the expression of different osteoblast related genes (Zhuang et al. 2015; Zhu., et al. 2013; Wang., et al. 2016).

Recently, it has been identified 231 lncRNA able to interact with HOX genes; in particular, one of those titled lncRNA HOX antisense intergenic RNA (HOTAIR) was identified in 2007. HOTAIR is transcribed from HOXC locus located at chromosome 12q13.13, in intergenic and anti-sense position to the flanking HOXC11 and HOXC12 genes (Rinn, et al. 2007; Kogo et al. 2011; Tano and Akimitsu 2012; He., et al. 2011). HOTAIR is the first lncRNA that control the gene expression with trans mechanism. In details, it is able to block the transcriptional control of the HOXD locus on the chromosome-2, including the unique Polycomb Responsive Element (PRE) to date identified in the genome. Moreover, the lncRNA HOTAIR and HOX genes are considered key factors in the control of the epithelial cell memory program (Procino, 2016). The HOTAIR is involved in pathogenesis and progress of multiple tumors. Recently, HOTAIR has been identified in Osteosarcoma (OS) and its upregulation confirmed that HOTAIR play a crucial role in the control of bone tissue deregulation (Zhou, et al. 2016)

The aim of this research was to study the osteoblasts differentiation considering the potential involvement of lncRNA HOTAIR and two HOX gene family members (HOXC11 and HOXC12) in mesenchymal stem cells (MSCs) treated with adipose tissue, obtained with Lipogems method.

Materials and Methods

Processing of Adipose Tissue with the Lipogems Device

Human subcutaneous adipose tissue samples were obtained with patient informed consent from abdominal lipoaspiration procedures performed before arthroscopic rotator cuff repair and processed by using the Lipogems device, according to the manufacturer's instructions previously described. Avoiding the presence of air, the lipoaspirate tissue was subjected to a first cluster reduction, obtained by pushing the lipoaspirate from the syringe into the device through the first large filter (blue cap), and allowing the corresponding quantity of saline to exit towards the wasting bag. The five stainless steel beads contained in the device were essential to obtain a temporary emulsion between oil, blood, and saline, which could be washed away against density following the current of saline moved by gravity in the wasting bag. After this washing step (the following solution appears clear and the lipoaspirate yellow), the saline flux was stopped and the device was reversed (gray cap up), leading to the second adipose cluster reduction. Such reduction was obtained by pushing the floating adipose clusters through the second cutting hexagonal filter (grey cap), pushing fluid from below with a 60mL syringe. The reduced cluster was collected in another 60mL syringe placed above and positioned to gently decant the Lipogems product by gravity in order to remove the excessive saline solution. A mean of 50mL of lipoaspirate was collected and processed by the Lipogems device to obtain an amount of 20mL of the final Lipogems product, which was transferred to several 10mL syringes to be injected in the same patient or used in our experiments. (Randelli, et al. 2016).

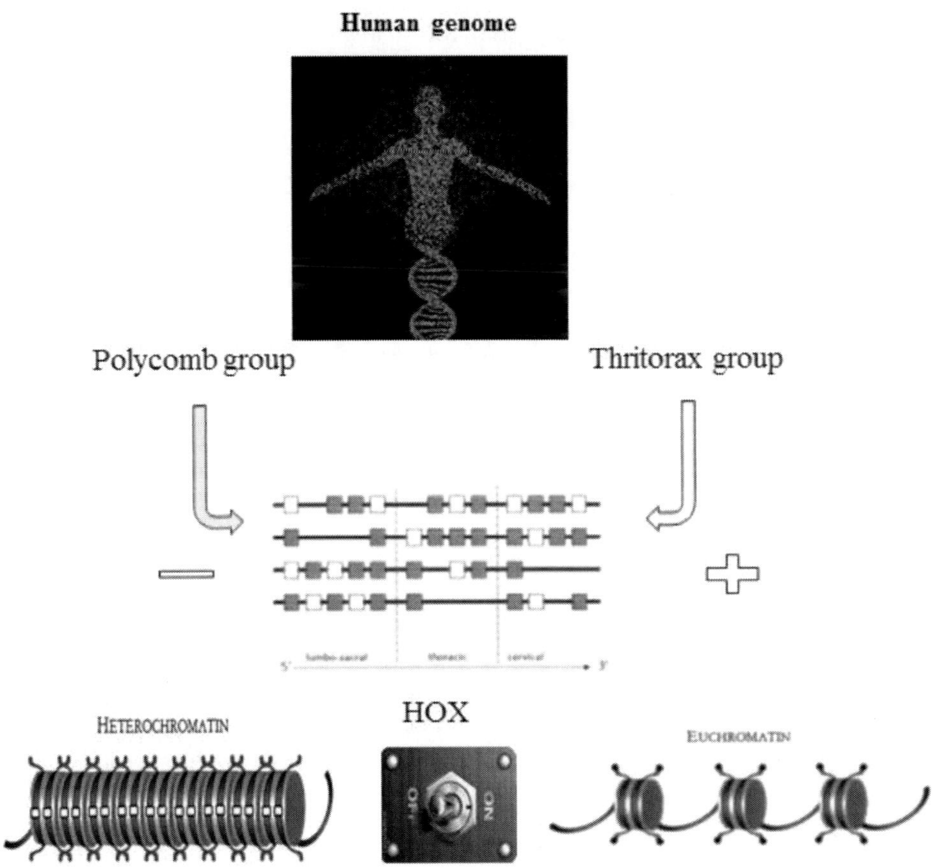

Figure 2. The HOX genes and Cell Memory Program control (see the text).

Preparation of Conditioned Medium (CM) from Lipogems (LG)

LG was washed in PBS at least three time by centrifugation at 250 g for 10.' LG was then aspirated and seeded (around 2 ml) in T25 flask in5 ml of IMDM serum free medium (a similar amount of LG was used for cell extraction with collagenase). The flasks (five for each LG preparation) were incubated for 5 days at 37°C in 5% CO_2. At the end of the incubation, the medium was aspirated and centrifuged at 250 x g for 10' LG-CM was filtered 0.22 μ, aliquoted and stored at -20°C until use (Cesarani, et al. 2016).

Cells Culture and Culture Medium

Osteoblasts culture medium was prepared using DMEM-low glucose (Sigma-Aldrich), 10% FBS (BioWest), 4mM L-glutamine (Euroclone), and 1% antibiotic-antimycotic mixture (Euroclone), supplemented with 0.1 □M dexamethasone, 50 □g/mL L-ascorbic acid-2-phosphate, and 10mM □-glycerophosphate (all reagents from Sigma-Aldrich). Mesenchyml Stem Cells Medium was constituted with alpha-MEM without L-glutamine (BioWhittaker), FBS (BioWest), 1% antibiotic-antimycotic mixture (Euroclone) and 4mM L-glutamine (Euroclone). The MSCs (LGC Milan, Italy) were seeded in Multiweel™ 6 wells. We seeded 3 x 10^4 for each well. In order to study the induction of osteogenesis-related genes, we perform three different multiwell: i) MSCs treated with alpha-MEM medium; ii) MSCs treated with Osteoblast medium; iii) MSCs treated with Lipogems Conditioned Medium (LG-CM). The cell were stimulated for 20 days and then we refresh the medium each three days (Randelli, et al. 2016).

MTT Assay

MTT, measuring the mitochondrial dehydrogenase activity. MSCs cells were seeded into 96-well flat-bottomed tissue culture plates respectively at 10^4 cells/well and treated as described in the paragraph 3. After 24 hours of incubation, the culture medium was replaced with 200□L/well of dilutions. Elapsed further 24 hours, the medium was substituted with 100□L/well of MTT solution (1 mg/ml) in PBS, and the cells were incubated for an additional hour at 37°C in a 5% CO_2 atmosphere. After the solution was removed, 100 □L/well of DMSO were added and the plates were swirled gently for 10 min. The optical density of each well was immediately measured in a spectrophotometer (Sunrise ™, Tecan, Mannederf/Zurich, Switzerland) at 590 nm. The optical density of cells cultured in the medium plus saline solution without the mini screw extracts was used as control for 100% of cell viability and as a reference for the determination of the cytotoxicity (%) in the assay. At least four independent experiments were performed in quadruplicate (Borelli, et al. 2017).

Alamar Blue Assay

MSCs (10_4 cells/well) were seeded in 96-multiwell plates. Cell proliferation of four replicate samples was assessed by alamarBlue assay (AB) at 7days and 14 days. AlamarBlue assay uses a visible blue fluorogen probe resazurin, which is reduced to a red fluorescent compound (resorufin) by cellular redox enzymes of the mitochondrial respiratory chain. Viable cells continuously convert resazurin to resorufin, thereby generating a quantitative measure of viability and proliferation [28]. The AB assay was performed according to the manufacturer's protocol (BioSource International, Camarillo, CA, USA). At predetermined culture intervals, 200 µL of alamar Blue dye were added directly into culture media at a final concentration of 10% and the plates were returned to the incubator for 4 h. As a negative control, AB was added to the medium without cells. The percentage AB reduction was calculated from the values of optical density at 540 and 590 nm using the manufacturer's formula (Gandolfi M.G. et al. 2014).

RNA Extraction and Analysis

Total RNA was isolated from MSCs culture cells, using RNeasy Micro-Kit (Qiagen GmbH, Hilden, Germany) following the manufacturer's instructions. Samples were treated with RNase-free DNase (Qiagen GmbH) to prevent amplification of genomic DNA. One microgram RNA was subjected to cDNA synthesis for 1 h at 37°C using the Ready to go You-Primer First-Strand Beads kit (Amersham Biosciences cod. 27-9264-01) in a reaction mixture containing 0,5 mg random hexamers (GeneAmp RNA PCR Random Hexamers Set N808-0127 Applied Biosystems, Foster City, CA).

Real Time-PCR (QRT-PCR)

QRT-PCR will be performed using Taq-Man or technology (QRT-PCR StepOne). This assay uses a specific oligonucleotide probe, annealing between the two primer sites, which is labelled with a reporter fluorophore and a quencher. Cleavage of the probe by the exonuclease activity of Taq polymerase during strand elongation releases the reporter from the probe resulting in an increase in reporter emission intensity owing to its separation from the quencher. This increment in net fluorescence is monitored in real-time during each PCR amplification. The cDNA, will be used for real-time PCR performed in 48-well optical reaction plates with cDNA equivalent to 100ng RNA in a volume of 25 ml reaction containing Taqman Universal Master Mix (Applied Biosystem 4304437). Optimized concentrations of FAM-labelled probe and specific forward and reverse primers for HOXC11, HOXC12, SATB2, Osx and lncRNA HOTAIR (Applied Biosystem) from Assay on Demand. The results will be analyzed using a comparative method, and the values will be normalized to the GAPDH expression as endogenous controls (Borelli, et al. 2017).

RESULTS

MMT Assay

The data of the MTT assay in MSCs treated with Osteoblast Medium (OsteoB), Lipogems Conditioned Medium (LG-CM) and LG-CM diluted with OsteoB (LGCM-OsteoB 1:1) are shown in the Figure 3. The viability of MSCs, was assessed considering the medium used in the different wells; No differences in the cytotoxic effect have been demonstrated in MSCs treated with OsteoB, LG-CM and LGCM+OsteoB compared with control.

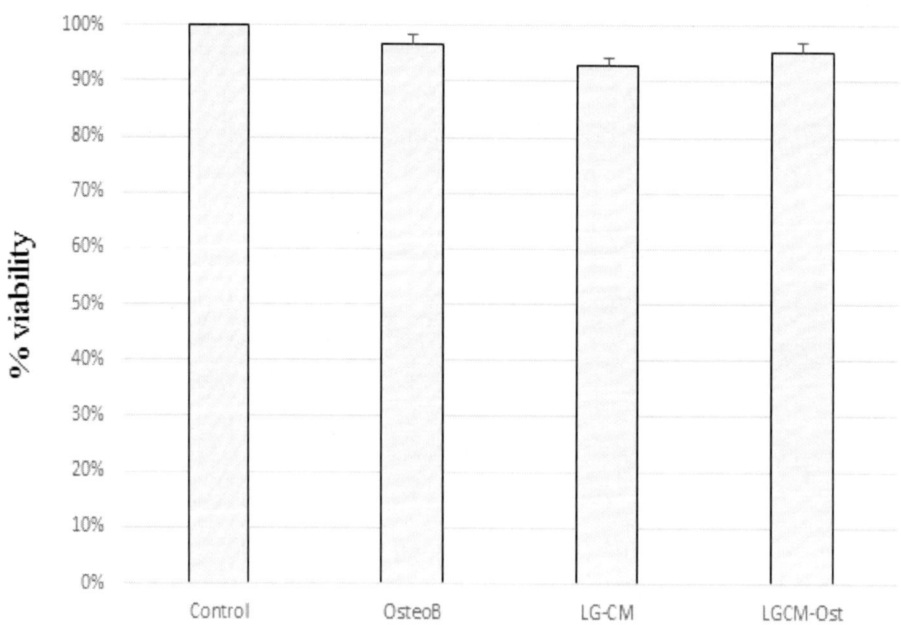

Figure 3. MTT assay did not show cytotoxic effects in MSCs treated for 24h with OsteoB, LG-CM and LGCM-OsteoB medium.

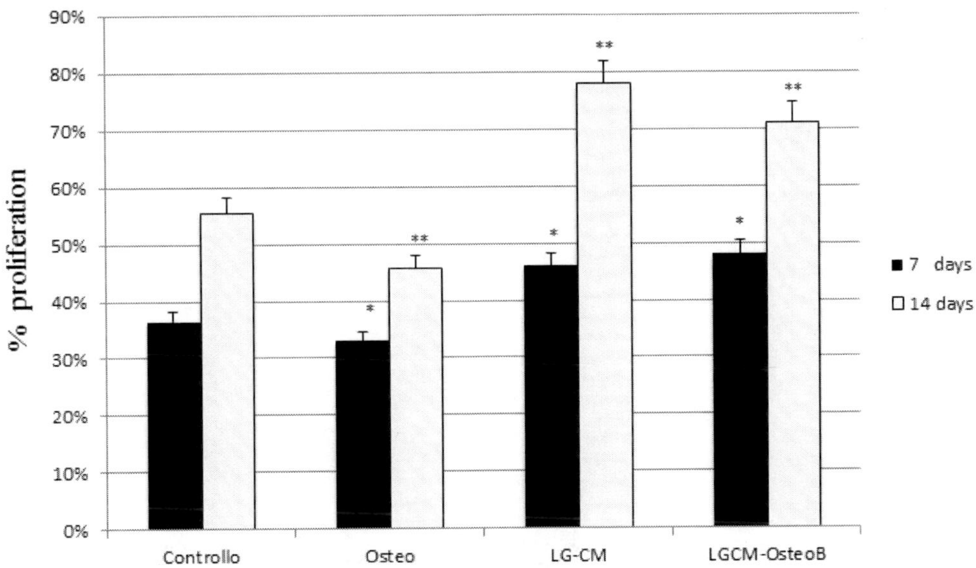

Figure 4. The proliferation of MSCs treated with OsteoB, LG-CM and LGCM-OsteoB medium for 7 and 14 days, was always increased during the experiments (see the text).

alamarBlue

The cell proliferation has been studied with the alamarBlue assay. The MSCs cells were treated with three different medium sorted out for the experiments: OsteoB, LG-CM and LGCM-OsteoB (see paragraph 1) and the cells proliferation was followed at 7 and 14 days. The cells viability was always increased in the experiment with OsteoB, LG-CM, LGCM-OsteoB, compared with control. The rate of proliferation in MSCs treated with LG-CM and LGCM-OsteoB at 7 days, was higher 10% than control and OsteoB; at 14 days the proliferation increased more than of 30% (Figure 4).

HOXC11, C12, Osx, SATB2 and lncRNA HOTAIR Gene Expression

qRT-PCR was used to analyze gene expressions of HOXC11, HOXC12, SATB2, Osx and HOTAIR in MSCs using specific medium. The time and condition of the experiments were performed following the material and methods described above. Both, OsteoB and LGCM-OsteoB medium were able to regulate the expression of all genes object of the study. Moreover, the rate of hyper-expression was related to the medium used during the experiments; in MSCs stimulated with LGCM+OsteoB the up-regulation of the HOXC11, SATB2, Osx and HOTAIR genes was always higher than MSCs treated with OsteoB medium. The differences in the rate of the gene expression were related to the specific characteristic of the single gene studied. Conversely, the HOXC12 gene expression did not show appreciable differences in cells exposed to different treatment, compared with MSCs control (Figure 5-6).

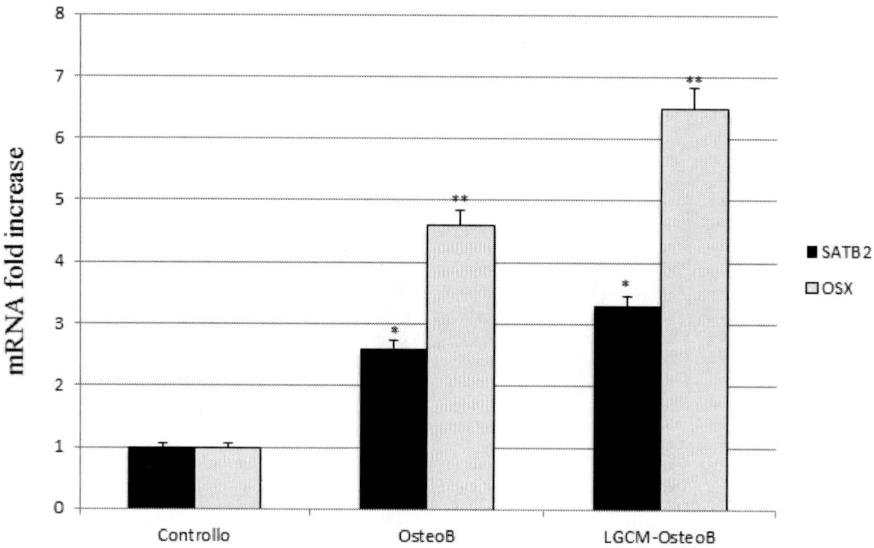

Figure 5. SATB2 and Osx gene expression was higher in MSCs treated with OsteoB and LGCM-OsteoB, compared with control (see the text).

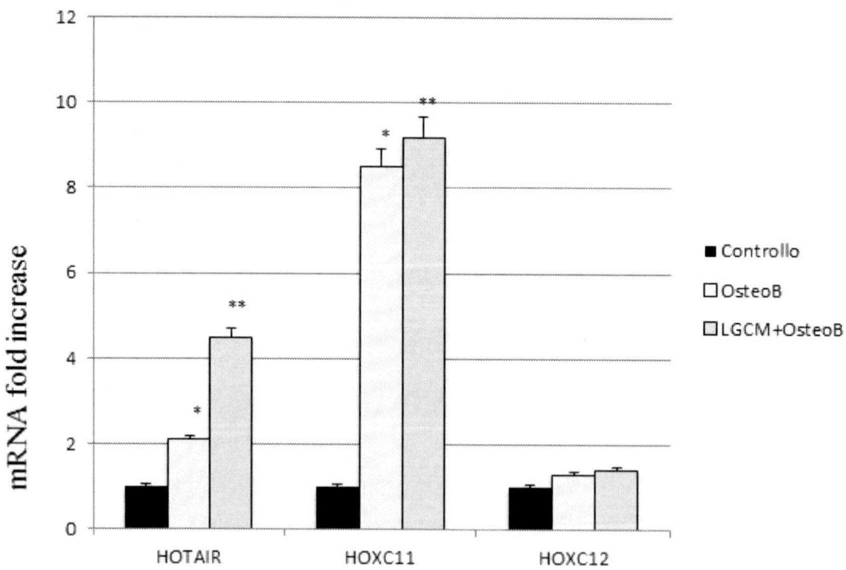

Figure 6. HOTAIR, HOXC11 gene expression was higher in MSCs treated with different medium than the control, while no HOXC12 gene expression was not influenced by OsteoB and LGCM-OsteoB medium.

Discussion

In the last years, the information about the role of Osteoblasts in the sorting out of bone tissue are increased. The bone remodeling process, is constantly controlled by osteoblastic lineage cells (Osteoclasts, Osteocytes). It has already demonstrated the commitment of osteoblastic lineage cells in the regulation of bone development, that play a key role like mechanic support for skeletal muscle; furthermore, bone matrix is a crucial reserve of Calcium and Phosphorus (Razzaque., 2009).

The osteoblasts development takes place through the transition from MSCs to pre-osteoblasts and, finally, osteoblasts. However, has not been clarified the molecular identity of the different stages; the characterization of these three steps provides a model to study the Osteoblasts differentiation. The molecular markers of mature Osteoblasts were already identified (Osteocalcin, Alkaline phosphatase, etc.); despite this, it is not clear which gene network regulates the transition from MSCs to pre-osteoblasts. Finally, the pre-osteoblast include a series of heterogeneous cells with intermediate characteristics between MSCs and osteoblasts, in this case is a lot of complex the identification of molecular markers able to lead the differentiation and development of mature Osteoblasts (Pritchard., et al. 1952; Owen., 1995; Owen and Machperson 1963). It has been shown the crucial role played by the specific transcription factors able to characterize different stages of Osteoblasts differentiation; these proteins work in a specific way during Osteoblasts lineage cell development and their activation is time-dependent.

Some of these Transcription factors have been already identified. Osx belong at family of homeobox genes located on chr 12q13.13, Osterix is one of the most regulator of osteoblasts differentiation (33 Long., 2011); this protein exerts its action inhibiting SATB2 gene expression. The SATB2 gene is a homeobox family member located on chr 2q32-33 and take part in the bone formation; its deregulation is related to several malformation like generalized osteoporosis, craniofacial dysmorphisms (cleft palate, mandibular hypoplasia and protruding incisor) (Leoyklang., et al. 2007; Rosenfeld., et al. 2009) .

Moreover, it has been demonstrated the role HOX gene family like must regulator of osteogenesis. Deregulation of Hoxd locus is related to the morphological anomaly of long bone development (Gonzalez-Martin., et al. 2014). The HOXA10 gene controls the Osteoblasts differentiation mainly by means of the Osteoblast related genes activation like Runx2, Alkaline phosphatase (ALP), Bone Sialoprotein (Spp) (Hassan, et al. 2007). Conversely, HOXA2 is an HOX protein inhibited by SATB2 and this regulation is related to a negative control of the osteogenesis (Zhang, et al. 2012).

Several studies suggested that HOTAIR is able to control the epigenetic regulation of DNA through a deep relationship with chromatin, involving the recruitment of complexes, which are able to identify specific sequences and induce silencing of several target genes. The HOTAIR epigenetically promotes Epithelial Mesenchymal Transition

(EMT) mainly by means of silencing miRNA34a and interacting with Polycomb-Responsive-Element-2 (PRC2). In conclusion, based on the literature and the results obtained, it has been hypothesized that HOTAIR and HOX genes might induce the alteration of the epithelial cell memory program (Pdua AC et al. 2013; Tano and Akimitsu et al. 2013; Liu, et al. 2015; Procino et al. 2016)

In this study, we have demonstrated, for the first time, the involvement of HOXC locus and lncRNA HOTAIR during MSCs differentiation towards osteoblastic phenotypes. In general, in MSCs treated with LGCM-OsteoB and OsteoB medium obtained by adipose tissue with indirect method, we have always observed an up-regulation of HOTAIR and HOXC11 genes, but no differences in HOXC12 mRNA; the fold increase of HOTAIR and HOXC11 was higher in Mesenchymal Stem Cells treated with LGCM-OsteoB medium than MSCs treated with OsteoB (Figure 5-6). According with the literature, Osx and SATB2 genes expression were increased in MSCs treated with OsteoB and LGCM-OsteoB medium, compared with control. The hyper-expression of Osterix and SATB2 were always higher in MSCs treated with LGCM-OsteoB medium than MSCs exposed to OsteoB medium. Moreover, the positive regulation of Osterix gene was always related to the deregulation of SATB2 gene.

In our opinion, HOTAIR is a good candidate in the control of osteogenesis, on the basis of its position within the locus and the ability to regulate the HOXC locus. In this paper we have shown, like HOTAIR is able to activate HOXC11 gene in MSCs treated with LGCM-OsteoB and OsteoB, conversely we did not observe none change in HOXC12 gene expression. On the other hand, we believe (on the basis of our results) that HOXC11 mRNA fold increase, is related to the up-regulation of Osx gene in MSCs exposed to a different medium; In turn, the hyper-expression of Osterix gene down-regulates SATB2 mRNA. Our theory is supported by the position within the genome of Osx (chr 12q13.13) close to HOXC cluster and SATB2 (chr 2q31.1) close to HOXD cluster, two area controlled by of HOX network and in particular by HOTAIR.

In conclusion, the data demonstrated the involvement of lncRNA HOTAIR, HOXC11 but not HOXC12, in the regulation of osteoblast related genes, although this is a preliminary study and further experiments will be useful in order to understand the role of HOX genes in the control of human osteogenesis.

REFERENCES

[1] Attanasio, C., Nord, A. S., Zhu, Y., Blow, M. J., Li, Z., Liberton, D. K., Morrison, H., Plajzer-Frick, I., Holt, A., Hosseini, R., Phouanenavong, S., Akiyama, J. A., Shoukry, M., Afzal, V., Rubin, E. M., FitzPatrick, D. R., Hallgrímsson, B., Pennacchio, L. A. & Visel, A. (2013). Fine tuning of craniofacial morphology by distant-acting enhancers *Science*., Oct 25, *342*(6157), 1241006.

[2] Baek, W. Y1., Lee, M. A., Jung, J. W., Kim, S. Y., Akiyama, H., de Crombrugghe, B. & Kim, J. E. (2009). Positive regulation of adult bone formation by osteoblast-specific transcription factor osterix. *J Bone Miner Res.*, Jun, *24*(6), 1055-65.

[3] Bantignies, F. & Cavalli, G. (2006). Cellular memory and dynamic regulation of polycomb group proteins. *Curr Opin Cell Biol*, *18*, 275-283.

[4] Barnes, T. M., Kohara, Y., Coulson, A. & Hekimi, S. (1995). Meiotic recombination, noncoding DNA and genomic organization in Caenorhabditis elegans. *Genetics*, *141*, 159-179.

[5] Borelli, B., Zarone, F., Rivieccio, V., Riccitiello, F., Simeone, M., Sorrentino, R., Rengo, S., Spagnuolo, G. & Procino A. (2016). Polyacrylic resins regulate transcriptional control of interleukin-6, gp80, and gp130 genes in human gingival fibroblasts. *J Oral Sci.*, Dec 28.

[6] Ceserani, V., Ferri, A., Berenzi, A., Benetti, A., Ciusani, E., Pascucci, L., Bazzucchi, C., Coccè, V., Bonomi, A., Pessina, A., Ghezzi, E., Zeira, O., Ceccarelli, P., Versari, S., Tremolada, C. & Alessandri, G. (2016). Angiogenic and anti-inflammatory properties of micro-fragmented fat tissue and its derived mesenchymal stromal cells. *Vasc Cell.*, Aug, *18*, 8, 3.

[7] Dobreva, G1., Chahrour, M., Dautzenberg, M., Chirivella, L., Kanzler, B., Fariñas, I., Karsenty, G. & Grosschedl, R. (2006). SATB2 is a multifunctional determinant of craniofacial patterning and osteoblast differentiation. *Cell.*, Jun 2, *125*(5), 971-86.

[8] Ferber, S., Halkin, A., Cohen, H., Ber, I., Einav, Y., et al. (2000). Pancreatic and duodenal homeobox gene 1 induces expression of insulin genes in liver and ameliorates streptozotocin-induced hyperglycemia. *Nat Med*, *6*, 568-572.

[9] Foucher, I., Volovitch, M., Frain, M., Kim, J. J., Souberbielle, J. C., et al. (2002). Hoxa5 overexpression correlates with IGFBP1 upregulation and postnatal dwarfism: evidence for an interaction between Hoxa5 and Forkhead box transcription factors. *Development*, *129*, 4065-4074.

[10] Gandolfi, G., Spagnuolo, F., Siboni, A., Procino, V., Rivieccio, G. A., Pelliccioni, C. & Prati, and S. Rengo. (2015). Calcium silicate/calcium phosphate biphasic cements for vital pulp therapy: chemical-physical properties and human pulp cells response. *Clin Oral Investig*. Nov, *19*(8), 2075-89.

[11] Gehring, W. J. & Hiromi, Y. (1986). Homeotic genes and the homeobox. *Annu Rev Genet*, *20*, 147-173.

[12] González-Martín, M. C., Mallo, M. & Ros, M. A. (2014). Long bone development requires a threshold of Hox function. *Dev Biol.*, Aug 15, *392*(2), 454-65.

[13] Gordon, J. A1., Hassan, M. Q., Koss, M., Montecino, M., Selleri, L., van Wijnen, A. J., Stein, J. L., Stein, G. S. & Lian, J. B. (2011). Epigenetic regulation of early osteogenesis and mineralized tissue formation by a HOXA10-PBX1-associated complex. *Cells Tissues Organs*, *194*(2-4), 146-50.

[14] Hassan, M. Q1., Tare, R., Lee, S. H., Mandeville, M., Weiner, B., Montecino, M., van Wijnen, A. J., Stein, J. L., Stein, G. S. & Lian, J. B. (2007). HOXA10 controls osteoblastogenesis by directly activating bone regulatory and phenotypic genes. *Mol Cell Biol.*, May, *27*(9), 3337-52.

[15] He, S., Liu, S. & Zhu, H. (2011). The sequence, structure and evolutionary features of hotair in mammals. *BMC Evolut. Biol.*, *11*, 102.

[16] Kogo, R., Shimamura, T., Mimori, K., Kawahara, K., Imoto, S., Sudo, T., Tanaka, F., Shibata, K., Suzuki, A. & Komune, S. (2011). Long noncoding RNA hotair regulates polycomb-dependent chromatin modification and is associated with poor prognosis in colorectal cancers. *Cancer Res.*, *71*, 6320–6326.

[17] Leoyklang, P., Suphapeetiporn, K., Siriwan, P., Desudchit, T., Chaowanapanja, P., et al. (2007). Heterozygous nonsense mutation SATB2 associated with cleft palate, osteoporosis, and cognitive defects. *Hum Mutat*, *28*, 732–738.

[18] Liu, Y. W., Sun, M., Xia, R., et al. (2015). LincHOTAIR epigenetically silences miR34a by binding to PRC2 to promote the epithelial-tomesenchymal transition in human gastric cancer. *Cell Death Dis*, *6*, e1802.

[19] Long, F. (2011). Building strong bones: molecular regulation of the osteoblast lineage. *Nat Rev Mol Cell Biol.*, Dec 22, *13*(1), 27-38.

[20] Nakashima, K., Zhou, X., Kunkel, G., Zhang, Z., Deng, J. M., Beh-ringer, R. R. & de Crombrugghe, B. (2002). The novel zinc finger-containing transcription factor osterix is required for osteo-blast differentiation and bone formation. *Cell.*, *108*, 17–29.

[21] Owen, M. (1995). Cell population kinetics of an osteo-genic tissue. I. *Clin Orthop Relat Res.*, Apr, (313), 3-7

[22] Owen, M. & Machpherson, S. (1963). Cell Population Kinetics o fan Osteogenic Tissue. *II J Cell Biol.*, Oct, *19*, 33-44.

[23] Pdua Alves, C., Fonseca, A. S., Muys, B. R., et al. (2013). The lincRNA Hotair is required for epithelial-to-mesenchymal transition and stemness maintenance of cancer cell lines. *Stem Cells*, *31*(12), 2827-32.

[24] Pritchard, J. J. (1952). A cytological and histochemical study of bone and cartilage formation in the rat. *J Anat.*, Jul, *86*(3), 259-77.

[25] Procino, A. (2016). The HOX network leads the neoplastic transformation in human solid tumors International. *Journal of Innovations in Biological and Chemical Sciences*, Volume 9, 8-13

[26] Procino, A. (2016). lncRNA HOTAIR, HOXC11 and HOXC12 Gene Expressions are Upregulated in CaCo2 Cells Treated with P31-43 Toxic Peptide. *Current Signal Transduction Therapy*, *11*, 33-36.

[27] Procino, A. (2014). The Paralogous Group HOX 13 Discriminates between Normal Colon Tissue and Colon Cancer. *J Mol Genet Med*, *8*, 3.

[28] Randelli, P., Menon, A., Ragone, V., Creo, P., Bergante, S., Randelli, F., De Girolamo, L., Alfieri Montrasio, U., Banfi, G., Cabitza, P., Tettamanti, G. & Anastasia, L. (2016). Lipogems Product Treatment Increases the Proliferation Rate of Human Tendon Stem Cells without Affecting Their Stemness and Differentiation Capability. *Stem Cells Int.*, 4373410.

[29] Razzaque, M. S. (2009). The FGF23-Klotho axis: endocrine regulation of phosphate homeostasis. *Nat Rev Endocrinol.*, Nov, *5*(11), 611-9.

[30] Rinn, J. L., Kertesz, M., Wang, J. K., Squazzo, S. L., Xu, X., Brugmann, S. A., Goodnough, L. H., Helms, J. A., Farnham, P. J., Segal, E., et al. (2007). Functional demarcation of active and silent chromatin domains in human hox loci by noncoding RNAs. *Cell*, *129*, 1311–1323.

[31] Rosenfeld, J. A.1, Ballif, B. C., Lucas, A., Spence, E. J., Powell, C., Aylsworth, A. S., Torchia, B. A. & Shaffer, L. G. (2009). Small deletions of SATB2 cause some of the clinical features of the 2q33.1 microdeletion syndrome. *PLoS One.*, Aug 10, *4*(8), e6568.

[32] Shen, J., Nair, A., Saxena, R., Zhang, C. C., Borrelli, J. Jr. & Tang, L. (2014). Tissue engineering bone using autologous progenitor cells in the peritoneum. *PLoS One.*, Mar 28, *9*(3), e93514.

[33] Tano, K. & Akimitsu, N. (2012). Long non-coding RNAs in cancer progression. *Front. Genet.*, *3*, 219.

[34] Tano, K. & Akimitsu, N. (2013). Long non-coding RNAs in cancer progression. *Front Genet*; *3*, 219.

[35] Wang, L., Wu, F., Song, Y., Li, X., Wu, Q., Duan, Y. & Jin, Z. (2016). Long noncoding RNA related to periodontitis interacts with miR-182 to upregulate osteogenic differentiation in periodontal mesenchymal stem cells of periodontitis patients. *Cell Death Dis.*, Aug 11, *7*(8), e2327.

[36] Weiss, K. M., Ruddle, F. H. & Bollekens, J. (1995). Dlx and other homeobox genes in the morphological development of the dentition. *Connect Tissue Res.*, *32*(1-4), 35-40.

[37] Wozney, J. M., Fujisawa-Sehara, A., Suda, T., Katagiri, T., Yamaguchi, A., Komaki, M., Abe, E., Takahashi, N., Ikeda, T. & Rosen, V. (1994). Bone morphogenetic protein-2 converts the differentiation pathway of C2C12 myoblasts into the osteoblast lineage. *J Cell Biol.*, Dec, *127*(6 Pt 1), 1755-66. Erratum in: *J Cell Biol*, 1995 Feb, *128*(4), following 713.

[38] Yamaguchi, A1., Katagiri, T., Ikeda, T., Wozney, J. M., Rosen, V., Wang, E. A., Kahn, A. J., Suda, T. & Yoshiki, S. (1991). Recombinant human bone morphogenetic protein-2 stimulates osteoblastic maturation and inhibits myogenic differentiation *in vitro*. *J Cell Biol.*, May, *113*(3), 681-7.

[39] Zhao, X., Qu, Z., Tickner, J., Xu, J., Dai, K. & Zhang, X. (2013). The role of SATB2 in skeletogenesis and human disease. *Cytokine Growth Factor Rev.*. 2014 Feb, *25*(1), 35-44.

[40] Zhang. P1., Men, J., Fu, Y., Shan, T., Ye, J., Wu, Y., Tao, Z., Liu, L. & Jiang, H. (2012). Contribution of SATB2 to the stronger osteogenic potential of bone marrow stromal cells from craniofacial bones. *Cell Tissue Res.*, Dec, *350*(3), 425-37.

[41] Zhou, Q., Chen, F., Fei, Z., Zhao, J., Liang, Y., Pan, W., Liu, X. & Zheng, D. (2016). Genetic variants of lncRNA HOTAIR contribute to the risk of osteosarcoma. *Oncotarget.*, Apr 12, *7*(15), 19928-34.

[42] Zhuang, W., Ge, X., Yang, S., Huang, M., Zhuang, W., Chen, P., et al. (2015). Upregulation of lncRNA MEG3 promotes osteogenic differentiation of mesenchymal stem cells from multiple myelomapatients by targeting BMP4 transcription. *Stem Cells*, *33*, 1985–1997.

[43] Zhu, L. & Xu, P. C. (2013). Downregulated LncRNA-ANCR promotes osteoblast differentiation by targeting EZH2 and regulating Runx2 expression. *Biochem Biophys Res Commun*, *432*, 612–617.

In: Stem Cells Origins of Cytokine Interactions
Editor: Prasad S. Koka
ISBN: 978-1-53615-302-6

Chapter 9

IN VITRO DIFFERENTIATION AND CHARACTERIZATION OF HUMAN HEMATOPOIETIC CD34+ STEM CELLS INTO ERYTHROCYTES

Pasupuleti Santhosh Kumar[1], Chodimella Chandrasekhar[2], Lokanathan Srikanth[1], Manne Mudhu Sunitha[1] and Potukuchi Venkata Gurunadha Krishna Sarma[1,*]

[1]Department of Biotechnology
[2]Department of Haematology
Sri Venkateswara Institute of Medical Sciences,
Tirupati, Andhra Pradesh, India

ABSTRACT

The differentiation capabilities of human CD34+ stem cells can be exploited in understanding the development of erythrocytes and its related erythroid lineage disorders. Hence, in the present study, human CD34+ stem cells were isolated from a donor and cultured in DMEM medium. Immunocytochemistry (ICC), Scanning electron microscopy (SEM) and flow cytometry were used to assess the purity and characteristic features of CD34+ stem cells. This pure culture was induced with erythrocyte differentiating medium (EDM) containing erythropoietin (EPO), stem cell factor (SCF), interleukin 3 (IL-3) and Granulocyte-macrophage colony-stimulating factor (GM-CSF) for 21 days, the morphological changes of erythrocytes were assessed with Leishman's stain and hematopoietic colony forming unit (CFU) assay. The expression of erythrocytes-specific markers ζ-globin, ε-globin, β-globin, *Glycophorin A* (GPA) and *BGP1* genes were evaluated by quantitative Real-time PCR. The ICC and FACS results showed (12×10^5 cells/mL) purity of the culture expressing CD34 marker. These pure cells on culturing

* Corresponding Author Email: sarmasvims@gmail.com.

with EPO, SCF, IL-3 and GM-CSF for 21 days showed conspicuous biconcave morphology when stained with the Leishman's stain. The results of the CFU-E assay indicated majority of the differentiated cells are erythroid progenitors. The qRT-PCR analysis showed distinct expression with ζ-globin, ε-globin, β-globin, *GPA* and *BGP1* genes in differentiated erythrocytes compared to $CD34^+$ cells. Further, GPA^+ expression was also confirmed by ICC. All these results conclusively elucidate that the $CD34^+$ stem cells were successfully differentiated into erythrocytes and these differentiated cells may be useful in the treatment of various erythrocytosis and erythroid lineage disorders.

Keywords: $CD34^+$, erythrocytes, RBC, Immunocytochemistry, CFU-E, β-globin, GPA

INTRODUCTION

Erythropoiesis defines the process of differentiation and proliferation from hematopoietic stem cells (HSCs) to mature red blood cells (RBCs) [1]. New erythrocytes are constantly produced in the bone marrow, which provides a niche consisting of endothelial cells of the vascular system, osteoblasts, stromal cells, hematopoietic cells and the extracellular matrix [2, 3]. This complex niche supports direct cell–cell contact and exposure of the developing hematopoietic cells to cell adhesion molecules, growth factors, and cytokines [4]. Erythroid cells at the terminal stages of differentiation have shed their nucleus, endoplasmic reticulum, mitochondria and consequently, they are no longer able to proliferate. To maintain the red blood cell count in the ~5 L of blood of an adult individual, $\sim 2.4 \times 10^6$ new erythrocytes have to be produced each second [4, 5, 6]. In adult humans, erythroid differentiation produces about 2×10^{11} red cells per day. Erythroid lineage development requires a delicate balance between the opposing effects of proliferation promoting factors and differentiation-inducing factors. The earliest erythroid progenitors are responsive to cytokines including EPO, SCF, IL 3 and GM-CSF [1, 4, 7, 8].

EPO plays a key role for activating the anti apoptotic proteins, protects erythroid progenitor cells from apoptosis and stimulates the haemoglobin (Hb) synthesis, which is essential for terminal differentiation [6, 8, 9]. The newly synthesized cells enter into the circulation as reticulocytes and are engaged in protein translation and the SCF binds to its c-KIT receptor (CD117), a tyrosine kinase that signals through several pathways including PI-3 kinase, Src kinases, and PLC-g. SCF acts synergistically with this lineage specific factor by promoting proliferation of erythroid progenitor cells [1, 8]. At subsequent stages, SCF acts synergistically with EPO in the proliferation and expansion of the developing erythroid progenitors and may play a key role in phosphorylating the EPO receptor [5, 6]. EPO production in the kidneys is directly regulated by tissue oxygen tension through the activity of the hypoxia-inducible transcription factor complex. Although EPO is the main regulator of red cell production other factors have an auxiliary role in the expansion of the erythroid progenitor compartment through their support of progenitor self-renewal. Development of reticulocyte from pro-erythroblast involves four to five rapid cell divisions, resulting in the decrease in cell size [1, 5]. Mature erythrocytes have a diameter of 6-8 μm in size. Their small size and biconcave shape creates a large surface area for gas exchange and allows the cells to enter the micro capillaries in the tissues. In the bone marrow, a series of intermediate erythroid precursors can be recognized that progressively gains erythroid characteristics [5, 8, 10].

The development of RBC mechanisms underlying *in vitro* erythroid differentiation and maturation, both at cellular and molecular levels, still remains unknown. In addition, the mechanisms of enucleation phenomenon of human erythropoiesis are also still unknown [10]. Thus, the availability of an *in vitro* erythropoiesis model, which exhibits all stages of RBC development, would be a powerful tool for investigating the factors and molecular mechanisms involved in proliferation and differentiation of human erythroid cells under normal and pathophysiological conditions [4, 8, 10]. The human haematopoietic stem cells (HSCs) reside in hypoxic bone marrow niche with the help of articulate HIF1α to maintain the anaerobic glycolysis as an energy source and sustain their quiescent nature [11]. These cells are mobilized by using cytokines like Granulocyte colony stimulating factor (G-CSF) which stimulates HSC mobilization by decreasing SDF-1α expression with increasing protease levels resulting in the cleavage of interaction between HSC and bone marrow niche [12, 13]. Thus mobilized haematopoietic stem cells were separated by apheresis technique [14]. Induction of these cells with extrinsic signaling molecules mimics specific signaling, leading to specific response for stem cell differentiation. The HSCs have the ability to correct damaged tissues. To understand the capability of these stem cells in the regeneration of RBC is to correct the various erythroid lineage disorders. Therefore, the present study is focused on *in vitro* differentiation of erythrocyte from $CD34^+$ stem cells.

Materials and Methods

Isolation of $CD34^+$ Stem Cells

Human haematopoietic $CD34^+$ stem cells were previously isolated from an apparently healthy donor peripheral blood by mobilizing with G-CSF 5μg/kg/day for up to three consecutive days, and collecting by apheresis technique using Rvy kit fitted in a Fresenius-kabi ASTec.204 cell separator (Bad Homburg, Germany) and cryopreserved. The same $CD34^+$ stem cells were used in the present study [14]. The protocol was approved by Institutional Ethics Committee, Sri Venkateswara Institute of Medical Sciences and University, Tirupati (IEC No: 31/06/2006 and IEC No: 565/15.07.2016).

Culturing and Characterization of CD34⁺ Stem Cells

The viability and assessment of human haematopoietic $CD34^+$ cells were carried out by using fluorescence activated cell sorting (FACS) (Becton Dickinson FACS Calibur™, USA), for this, monoclonal antibodies of CD45 coupled with Atto and CD34 tagged with PE were used following the ISHAGE guidelines [15]. These pure $CD34^+$ stem cells were cultured in T_{25} plastic culture flasks containing Dulbecco's Modified Eagle's Medium (DMEM) in 95% humidified atmosphere with 5% CO_2, at 37°C. Thus, obtained monolayer culture of $CD34^+$ cells were assessed by ICC using anti-human CD34 monoclonal antibody (QBEnd 10 clone) (Dako, Denmark) [15]. The morphology was observed by using Giemsa staining and SEM analysis (EVO MA15, Carl Zeiss, Germany) at a magnification of 1.00K X.

In Vitro Differentiation of CD34⁺ Stem Cells into Erythrocytes

Monolayer culture of $CD34^+$ stem cells was seeded at a density of 1×10^3 cells/Cm^2 and induced with Erythrocyte Differentiating Medium (EDM) containing 2U/mL EPO (Life Technologies), 100 ng/mL SCF (Life Technologies), 0.01U/ml IL-3 (Invitrogen) and 0.001ng/ml GM-CSF (Invitrogen) and incubated at 37°C, 5% CO_2 with 95% humidity for 21 days. Changes in the culture were observed at regular intervals of day 1 (D1), day 8 (D8), day 12 (D12), day 16 (D16) and day 21 (D21) and the morphological changes were recorded using microscope image processing system (MIPS) (Magnus analytics, India) [4].

Erythrocytes Culture Maintenance and Cryopreservation

On D21 the cells in the erythrocyte differentiating medium (EDM) showing distinct morphology were purified by limiting dilution method (1cell/well) with DMEM and sub-cultured with same induction medium in a T_{25} culture flask and the grown cells were cryopreserved as previously described [14].

Immunocytochemistry

Smear of differentiated culture and control $CD34^+$ cells were fixed with 4% paraformaldehyde at 37°C for 1hour. The internal peroxidase activity was blocked with 3% H_2O_2, and after further blocking with 2% gelatin, the cells were incubated with anti-human Glycophorin-A monoclonal antibody (Life technologies). For the detection of the primary antibody, the HRP conjugated rabbit anti-mouse IgG (Merck Biosciences) secondary antibody was used and finally the slides were counterstained with H_2O_2 and diaminobenzidine (DAB) (SRL) for 30 min at room temperature. The slides were washed with distilled water and observed under microscope with 100X magnification (Olympus, India) [11, 14].

Characterization of Differentiated Erythrocytes

The Leishman's and Giemsa staining were used to identify the morphological changes in differentiated cultured cells analyzed by light microscopy. The differentiated cultured cells on D21 were evaluated for the presence of $CD34^+$ marker by using monoclonal antibodies of CD45 coupled with Atto and CD34 tagged with PE against cell surface antigen, glycophorin A (GPA)-PE (Invitrogen) using fluorescence activated cell sorting (FACS) (Becton–Dickinson FACS Calibur™, US) following the ISHAGE guidelines [11, 15].

Erythroid Colony Forming Unit Assay (CFU-E)

Hematopoietic CFU-E assay was performed on D21 using a modification of the technique of Masters et al., 1990 [16] and these cells were plated in 35 mm plastic culture dishes containing DMEM agar-based medium. Cells were incubated for 14 days at 37°C and 5% CO_2 and colonies were enumerated by inverted microscope.

Enucleation of $CD34^+$ Derived Erythroid Cells

To induce terminal maturation, proliferating erythroid progenitor cells were co-cultured with EPO (10U/mL) and iron-saturated human transferrin (400μg/mL). Differentiating erythroblasts were maintained at 2×10^6 cells/mL for 8 days and the morphological changes were observed by Giemsa staining [4].

Quantitative Real-Time PCR (qRT-PCR)

Total RNA Isolation and cDNA Synthesis

The total RNA was extracted from control $CD34^+$ and differentiated cells by using Medox easy spin kit (Medox) and then transcribed to cDNA by using reverse transcriptase PCR (Applied Biosystems) method in a Thermocycler (Eppendrof) using the following reaction conditions: initial step at 25°C for 10 min, followed by 37°C for 120 min and thereafter 85°C for 5 min to inactivate the enzyme. Thus obtained cDNA of $CD34^+$ and differentiated erythrocyte cells was stored at -20 °C until use for qPCR [11, 17].

qRT-PCR

The qRT-PCR assay was performed in the ABI 7300 Real-Time PCR System (Applied Biosystems, USA) using different primers listed in (Table 1). All the primers were designed using the Primer Express Software (Applied Biosystems, California, USA), purchased from Eurofins genomics India Pvt Ltd, (Bangaluru, India). Reaction volume of 20 µl contains 100 µM of each forward and reverse primer, 10 µl of 1x SYBR select mastermix (Gibco, invitrogen), 5 µl Nuclease free water and cDNA. In the no-template control (NTC) cDNA was substituted with nuclease free water. The qPCR reactions were carried out using an initial step of 2 min at 50°C, followed by 10 min at 95°C to activate the Taq polymerase, followed by 40 cycles consisting of 15 s at 95°C, 60 s at 60°C. The expression levels were normalized with endogenous control GAPDH gene. The results were presented as Threshold Cycle values (Ct-values). The Ct-values were used to calculate the amount of PCR product by subtracting the C_T-value of reference genes from the C_T-value of gene studied (ΔC_T). Relative mRNA expression was calculated as $2^{-\Delta\Delta CT}$ [18].

Table 1. Quantitative RT-PCR primers used in this study

Gene	Accession number	Forward primer (5'-> 3')	Reverse primer (5'-> 3')	Tm (°C)
ζ-globin	NM_005332.2	CGGTGAAGAGCATCGACG	GGATACGACCGATAGGAACTTGT	56
ε-globin	NM_005330.3	GCCTGTGGAGCAAGATGAAT	GCGGGCTTGAGGTTGT	56
β-globin	NM_000518.4	TACATTTGCTTCTGACACAAC	ACAGATCCCCAAAGGAC	56
GPA	NM_002099.7	CAAACGGGACACATATGCAG	TCCAATAACACCAGCCATCA	56
BGP1	NM_001712.4	ACCCACCTGCACAGTACTCC	TTGATTTGGGGCTTTGCTAC	56
GAPDH	NM_002046.4	GACCTGACCTGCCGTCTAGAAA	CCTGCTTCACCACCTTCTTGA	56

Statistical Analysis

Each value presents the average of 3 independent experiments (n = 3). Statistical analysis was performed using ANOVA and statistical significance was assigned at $p < 0.05$.

RESULTS

Characterization of CD34⁺ Cells

The Cultured human $CD34^+$ cells, identified morphologically by Leishman's stain and SEM analysis, showed prominent rounded nuclei. Further confirmation was done by ICC and the cells were positive for HSC marker CD34. Enumeration of $CD34^+$ cells by flow cytometry showed 150.86 cells/μl. Thus; obtained pure $CD34^+$ cells were further cultured in DMEM for the formation of adherent monolayer cells (Figure 1a-f).

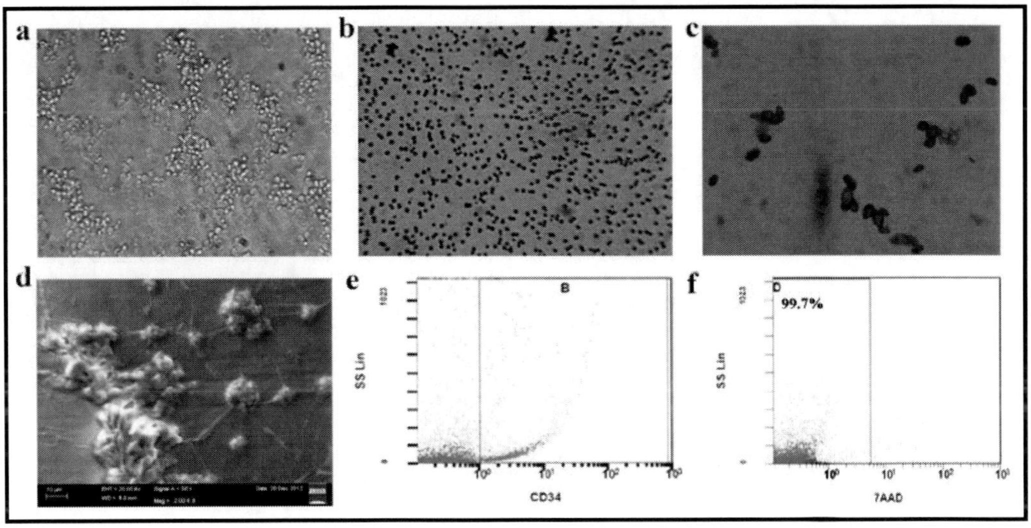

Figure 1. Culture and characterization of $CD34^+$ cells a) monolayer culture of $CD34^+$ cells in DMEM. b) Leishman's staining of cultured $CD34^+$ cells showing distinct morphology of round cells. c) ICC staining of $CD34^+$ stem cells showing positive against monoclonal mouse anti-human CD34 antibody. d) SEM analysis showed the round morphology of $CD34^+$ stem cells. e) FACS histogram showing enumerated absolute count of $CD34^+$ cells in the gated region 'B.' f) Percent viability of CD34+ cells with 7-AAD gated region 'D' showing 99.7% cells is viable.

Characterization of *In Vitro* Differentiated Erythrocytes

The monolayer cultured cells were directed towards the differentiation of Human CD34+ stem cells using the EDM containing 2U/mL EPO (Life Technologies), 100 ng/mL SCF (Life Technologies), 0.01U/mL IL-3 (Invitrogen) and 0.001ng/mL GM-CSF (Invitrogen). The formation of erythrocytes was observed on D21 (Figure 2a). The enucleation of RBC was obtained 8days later, when co-cultured with EPO (10U/mL) and iron-saturated human transferrin (100μg/mL). Thus attained differentiated cells were stained with leishman's staining, showed biconcave shape and the Giemsa staining confirmed the morphology of the cultured erythroid progenitor cells (Figure 2b, c). The cells were further confirmed by its positive reaction with anti-Glycophorin-A monoclonal antibody and visualized under microscope (Figure 2d). Further, the FACS analysis showed the absence of $CD34^+$ marker elucidating the erythrocytes differentiation (Figure 2g). Differentiation over time of $CD34^+$ cells was analyzed by FACS analysis of the cell surface markers. At D21, 89.1% of cells expressed GPA (Figure 2e,f). Hematopoietic CFU-E assay performed on D21 showed that most of the differentiated cells of our system were erythroid progenitors (Figure 2h).

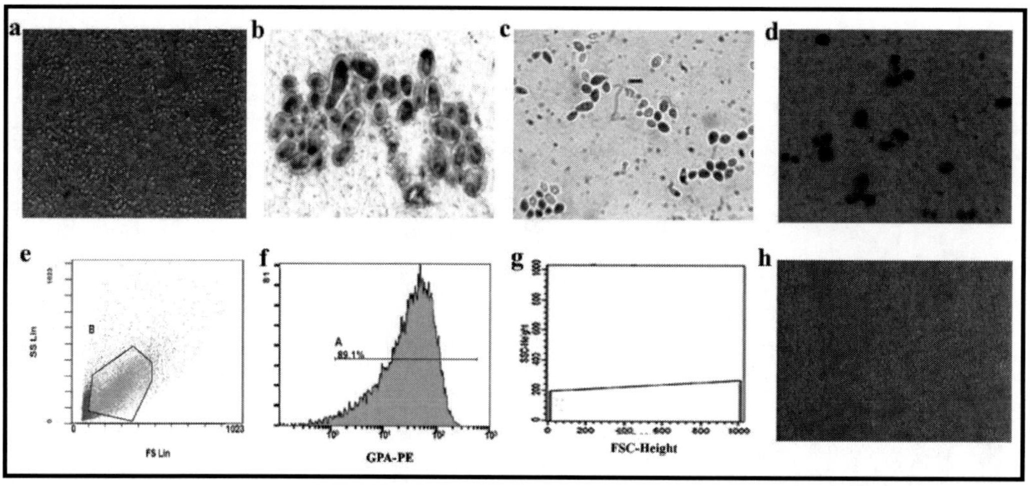

Figure 2. Culture and characterization of differentiated erythrocytes a) D21 differentiated erythrocytes. b) Leishman's staining of differentiated erythrocytes. c) Giemsa staining of differentiated erythrocytes. d) ICC staining of differentiated erythrocytes showing positive reaction against monoclonal mouse anti-human Glycophorin-A antibody. e,f) FACS analysis showing at D21, 89.1% of cells expressed GPA. g) FACS histogram showing differentiated cells showed less $CD34^+$ events. h) CFU-E assay showed most of the differentiated cells are erythroid progenitors.

Quantitative Gene Expression (qPCR) Analysis in Differentiated Erythrocytes

The quantitative gene expression analysis of ζ-globin, ε-globin, β-globins, GPA and BGP1 along the day D1, D8, D12, D16 and D21 were done for erythroid gene expression in which the D1 expression levels were taken as the calibrator. The early erythroid genes, ζ-globin and ε-globin revealed an upregulated expression at D8 and D12. Maturation at D16 showed increased expressions of β-globin mRNA. At D21 there was decreased expression observed in ζ-globin, ε-globin and constant expression of GPA and BGP1 genes in the differentiating cells confirming the erythrocytes formation (Figure 3). Statistical analysis of these qRT-PCR results using ANOVA shows significance at $p < 0.05$ value with its mean ± standard deviation.

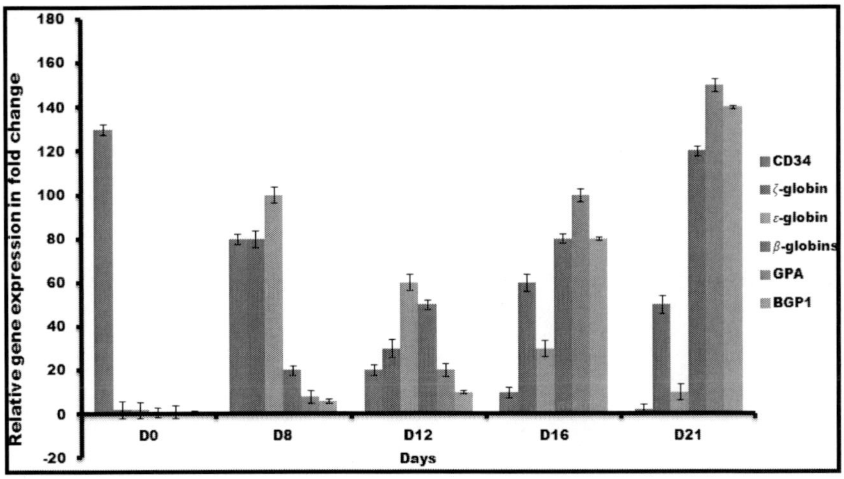

Figure 3. Relative quantitative expressions of marker genes. qRT-PCR expressions of erythrocyte gene specific markers in *ζ*-globin, *ε*-globin, *β*-globin, GPA and BGP1. The relative expression levels were indicated in fold change deduced by $2^{-\Delta\Delta CT}$ method.

Discussion

Human haematopotic CD34$^+$ cells have the ability to differentiate into different cell types of three germ layers when these cells were initiated with specified growth factors and cytokines [19]. *In vitro* erythrocyte development from hematopoietic CD34$^+$ stem cells would be especially useful for the basis of patient specific blood transfusion in patients suffering with sickle cell anaemia, β-thalasemia, erythrocytosis and other erythroid lineage disorders [4, 8]. For potential clinical application of *in vitro* erythrocyte production, a culture system using human cells is a prerequisite and we generated RBCs using human CD34$^+$ stem cells by inducing with erythrocytes differentiating medium (EDM) containing 2U/ml EPO, 100 ng/ml SCF, 0.01U/ml IL-3 and 0.001ng/ml GM-CSF [4, 20].

EPO is an anti-apoptotic protein essential for the terminal differentiation, which stimulates the hemoglobin synthesis [21]. SCF is a tyrosine kinase which binds the c-KIT receptor (CD117) that signals through several pathways including PI-3 kinase, Src kinases, and PLC-g, promotes the proliferation of erythroid progenitor cells and plays a key role in phosphorylating the EPO receptor. IL-3, GM-CSF are the cytokines which trigger the erythrocytes differentiation [1, 5, 22, 23].

Almost all the CD34$^+$ stem cells in the presence of EPO, SCF, IL-3, GM-CSF at D21 culture, were differentiated to erythroid progenitor cells and were positive to erythroid specific cell surface markers like GPA and *β*-globin (Figure 2). The distinct expressions of erythrocyte specific genes *ζ*-globin, *ε*-globin, *β*-globin, GPA and BGP1 were accessed (Figure 3). At D8 and D12 the early erythroid genes *ζ*-globin, *ε*-globin demonstrated high expressions and at D16 the β-globin displayed increased expressions. At D21, GPA and BGP1 genes were showing constant expression while, ζ-globin, ε-globin genes exhibited decreased expression. The GPA expression at D21 was very high, indicating cell cycle arrest and terminal maturation to erythrocytes and therefore confirming the successful differentiation of $CD34^+$ stem cells into erythrocytes [24] (Figure 3). The FACS analysis showed 89.1% of the differentiated cells were positive to GPA (Figure 2f, g). The ICC results displayed positive reaction with anti-Glycophorin-A monoclonal antibody clearly elucidating successful differentiation of $CD34^+$ stem cells to erythrocytes (Figure 2d). Further, the CFU-E assay results confirmed that most of the differentiated cells in our system were erythroid progenitors. *In vitro* erythrocytes production with the present method may help elucidate the process of understanding the RBC development, the required growth factors and receptor-mediated signal cascade, and the overall mechanism of enucleation and mature functional RBC formation (Figure 2).

In conclusion, the differentiation of human haematopoietic $CD34^+$ stem cells into the erythrocytes was successfully generated in the presence of EPO, SCF, IL-3, and GM-CSF. The GPA^+ expression indicates $CD34^+$ stem cells terminal maturation to erythrocytes. Therefore, the present study may be useful in the understanding of the development of RBC and help in understanding the erythroid lineage disorders like sickle cell anaemia, β-thalasemia, and erythrocytosis and provides a useful basis in the patient specific blood transfusion.

Acknowledgments

We sincerely acknowledge Sri Venkateswara Institute of Medical Sciences (SVIMS University), Tirupati, India for providing facilities to carry out this work, and this paper forms a part of PhD thesis work going to be submitted to SVIMS University, Tirupati, Andhra Pradesh, India. This work was carried out in the Department of Biotechnology with the liberal financial support given by Sri Balaji Arogya Varaprasadini Scheme [Grant No: SBAVP-RG/PhD/19] from Sri Venkateswara Institute of Medical Sciences, Tirupati.

References

[1] Muta K, Krantz SB, Bondurant MC, Wickrema A. Distinct roles of erythropoietin, insulin-like growth factor I, and stem cell factor in the development of erythroid progenitor cells. *J Clin Invest*, 1994, 94(1), 34-43.

[2] Baek EJ, Kim HS, Kim S, Jin H, Choi TY, Kim HO. *In vitro* clinical-grade generation of red blood cells from human umbilical cord blood CD34+ cells. *Transfusion*, 2008, 48(10), 2235-2245.

[3] Nakamura Y. *In vitro* production of transfusable red blood cells. Biotechnol *Genet Eng Rev* 2008, 25, 187-201.

[4] Xi J, Li Y, Wang R, Wang Y, Nan X, He L, Zhang P, Chen L, Yue W, Pei X. *In vitro* large scale production of human mature red blood cells from hematopoietic stem cells by co-culturing with human fetal liver stromal cells. *Biomed Res Int*, 2013, 2013, 807863.

[5] Dzierzak E, Philipsen S. Erythropoiesis: development and differentiation. *Cold Spring Harb Perspect Med*, 2013, 3(4), a011601.

[6] Malik P, Fisher TC, Barsky LL, Zeng L, Izadi P, Hiti AL, Weinberg KI, Coates TD, Meiselman HJ, Kohn DB. An *in vitro* model of human red blood cell production from hematopoietic progenitor cells. *Blood*, 1998, 91(8), 2664-2671.

[7] Baek EJ, Kim HS, Kim S, Jin H, Choi TY, Kim HO. *In vitro* clinical-grade generation of red blood cells from human umbilical cord blood $CD34^+$ cells. *Transfusion*, 2008, 48(10), 2235-2245.

[8] Dorn I, Lazar-Karsten P, Boie S, Ribbat J, Hartwig D, Driller B, Kirchner H, Schlenke P. *In vitro* proliferation and differentiation of human $CD34^+$ cells from peripheral blood into mature red blood cells with two different cell culture systems. *Transfusion*, 2008, 48(6), 1122-1132.

[9] Dolznig H, Habermann B, Stangl K, Deiner EM, Moriggl R, Beug H, Müllner EW. Apoptosis protection by the Epo target Bcl-X(L) allows factor-independent differentiation of primary erythroblasts. *Curr Biol*, 2002, 12(13), 1076-1085.

[10] Hyun Ok Kim. *In-Vitro* Stem Cell Derived Red Blood Cells for Transfusion: Are We There Yet? *Yonsei Med J*, 2014, 55(2), 304-309.

[11] Srikanth L, Sunitha MM, Venkatesh K, Kumar PS, Chandrasekhar C, Vengamma B, Sarma PV. Anaerobic Glycolysis and HIF1α Expression in Haematopoietic Stem Cells Explains Its Quiescence Nature. *J Stem Cells*, 2015, 10(2), 97-106.

[12] Petit I, Szyper-Kravitz M, Nagler A, Lahav M, Peled A, Habler L, Ponomaryov T, Taichman RS, Arenzana-Seisdedos F, Fujii N, Sandbank J, Zipori D, Lapidot T. G-CSF induces stem cell mobilization by decreasing bone marrow SDF-1 and up-regulating CXCR4. *Nat Immunol*, 2002, 3(7), 687-694.

[13] Lévesque JP, Hendy J, Takamatsu Y, Williams B, Winkler IG, Simmons PJ. Mobilization by either cyclophosphamide or granulocyte colony-stimulating factor transforms the bone marrow into a highly proteolytic environment. *Exp Hematol*, 2002, 30(5), 440-449.

[14] Sarma PV, Subramanyam G. *In vitro* cardiogenesis can be initiated in human $CD34^+$ cells. *Indian Heart J*, 2008, 60(2), 95-100.

[15] Sutherland DR, Anderson L, Keeney M, Nayar R, Chin-Yee I. The ISHAGE guidelines for $CD34^+$ cell determination by flow cytometry. International Society of Hematotherapy and Graft Engineering. *J Hematother*, 1996, 5(3), 213-26.

[16] Masters GS, Baines P, Jacobs A. Erythroid colony growth from peripheral blood and bone marrow in polycythaemia. *J Clin Pathol*, 1990, 43(11), 937-941.

[17] Sunitha MM, Srikanth L, Kumar PS, Chandrasekhar C, Sarma PV. *In vitro* differentiation potential of human haematopoietic $CD34^+$ cells towards pancreatic β-cells. *Cell Biol Int*, 2016, 40, 1084-1093.

[18] Livak KJ, Schmittgen TD. Analysis of relative gene expression data using real-time quantitative PCR and the 2(-Delta Delta C(T)) Method. *Methods*, 2001, 25(4), 402-408.

[19] Jiang X, Fujisaki T, Nicolini F, Berger M, Holyoake T, Eisterer W, Eaves C, Eaves A. Autonomous multi-lineage differentiation *in vitro* of primitive CD34+ cells from patients with chronic myeloid leukemia. *Leukemia*, 2000, 14(6), 1112-1121.

[20] van den Akker E1, Satchwell TJ, Pellegrin S, Daniels G, Toye AM. The majority of the *in vitro* erythroid expansion potential resides in $CD34^{(-)}$ cells, outweighing the contribution of $CD34^{(+)}$ cells and significantly increasing the erythroblast yield from peripheral blood samples. *Haematologica*, 2010, 95(9), 1594-1598.

[21] Dolznig H, Habermann B, Stangl K, Deiner EM, Moriggl R, Beug H, Müllner EW. Apoptosis protection by the Epo target Bcl-X(L) allows factor-independent differentiation of primary erythroblasts. *Curr Biol*, 2002, 12(13), 1076-1085.

[22] Panzenböck B, Bartunek P, Mapara MY, Zenke M. Growth and differentiation of human stem cell factor/erythropoietin-dependent erythroid progenitor cells *in vitro*. *Blood*, 1998, 92(10), 3658-3668.

[23] Leberbauer C, Boulmé F, Unfried G, Huber J, Beug H, Müllner EW. Different steroids co-regulate long-term expansion versus terminal differentiation in primary human erythroid progenitors. *Blood*, 2005, 105(1), 85-94.

[24] di Giacomo V, Matteucci A, Stellacci E, Battistini A, Di Baldassarre A, Capitani S, Alfani E, Migliaccio AR, Cocco L, Migliaccio G. Expression of signal transduction proteins during the differentiation of primary human erythroblasts. *J Cell Physiol*, 2005, 202(3), 831-838.

In: Stem Cells Origins of Cytokine Interactions
Editor: Prasad S. Koka
ISBN: 978-1-53615-302-6

Chapter 10

A RETROSPECTIVE STUDY OF *JAK-2* MUTATIONS, ENDOGENOUS ERYTHROID COLONY FORMATION ASSAY IN ABSOLUTE ERYTHROCYTOSIS PATIENTS IN A TERTIARY CARE CENTRE, SOUTH INDIA

Chandra Sekhar Chodimella[1,*], Pasupuleti Santhosh Kumar[2], Marthyaya Sumanth Kumar[1] and Potukuchi Venkata Gurunadha Krishna Sarma[2]

[1]Department of Hematology
[2]Department of Biotechnology
Sri Venkateswara Institute of Medical Sciences,
Tirupati, Andhra Pradesh, India

ABSTRACT

The Myeloproliferative disease that arises due to variation in the haematopoietic stem cell (HSC), and usually proliferates and dominates over normal HSC causing formation of enormous amount of erythrocytes (absolute erythrocytosis) in the blood is called Polycythemia vera (PV). Idiopathic erythrocytosis (IE) is absolute erythrocytosis of undefined etiology. In the present study, 54 patients with primary absolute erythrocytosis were thoroughly worked up including Endogenous erythroid colony forming assay (EEC), followed up for 2-4 years and analyzed retrospectively. The diagnoses were based on BCSH guidelines (2007) and WHO revision criteria (2016). We diagnosed 33% (18/54) patients as Polycythemia vera (PV) and 66.6% (36/54) patients as absolute erythrocytosis most probably Idiopathic Erythrocytosis (IE). Overall, hypertension (74%), dyslipidemia (55%), overweight (18%), consumption of alcohol

* Corresponding Author Email: drccsekhar@gmail.com.

(66.6%), smoking (55%) were seen and all these could have contributed to the pathology. Splenomegaly was seen in 14/18 (77.7%) in PV, 6/36 (16.6%) in IE. Serum EPO levels were below normal range 12/18 (66.6%) in PV, 14/36 (38.8%) in IE patients. EEC formation was observed 10/12 (83.3%) in PV, 14/22 (63.6%) in IE patients. Differences in WBC, platelet counts, EPO levels between PV and IE were statistically significant. In PV patients, *JAK-2* exon 14 mutation (V617F) was seen in 17/18 patients (94.4%). Only one patient was *JAK-2* (exon 12, 14) Negative and none were JAK2 exon 12 positive. IE was twice as common as PV in our study group and interestingly, most of the IE patients were EEC positive. The number of EEC colonies was significantly high ($p < 0.001$) in PV patients when compared with IE patients and normal positive controls.

Keywords: Erythrocytosis, Polycythemia vera, Idiopathic erythrocytosis, JAK-2; EEC

Introduction

The word 'polycythemia' is used for myeloproliferative neoplasia (MPN) when all the three cell lines are affected and 'erythrocytosis' if only red cell line is involved. These can be primary or secondary and congenital or acquired [1]. The diagnosis is largely based on the EPO levels, genetic mutation studies, bone marrow examination in addition to the detailed family history and clinical examination. The widely used criteria for the diagnosis of PV were the 2007 BCSH guidelines where persistently raised hematocrit (Hct) and *JAK-2* mutations were the major criteria which include nearly 99% of the cases of PV [2, 3]. The recent WHO guidelines (2016) [4] stressed on bone marrow examination also, in order not to miss the masked PV, essential thrombocythemia (ET) and other myeloproliferative disorders [5].

Absolute erythrocytosis is to be confirmed by red cell mass (RCM) estimation, for the given sex, weight, age and it may not be necessary if Hct >60% in males and >56% in females [6]. Erythrocytosis can be Primary (congenital and acquired), or Secondary (congenital and acquired). Primary congenital erythrocytosis (CE) usually have subnormal EPO levels (EPO independent erythroid proliferation), polyclonal hematopoiesis, normal WBC and platelets. e.g., Epo receptor (*EPOR*) mutation (autosomal dominant), *LNK* mutations, familial polyclonal MPN. On the contrary, Secondary congenital erythrocytosis is due to increased endogenous EPO production e.g., High O_2 affinity hemoglobins, reduced 2-3 DPG levels and mutations in prolyl hydroxylase domain proteins (*PHD2*), hypoxia inducible factor (*HIF1α*) and von Hipple Lindau (*VHL*) genes. These CE entities are seen extremely rarely [7]. Onset during childhood or at very young age and positive family history are suggestive of congenital familial erythrocytosis [1, 8-11]. EPOR is less than 15% of all the primary congenital varieties [12] and VHL mutation constitutes 10% of all IE cases [9]. All these are considered as rare entities [1, 7].

PV is a hematopoietic stem cell disorder with uncontrolled red cell production, and characterized by panhyperplastic and neoplastic marrow. The origin of stem cell transformation is not yet clear. The sensitivity of the PV progenitors to multiple cytokines suggests that the defect may lie in a common pathway downstream from multiple receptors. Decreased cellular apoptosis is suggested by the increased expression of BCLX. A unique valine to phenylalanine substitution (V617F) in the pseudokinase *JAK-2* domain has been identified. This mutation leads to deregulation of the kinase activity and thus to constitutive tyrosine phosphorylation activity resulting in gain of function mutation. Unlike *JAK-2* V617F, exon 12 mutations are also observed in *JAK-2* V617F negative PV cases only and are specific to patients with isolated erythrocytosis without concomitant leucocytosis and thrombocytosis. *JAK-2* V617F mutation is seen in 95% PV, *JAK-2* exon 12 mutation in 4% PV and *JAK-2* negative (exon 12 and 14) in less than 1% PV cases. In the absence of *JAK-2* mutations and subnormal EPO levels PV is less likely. *JAK-2* V617F mutation however can be seen in ET (40-50%) and PMF (50%) cases also. The prevalence of PV was 0.3/1000 persons [5, 13, 14].

IE is a collection of undefined, unidentified, heterogeneous and rare hematological disorders. It is different from PV in having variable EPO levels, low levels (in 1/3rd patients) to normal (inappropriately normal) levels depending on the various etiopathogenic factors [3]. It is classifiable from the already established entities mentioned in the above paragraphs and this separate entity is therefore a diagnosis by exclusion. The IE group further has two possible sub groups. In the first congenital group, the low EPO levels suggest molecular defect in the EPO signal transduction pathway. The second group has normal (inappropriately) or elevated EPO levels indicating, therefore, an unidentified secondary cause [1]. IE is the most common type of erythrocytosis seen in clinical practice. When dealing with a case of erythrocytosis, if secondary forms are excluded and no *JAK-2* mutations are found, the issue is only whether the patient has one of a) the rare *JAK-2* Negative (*JAK-2* wild type) forms of PV, b) the rare congenital erythrocytosis varieties, or c) the separately classifiable and more common IE [7]. The incidence of IE is 1.1 in 1000 persons [15].

EEC, which can be seen *in vitro* as the ability to form erythrocyte colonies, is considered as a surrogate marker for the EPO hypersensitive/independent erythropoiesis seen in certain PV/IE. Relative erythrocytosis is usually seen in dehydration and Apparent Erythrocytosis (AE) is the term used for the increased venous hematocrit without increase in Red Cell Mass. This entity has higher morbidity and mortality due to thrombotic events e.g., obesity, smoking, alcohol, hypertension etc. In general, all groups of erythrocytosis and polycythemia vera produce a variety of symptoms and signs due to increased thrombotic tendency and therefore should be managed appropriately [2].

MATERIALS AND METHODS

Patients

This is a retrospective study carried out between 2011-2015 where, sixty one (n = 61) patients with erythrocytosis reported to Department of Hematology, Sri Venkateswara Institute of Medical Sciences (SVIMS), Tirupati, the Southern state of Andhra Pradesh, India, out of which 7 patients refused to undergo further work up. Therefore, only fifty four (n = 54) asymptomatic individuals, found to have high Hct and Hb level on routine checkup during blood donor screening procedures and for some trivial complaints, were investigated. The study was cleared by the Institutional Ethics Committee (IEC.NO: 567). A written consent was obtained from all the patients studied.

Clinical Investigations

Hemoglobin (Hb), hematocrit (Hct), red blood cell (RBC) indices, total white blood cell (TWBC) count, differential WBC count, Platelet count were measured using auto hematology analyzer BC-5300 (MINDRAY, Shenzhen, China), and peripheral blood smear was done by using Leishman's stain. Liver function tests, renal function tests, lipid profile, serum iron, total iron binding capacity (TIBC) were measured by using (Beckman ante analyzer DXC 600). Serum ferritin (Chem well analyzer), serum erythropoietin (EPO) levels (Chemiluminesence immune assay CLIA), arterial blood gas analysis, ultrasonography abdomen, X-ray chest were done. The exon 14 and exon 12 of *JAK-2* gene were PCR amplified from the patients' and healthy individuals' genomic DNA by using the exon 14 primers: forward primer 5'-TGGCAGAGAGAATTTTCTGAACT-3' and reverse primer 5'-TGTTTGGGCATTGTAA CCTTC-3' and exon 12 primers: forward primer 5'-TCTGATGTACCAACCTCACCA-3' and reverse primer 5'-CATGAATGTAAATCAAGAAAACAGA TG-3' respectively [16] in a Mastercycler gradient Thermocycler (Eppendorf). The PCR products were then verified by running 2% agarose gel electrophoresis. The expression of *JAK-2* gene was assessed by quantitative Real-time PCR (Applied Biosystems ABI 7300, USA and Rotor Gene 3000, Corbett Research, Australia) [17]. Endogenous erythroid colony forming assay (EEC) method was performed to characterize the clonal ability in these patients [18, 19]. CT-scan chest/abdomen, MRI screening of CNS, pulmonary function tests were done where needed.

Investigative Pathway

- Only the patients with absolute erythrocytosis were considered in this retrospective study. Confirmation of the presence of absolute erythrocytosis was done on repeated blood counts, if needed, on the basis of a) the Hct >60% in males >56% in females b) when Hct >54% despite weekly phlebotomies in a month.
- Apparent Erythrocytosis (AE) was considered when the above criteria were not met and these patients were not included in the study.
- Individuals with Hb/Hct higher than normal but reverting to normal ranges without specific treatment were not studied but followed up [2].
- BCSH guidelines amended in 2007, WHO 2016 were followed. Bone marrow examination was done in all patients and morphological diagnosis was made as per the WHO guidelines. Karyotyping could not be done due to paucity of funds. Features like splenomegaly, TWBC/Platelet count were followed up as per the proposed review WHO criteria 2014 [5, 13].
- EPO levels were repeated after normalizing Hb/Hct levels to <15gm/<45% with periodic phlebotomies as tolerated, and the highest value of EPO was taken into account.
- When EPO levels were higher >26 units (4.3-26 IU/L), radiological work up was done to exclude tumors, organ enlargement, A-V malformations in CNS and these were considered as secondary polycythemic and not included in the study.
- To substantiate the low to normal levels of EPO, EEC forming ability was exploited, and could be performed only in 34 out of 54 patients.

Polycythemia vera: PV was considered if BCSH 2007, WHO 2016 (revised) criteria [4] were satisfied as given below:

a) *JAK-2* exon 14 or 12 mutation detection.
b) *JAK-2* Negative PV was considered on bone marrow exam, splenomegaly, EPO levels and EEC.
c) Major and minor criteria of BCSH guidelines seen.
d) Bone marrow trephine biopsy and aspiration cytology.

Congenital erythrocytosis (CE): *EPOR*, *LNK*, *VHL*, *PHD2*, *HIF1α* gene mutations could not be analyzed. However, the low probability of CE was considered because of:

1) Absence of positive family history
2) Older age of onset in our patient group

3) Lower normal and subnormal levels of EPO
4) Normal P_{50} values in blood gas analysis

High probability of Idiopathic Erythrocytosis was considered on the following basis:

a) *JAK-2* negative patient
b) Absence of splenomegaly
c) Splenomegaly attributable to liver diseases or recurrent malaria, or other known causes
d) Normal TWBC/platelet count throughout follow up
e) Bone marrow examination
f) Absence of disease in the family members and relatives
g) Older age of onset

Statistical Analysis

The association between two categorical variables was evaluated by Chi-square (χ^2) test, a p-value <0.05 was considered as significant by using statistical software IBM SPSS, version 20 (IBM SPSS statistics, Somers NY, USA).

Results and Discussion

In this retrospective study of 54 patients of absolute erythrosytosis, 18/54 (33.3%) patients were PV and 36/54 (66.6%) were probably IE and the clinical and laboratory investigations results were summarized in Tables 1 and 2.

Overall, the male/female ratio was 12.5:1 in our study. In PV group it was 3.5:1 [7]. It is interesting to note the male dominance in our study when globally the male: female ratio in PV was 1.1:1 [20] and the youngest patients in our PV group were 34 and 39 years old [18, 21] (Tables 1 and 2). All patients in the IE group were males [7].

In both the groups most of them were habituated to alcohol and smoking and were hypertensive and dyslipidemic. These factors must have contributed to the rise in viscosity in the already established absolute erythrocytosis thereby increasing the risk of thrombotic tendency [2].

In the 18 patients of PV group *JAK-2* (c.1849; V617F) mutation was seen in 17/18 (94.4%) (Figure 1A). No mutations were observed in healthy subjects (controls) in allele specific PCR (Figure 1B) [2]. Looking at them in the light of the proposed WHO (2015) and revised WHO (2016) criteria [4], the features were more suggestive of PV than essential thromocythemia (ET) and primary myeolofibrosis (PMF). The sustained high values in Hb (>18.5%), Hct (>56%) and the persistently normal platelet count at presentation or during the 4 years follow up in these patients, were against considering PMF and ET. Grade II reticulin status was seen in 22.2% (4/18) in this group of PV patients [5, 14, 22]. There were no *JAK-2* exon 12 mutations in our patients [13, 23, 24, 25]. *JAK-2* Negative (exon 14 and 12) PV was diagnosed in one patient, on the basis of low EPO levels, EEC positivity, splenomegaly, and bone marrow biopsy [2].

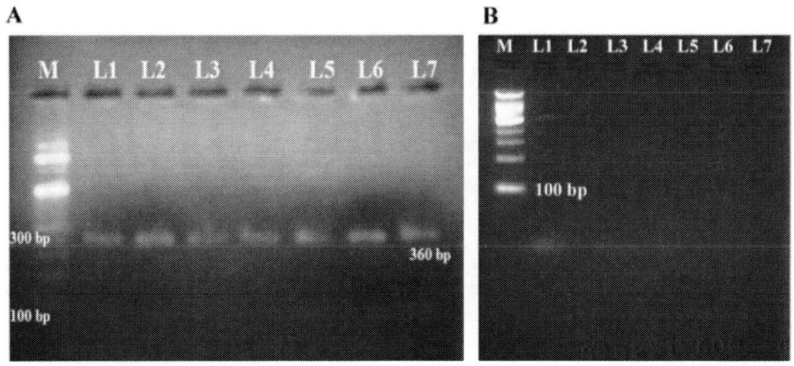

Figure 1. Electrophoretogram showing PCR products exon 14 of *JAK-2* gene. A) PCR amplification of PV patients, Lane M: 100bp ladder obtained from Merck Biosciences, Bangalore, India, Lane L1-L6: PCR product of (360 bp) indicates Positive to exon 14 (V617F) of *JAK-2* gene. B) PCR amplification of healthy individuals, Lane M: 100bp ladder obtained from Merck Biosciences, Bangalore, India, Lane L1-L6: showing no PCR product amplification indicates Negative to exon 14 (V617F) of *JAK-2* gene.

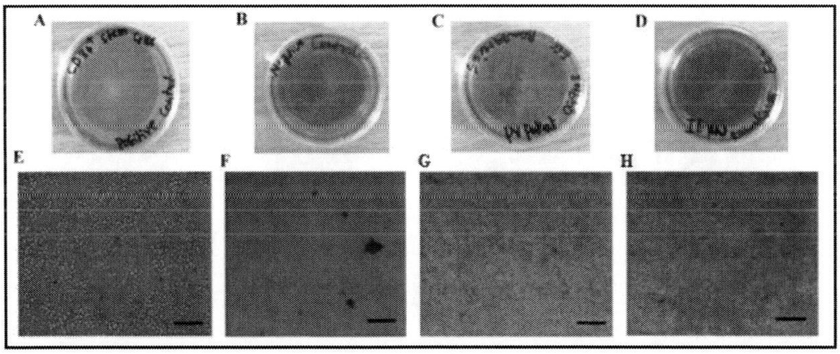

Figure 2. EEC formation assay. A-D. 35 mm Petri dishes having Methyl cellulose medium with A) Positive control, B) Negative control, C) PV patients and D) IE patients grown EEC colonies E. Positive control $CD34^{+}$ stem cells with EPO showing 8386 ± 200 EEC colonies [19] F. Negative control without EPO showing no EEC colonies G. PV patients showing without EPO 10900 ± 489 EEC colonies and H. IE patients showing without EPO 4879 ± 356 EEC colonies.

EPO levels were lower than normal in 12/18 (66%) of PV [7, 20]. EEC was done in 12/18 (66%) of PV patients. Ten (10) out of these 12 PV patients (83.3%) were found to be having subnormal EPO levels and EEC positivity (Figure 2) [7, 26] while, two patients with *JAK-2* V617F mutation and normal EPO levels did not form EEC. Splenomegaly was seen in 63.3% PV patients [1, 2, 7, 13]. The relatively higher (than in IE group) Hct,

TWBC, absolute neutrophil count (ANC) with or without splenomegaly were indicative of panmy-eloproliferative potential in the PV patients [3, 7].

Table 1. Characteristics of Idiopathic Erythrocytosis (IE) (n = 36) patients

	Total cases (n = 36)	Percentage	Range	Average	Median
Age (years)			25-60	38.5	38
Sex-M/F	36 M	100%			
EEC test done	22 (out of 36)	14 (63.6% Pos), 8 (36.3% Neg)			
Serum EPO (<4.3 IU/L)	14 (out of 36)	(38.8%)			
EEC done	8 (out of 14)	6 (75% Pos), 2 (25% Neg)			
Serum EPO (>4.3 IU/L)	22 (out of 36)	(61.1%)			
EEC done	14 (out of 22)	8 (57.1% Pos), 6 (42.9% Neg)			
WBC, mean (range) x 10^9/L			6300-13200	8972.22	8418
Absolute neutrophil count (ANC)			3402-10032	5559.94	5359
Hb (g/dl), mean (range)			17.8-23	19.86	19
Packed cell Volume (PCV) %			58-65.2	61.67	60.5
Mean Corpuscular Volume (MCV) fL			83-108	95.55	94
Platelets, mean (range) x 10^{11}/L			1.53-3.5	2.49	2.1
Splenomegaly	6 (out of 36)	(16.6% Pos), (83.3% Neg)			
JAK 2 exon 14	0 (out of 36)	0%			
JAK 2 exon 12	0 (out of 36)	0%			
Hypertension	24 (out of 36)	(66.6% Pos), (33.3% Neg)			
Dislipidemia	18 (out of 36)	(50% Pos), (50% Neg)			
Alcohol	28 (out of 36)	(77.7% Pos), (22.2% Neg)			
Smoking	22 (out of 36)	(61.6% Pos), (38.8% Neg)			
Overweight	6 (out of 36)	(16.6% Pos), (83.3% Neg)			
Fatty liver	8 (out of 36)	(22.2% Pos), (77.7% Neg)			
Gr 0 reticulin on BM biopsy	36 (out of 36)	-			
Ferritin, mean (range) ng/dl			5-480	156	285
Transferrin Saturation, mean (range) %			20-65	33.8	31

Table 2. Characteristics of Polycythemia vera (n = 18) patients

	Total cases (n = 18)	Percentage	Range	Average	Median
Age (years)			34-71	53	53
Sex-M/F	14 M, 4 F	77.7% M, 22.2% F			
EEC test done	12 (out of 18)	5 (83.3% Pos), 1 (16.6% Neg)			
Serum EPO (<4.3 IU/L)	12 (out of 18)	(66.6%)			
EEC vs EPO (≤4.3 IU/L)	10 (out of 12) EEC done	10/10 (100% Pos)			
Serum EPO (>4.3 IU/L)	6 (out of 18)	(33.3%)			
EEC vs EPO (≥4.3 IU/L)	2 (out of 6) EEC done	2/2 (100% Neg)			
WBC, mean (range) x 10^9/L			6800-20400	12355.55	11170
Hb (g/dl)			19-21.5	20.4	20.7
Packed cell Volume (PCV) %			61-69.5	62.94	62
Mean Corpuscular Volume (MCV) fL			79-106	96	87
Absolute neutrophil Count (ANC)			4280-17740	9617.88	9001
Platelets x 10^{11}/L			2.5-4.2	3.51	3.5
Splenomegaly	14 (out of 18)	(63.3% Pos), (18.18% Neg)			
JAK-2 exon 14	17 (out of 18)	(94.4% Pos), (5.6% Neg)			
JAK-2 exon 12	0 (out of 18)	-			
Hypertension	16 (out of 18)	(88.8% Pos), (11.1% Neg)			
Dislipidemia	12 (out of 18)	(66.6% Pos), (33.3% Neg)			
Alcohol	8 (out of 18)	(44.4% Pos), (55.5% Neg)			
Smoking	8 (out of 18)	(44.4% Pos), (55.5% Neg)			
Overweight	4 (out of 18)	(22.2% Pos), (77.7% Neg)			
Fatty liver	4 (out of 18)	(22.2% Pos), (77.7% Neg)			
Gr II reticulin on BM biopsy	4 (out of 18)	(22.2% Pos)			
Ferritin, mean (range) ng/dl			15-941	139.55	
Transferrin Saturation, mean (range) %			12-75	29.26	

The limitation of the study has been that the known congenital varieties of erythrocytosis, though very rare, were not ruled out with definiteness by mutation analysis. Congenital primary erythrocytosis e.g., *EPOR*, *LNK* mutations are less likely due to the rarity of these entities and the absence of the positive family history in our patients. In the earlier studies, *EPOR* gene mutation was seen 1 out of 181 cases [25] and 1 out of 99 cases [7] of IE respectively.

Congenital secondary erythrocytosis e.g., *VHL, HIF1α, PHD2* mutations, should be having higher than normal or upper normal levels of EPO in addition to the other features like positive family history, consanguinity and younger age of onset [7], features, which were not seen in our study group. Since the P_{50} values and the EPO values of IE patients in our study were normal in range, conditions with increased oxygen affinity to Hb (e.g., high affinity Hb, 2-3 DPG deficiencies) were not considered as the causes for the erythrocytosis. In a study, 14 patients out of 181cases from IE registry (7.7%) had *EPOR, VHL* and *PHD2* mutations [25]. Further, in a follow up study for 20 years, twenty one out of 99 (21.2%) patients of erythrocytosis were diagnosed to have CE [7] and all of them had strong family history.

Therefore, IE was the highly probable diagnosis after excluding PV. All the 36 patients of IE were men (100%) in our study, and negative for *JAK-2* mutations [15]. Most of them were hypertensives and dyslipidemics and used to alcohol and smoking as observed in the PV group. Six out of thirty six (6/36; 16.6%) patients [7, 15] showed splenomegaly without abnormal values in TWBC and platelets at presentation and their bone marrow biopsies ruled out myeloproliferative neoplasia (MPN). All the 36 patients were followed up for changes in blood counts and splenic size to exclude any evolving MPN. Considering the features a) absence of *JAK-2* mutation (exon 14, 12), b) high (Hct >60%) and persistent (requiring repeated phlebotomies) absolute erythrocytosis, c) normal splenic size and normal bone marrow, d) normal TWBC and platelet counts and e) subnormal or low normal EPO levels at presentation and during follow up, the diagnosis of IE was more probable and PV was unlikely.

In the IE group of present study, the TWBC, absolute neutrophil count (ANC), platelet counts were relatively lower than what were observed in the PV group and the differences were statistically significant (Table 3). 38.8% patients of IE group [7, 27, 28], 66.6% of PV group showed subnormal serum EPO levels and 61.1% IE group, 33.3% PV group [14] showed normal serum EPO levels, the differences were again statistically significant (Table 3).

Interestingly, high EEC positivity was noticed among IE patients of our study group. EEC was done (Figure 2) on 22/36 (61.1%) of IE patients and EEC positivity was seen in 14/22 (63.6%) of which 6 (42.8%) patients had low and 8 patients had normal levels of EPO respectively. In our study, 8/14 (57.1%) EEC positive IE patients had one or two minor features of polycythemia vera of WHO 2008 criteria and 6 patients had none (*data not shown*). In a study [26], 8/13 (61.5%) patients of unclassified erythrocytosis and 43/43 (100%) of PV patients (as per polycythemia vera study group PVSG criteria) exhibited spontaneous EEC formation. Another study showed the incidence of EEC in IE as 40% (8/28) [29]. In another recent study none of the IE patients (18/78 (23%)) showed EEC positivity [7].

Table 3. Clinical and laboratory diagnosis data of idiopathic erythrocytosis (IE), compared to polycythemia vera (PV)

	IE	PV	P value IE vs PV
No of patients	36	18	-
Male (%)	36 (100)	14 (77.7)	0.01
Mean age (y)	38	53	0.00735
Mean WBC	8972	12355	0.0001
Mean ANC	5559	9617	0.0001
Mean Platelets (x10^9/L)	2.1	3.5	0.0057
Median serum EPO (<4.3 UI/L) (range)	14 (38.8)	12 (66)	0.017
Median serum EPO (>4.3 UI/L) (range)	22 (61.1)	6 (33.3)	0.014
EEC yes/no (%)	14/22 (63.6)	10/12 (86.6)	0.45
Splenomegaly patients (%)	6 (16.6)	14 (63.3)	0.27

The number of EEC colonies from the $CD34^+$ stem cells incubated with the 1U/ml of EPO served as positive control and the mean value observed in our study was 8386 ± 200 colonies whereas, no colonies were observed in negative controls (without EPO).

The PV and IE patients without EPO in the culture exhibited 10900 ± 489 colonies and 4879 ± 356 colonies respectively and the differences in the growth potential was statistical significant ($p < 0.001$) (Figure 2 and Table 4). Thus, in the pathological states like PV and IE, variations in the EEC formation without the addition of EPO clearly validates the differential growth potential of the haematopoietic stem cell, and therefore, the EEC assay can be a good diagnostic tool for the analysis of absolute erythrocytosis. This can help differentiate the malignant PV from the possibly benign IE (Table 4).

Table 4. Endogenous erythroid colony (EEC) formation assay

	No. of EEC colonies (per 35 mm dish) mean value
Positive Control (with EPO)	8386 ± 200
Negative Control	0
PV (without EPO)	10900 ± 489
IE (without EPO)	4879 ± 356

SD ± three determinations.

The results of the present study strongly suggest that allogeneic stem cell transplantation could be the possible therapeutic approach in the treatment and management of such patients [30].

ACKNOWLEDGMENTS

We sincerely acknowledge Sri Venkateswara Institute of Medical Sciences and University, Tirupati, Andhra Pradesh, India for providing facilities to carry out this work. We thank Ms Vijayalakshmi for her secretarial assistance in this study.

CONFLICT OF INTEREST

All authors declare that there is no conflict of interest.

REFERENCES

[1] McMullin MF. Diagnosis and management of congenital and idiopathic erythrocytosis. *Ther Adv Hematol*, 2012, 3(6), 391–398.

[2] McMullin MF, Bareford D, Campbell P, Green AR, Harrison C, Hunt B, Oscier D, Polkey MI, Reilly JT, Rosenthal E, Ryan K, Pearson TC, Wilkins B; General Haematology Task Force of the British Committee for Standards in Haematology. Guidelines for the diagnosis, investigation and management of polycythaemia/ erythrocytosis. *Br J Haematol*, 2005, 130(2), 174-195.

[3] McMullin MF. Idiopathic erythrocytosis: a disappearing entity. *Hematology Am Soc Hematol Educ Program*, 2009, 629-635.

[4] Arber DA, Orazi A, Hasserjian R, Thiele J, Borowitz MJ, Le Beau MM, Bloomfield CD, Cazzola M, Vardiman JW. The 2016 revision to the World Health Organization classification of myeloid neoplasms and acute leukemia. *Blood*, 2016, 127(20), 2391-2405.

[5] Barbui T, Thiele J, Vannucchi AM, Tefferi A. Rationale for revision and proposed changes of the WHO diagnostic criteria for polycythemia vera, essential thrombocythemia and primary myelofibrosis. *Blood Cancer J*, 2015, 5, e337.

[6] Pearson TC. Apparent Polycythaemia, *Blood Reviews*, 1991, 5, 205-213.

[7] Randi ML, Bertozzi I, Cosi E, Santarossa C, Peroni E, Fabris F. Idiopathic erythrocytosis: a study of a large cohort with a long follow-up. *Ann Hematol*, 2016, 95(2), 233-237.

[8] Kralovics R, Prchal JT. Genetic Heterogenity of Primary familial and congenital polycythemia. *Am J Hematol.* 2001, 68(2), 115-121.

[9] Percy MJ, Jones FGC, Lappin TRJ, McMullin MF. Diverse molecular defects associated with idiopathic erythrocytosis reflect the heterogeneity of this disorder. *Haematologica*, 2006, 91(Suppl 1), 209-210.

[10] Cario H, Rives Sola S, Neusuess A, Kohne E, Schwarz K. Molecular analyses in familial and sporadic congenital primary erythrocytosis. *Haematologica*, 2006, 91(Suppl 1), 352-353.

[11] Jedlickova K, Stockton DW, Prechal JT. Possible primary familial and congenital polycythemia locus at 7q22.1 – 7q22.2. *Blood Cells Mol Dis*, 2003, 31, 327-331.

[12] Rives S, Pahl HL, Florensa L. Molecular genetic analyses in familial and sporadic congenital primary erythrocytosis. *Haematologica*, 2007, 92(5), 674-677.

[13] Vannucchi AM. How I treat polycythemia vera. *Blood*, 2014, 124(22), 3212-3220.

[14] Srikanth N, Besa EC. Polycythemia vera. *Medscape updated*, 2016.

[15] Ruggeri M, Tosetto A, Frezzato M, Rodeghiero F. The rate of progression to polycythemia vera or essential thrombocythemia in patients with erythrocytosis or thrombocytosis. *Ann Intern Med*, 2003, 139(6), 470-475.

[16] Jaradat SA, Khasawneh R, Kamal N, Matalka I, Al-Bishtawi M, Al-Sweedan S, Ayesh MH. Analysis of JAK2V617F mutation in Jordanian patients with myeloproliferative neoplasms. *Hematol Oncol Stem Cell Ther*, 2015, 8(4), 160-166.

[17] Sazawal S, Bajaj J, Chikkara S, Jain S, Bhargava R, Mahapatra M, Saxena R. Prevalence of JAK2 V617F mutation in Indian patients with chronic myeloproliferative disorders. *Indian J Med Res*, 2010, 132, 423-427.

[18] Masters GS, Baines P, Jacobs A. Erythroid colony growth from peripheral blood and bone marrow in polycythaemia. *J Clin Pathol*, 1990, 43(11), 937-941.

[19] Kumar PS, Chandrasekhar C, Srikanth L, Sunitha MM, Sarma PV. In vitro differentiation and characterization of Human hematopoietic $CD34^+$ stem cells into Erythrocytes. *J Stem Cells*, 2017, 12(2), (In press).

[20] Mossuz P, Girodon F, Donnard M, Latger Cannard V, Dobo I, Boiret N, Lecron JC, Binquet C, Barro C, Hermouet S, Praloran V. Diagnostic value of serum erythropoietin level in patients with absolute erythrocytosis. *Haematologica*, 2004, 89(10), 1194-1198.

[21] MeeraV. Clinical Profile of Polycythaemia patients from a tertiary centre in South India. *Int Journal of Clinical Cases and Invest*, 2015, 6(2), 31-39.

[22] Passamonti F. How I treat polycythemia vera. *Blood*, 2012, 120(2), 275-284.

[23] Scott LM. The*JAK-2* exon 12 mutations: a comprehensive review. *Am J Hematol*, 2011, 86(8), 668-676.

[24] Strnad M, Todoric Zivanovic B, Tatomirovic Z, Tarabar O, Elez M, Radic O, Stamatovic D. JAK2V617F mutation and endogenous erythroid colony formation in patients with polycythaemia vera. *J BUON*, 2014, 19(4), 985-991.

[25] Percy MJ, Scott LM, Erber WN. The frequency of JAK2 exon 12 mutations in idiopathic erythrocytosis patients with low serum erythropoietin levels. *Haematologica*, 2007, 92(12), 1607-1614.

[26] Partanen S, Juvonen E, Ikkala E, Ruutu T. Spontaneous erythroid colony formation in the differential diagnosis of erythrocytosis. *Eur J Haematol*, 1989, 42(4), 327-330.

[27] Usman M, Bilwani F, Kakepoto GN, Adil SN, Sajid R, Khurshid M. Polycythemia vera and idiopathic erythrocytosis: comparison of clinical and laboratory parameters. *J Pak Med Assoc*, 2004, 54(5), 249-251.

[28] Goerttler PS, Steimle C, März E, Johansson PL, Andreasson B, Griesshammer M, Gisslinger H, Heimpel H, Pahl HL. The Jak2V617F mutation, PRV-1 overexpression, and EEC formation define a similar cohort of MPD patients. *Blood*, 2005, 106(8), 2862-2864.

[29] Tey SK, Cobcroft R, Grimmett K, Marlton P, Gill D, Mills A. A simplified endogenous erythroid colony assay for the investigation of polycythaemia. *Clin Lab Haematol*. 2004, 26(2), 115-121.

[30] Kerbauy DM, Gooley TA, Sale GE, Flowers ME, Doney KC, Georges GE, Greene JE, Linenberger M, Petersdorf E, Sandmaier BM, Scott BL, Sorror M, Stirewalt DL, Stewart FM, Witherspoon RP, Storb R, Appelbaum FR, Deeg HJ. Hematopoietic cell transplantation as curative therapy for idiopathic myelofibrosis, advanced polycythemia vera, and essential thrombocythemia. *Biol Blood Marrow Transplant*. 2007, 13(3), 355-365.

In: Stem Cells Origins of Cytokine Interactions
Editor: Prasad S. Koka
ISBN: 978-1-53615-302-6

Chapter 11

MULTIPLE SCLEROSIS: IT IS OUR TURN TO ATTACK

Mohamed S. A. Mohamed*, MD
Cologne, Germany

ABSTRACT

Multiple sclerosis is an autoimmune disease characterized by inflammatory demyelination of the nerves in the central nervous system (brain and spinal cord). Depending on the attacked sites, the disease manifestations vary from limited sensory and or motoric deficits to wide spread neurological and cognitive deficits. Similar to other autoimmune neurological and non-neurological diseases, the available therapeutic strategies rely on the pharmacological modulation of the immunity, which could be associated with significant side effects.

In this paper, the light will be shed on another innovative therapeutic strategy that relies on the biological modulation of the immune response, together with the inhibition of the recruitment of the immune cells into the central nervous system after distortion of the blood brain barrier, as well as the enhancement of the regenerative ability towards the restoration of efficient blood brain barrier and the regenerative recovery of the neurological deficits.

Keywords: multiple sclerosis, autoimmunity, regenerative therapy, mesenchymal stem cells and fibrinogen

INTRODUCTION

Multiple sclerosis (MS) is a neurological disease characterized by disseminated encephalomyelitis. The dissemination in MS is related to the time and the affected site.

* Corresponding Author Email: MBBCh, MSc. Email: mohammed.shehatta1@gmail.com.

Females are usually more affected than males, where the usual age of onset (up to 80%) lies between 20-40 years. The clinical presentation can vary dramatically, however, sensory, motoric, visual, urinary bladder, sexual, and cognitive complaints are common in most of the cases. In addition, the natural history of the disease may take one of the following forms [1] (Figure 1):

- Clinically isolated syndrome (CIS)
- Relapsing-remitting MS (RRMS)
- Secondary progressive MS (SPMS)
- Primary progressive MS (PPMS)

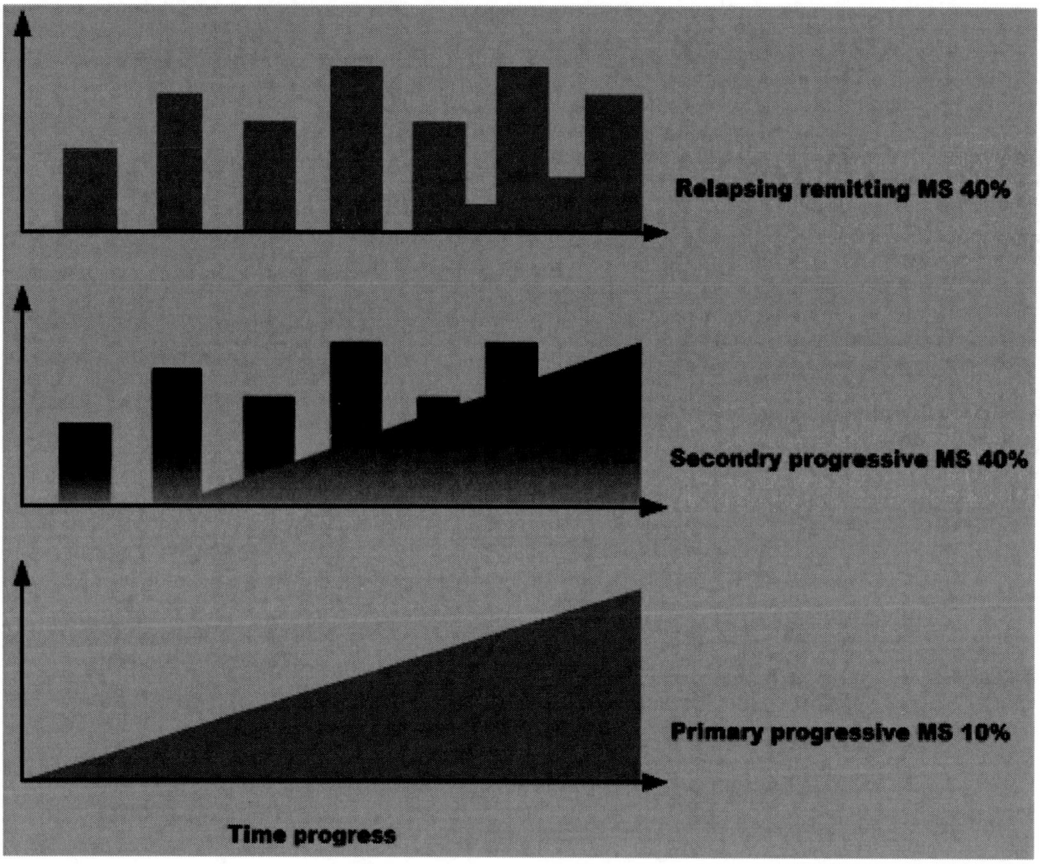

Figure 1. Diagrammatic representation of the common forms of multiple sclerosis.

The manifestations of MS can vary depending on the affected site of the central nervous system. Sensory deficits may appear in the form of positive Lhermitte sign, parathesia and or neuralgia. Motoric deficits such as weakness, wasting, or spastic paralysis could be manifested. Optic neuritis, double vision and other visual disturbances are also common manifestations. Ataxia, nystagmus, dysartheria and or intention tremors represents the cerebellar involvement. Urinary bladder and intestinal disturbances are frequently seen in MS, in the form of retention, urgency, incontinence and or diarrhea versus constipation. Swallowing and or respiratory complaints can indicate brain stem involvement. Erectile dysfunctions and or epilepsy could also occur. Psychological manifestations, such as depression, fatigue, memory affection, concentration problems and or emotional affection are frequently seen consequences of MS [2].

Table 1. Currently approved Anti-MS medications

Drug	Mechanism of action	Unfavorable effects
First line medications		
Interferon beta	Shifts the cytokine profile towards the anti-inflammatory side	The development of neutralizing antibodies. No efficiency against progressive forms of the disease. Elevated liver enzymes and depressed bone marrow function
Glatiramer acetate	Probably related to enhancing the activity of Th2 cells, which gives the anti-inflammatory cytokines the upper hand	Skin reaction at the site of injection. Systemic reaction. Lymphadenopathy, dyspnoea and lipoatrophy
Teriflunomide	May inhibit the proliferation of lymphocytes	Serious infections, elevated alanin transaminase, hypertension and decreased leukocyte count, up to pancytopenia
Dimethyl fumarate	Modulates the immune response towards the anti-inflammatory properties	Abdominal pain, diarrhea, leukopenia, lymphocytopenia and elevated liver transaminases
Second line medications		
Fingolimod	Reduces the mobilization of lymphocytes from the lymph nodes	Serious infections, transient bradycardia, atrioventricular block, elevated liver enzymes and macular odema
Natalizumab	Blockes the migration of the inflammatory cells into the CNS	Increased risk of progressive multifocal leucoencephalopathy. Prolonged severe lymphopaenia. Deveopment of neutrilizing antibodies
Alemtuzumab	Mediates the cytolysis and the complement-mediated lysis of T and B lymphocytes	May induce secondary autoimmunity
Mitoxantrone	Functions as anti-proliferative and apoptosis-inducing in: T-lymphocytes, B-lymphocytes, macrophages & other antigen-presenting cells	Urinary tract infections, elevated liver enzymes, leucopenia, amenorrhoea, acute promyelocytic leukaemia and cardiotoxicity with overdoses

The pathogenesis of MS involves immunological reaction, where the immune cells (lymphocytes) attack the blood brain barriers and the central nervous system, leading to demyelinating lesions that could be detected in magnetic resonance imaging (MRI). Antigen-presenting cells, macrophages, T lymphocytes, and B lymphocytes are involved in the pathogenesis of MS. Thus, almost all the therapeutic strategies of the disease rely on the control of the functions of those cells (Table 1). However, the treatment of MS might be more effective and less side effects-associated, if the homing of the immune cells to the nervous system is blocked [3].

First-Line Medications against MS

Interferon beta (IFN-β), which is predominantly produced by fibroblasts and shifts the cytokine profile towards the anti-inflammatory side. IFN-β also inhibits the proliferation and migration of lymphocytes into the CNS. For MS therapy, recombinant IFN-β-1b in a dose of 250 μg subcutaneously every other day or IFN-β-1a in a dose of 30 μg intramuscularly once weekly, or subcutaneously at doses of 22 or 44 μg three times a week, could be used. The application of IFN-β therapy has proved efficiency in decreasing the relapse rate by 30-34%. However, the use of IFN-β is associated with the development of neutralizing antibodies and has no efficiency against progressive forms of the disease. Up to 50-75% of the patients treated with IFN-β develope flu-like symptomes. In addition, IFN-β therapy has been associated elevated liver enzymes and depressed bone marrow function, which may increase the risk of cancer and serious infections [4-8].

Glatiramer acetate is a synthetic polypeptide, which resembles the myelin basic protein and is administered through subcutaneous injections at 20 mg per day. The mechanism of action of glatiramer acetate is not fully understood, but it is probably related to the enhancement of the activity of Th2 cells, which gives the anti-inflammatory cytokines the upper hand. Glatiramer acetate therapy has showed 29% reduction of the annual relapsing rate, however, no evidence of efficiency in the secondary progressive or primary progressive forms of the disease. The most side effects reported with the use of glatiramer acetate are skin reaction at the site of injection in 65% of cases, systemic reaction in 15% of cases, lymphadenopathy, dyspnoea and lipoatrophy, which is irreversible [9-13].

Teriflunomide is an immunomodulatory agent that is selectively and reversibly able to inhibit the mitochondrial enzyme dihydroorotate dehydrogenase, which is required for *de novo* pyrimidine synthesis. In this way, it may inhibit the proliferation of lymphocytes. It is used as tablets, in a dose of 14 mg once daily. Teriflunomide therapy in MS has the ability to reduce the relapse rate by 31-36% and the rate of disability progression by 26-27%. While no data is currently available regarding the efficiency of teriflunomide with progressive MS forms, its major side effects include serious infections, elevated alanin transaminase, hypertension and decreased leukocytic count, up to pancytopenia [14-16].

Dimethyl fumarate modulates the immune response towards the anti-inflammatory properties, mainly through the activation of the nuclear factor (erythroid-derived 2)-like 2 (Nrf2) transcriptional pathway. The drug is administered in a dose of 240 mg twice daily. The use of dimethyl fumarate can reduce the remission rate by 44-53%, the rate of disability progression by 22-32% and the MRI gadolinium-enhancing lesions by about 75-94%. The major side effects associated with the use of dimethyl fumarate are abdominal pain, diarrhea, leukopenia, lymphocytopenia and elevated liver transaminases [17-21].

Second-Line Medications

Fingolimod 5 mg/day is a modulator of sphingosine 1-phosphate receptor (S1PR), which has the ability to reduce the mobilization of the lymphocytes from the lymph nodes. Though fingolimod can reduce the relapsing rate by about 55%, it is not reported to be effective against the primary or the secondary progressive forms of the disease. It is main adverse effects are serious infections, transient bradycardia, atrioventricular block, elevated liver enzymes and macular odema [22-24].

Natalizumab is a blocking monoclonal antibody against α4-integrin, thus, blocking the migration of the inflammatory cells into the CNS. The dose of natalizumab is 300mg I.V. Infusion every 4 weeks. Natalizumab therapy has been associated with 68% reduction of the annualized relapse rate and 54% reduction of the disability progression rate. However, its major side effects are the increased risk of progressive multifocal leucoencephalopathy, especially in anti-John Cunningham virus antibody positive patients, and prolonged severe lymphopaenia. In addition, natalizumab is immunogenic in some patients, which leads to the development of neutralizing antibodies, decreasing the efficacy of the therapy [25-28].

Alemtuzumab, an anti-CD52 recombinant monoclonal antibody that mediates the cytolysis and the complement-mediated lysis of T and B lymphocytes. Thus, it reduces the relapse potential and disease progression in MS. Alemtuzumab is used as an intravenous infusion of 12 mg/day for five consecutive days followed by 12 months pause, before the infusion of another 12 mg/day for 3 successive days. Additional doses could be added, however, the drug has the potential to induce secondary autoimmunity, which limits its usability as a first line therapy [29-34].

Mitoxantrone is a synthetic anthracenedione derivative that functions as anti-proliferative and apoptosis-inducing in T lymphocytes, B lymphocytes, macrophages and other antigen-presenting cells. The treatment with mitoxantrone showed significant reduction in the relapsing rate (up to 60-70%). However, major side effects associated with its use include urinary tract infections, elevated liver enzymes, leucopenia, amenorrhoea and acute promyelocytic leukaemia. In addition, its maximum cumulative dose should not exceed 120–140 mg/m^2 of body surface, in order to avoid cardiotoxicity [35-39].

Innovative Thinking

The blood brain barrier (BBB) is a mechanical and functional barrier between the systemic circulation and the central nervous system that is composed of three cellular elements; the endothelial cells connected with their tight junctions, astrocyte end-feet, and pericytes. A major step of the pathogenesis of MS and any other autoimmunity-based CNS inflammatory disease, is the recruitment of the inflammatory cells across the injured or activated BBB [40].

Fibrinogen (FB) is a gylcoprotein, which is a hexamer, containing two sets of three different chains (α, β, and γ), linked to each other by disulfide bonds. FB plays an important role in the coagulation cascade, where it can form bridges between platelets by its binding to their GpIIb/IIIa surface membrane proteins, in addition to the major role, where prothrombin is converted into thrombin, which then converts the soluble FB (sFB) into insoluble fibrin strands that are then cross-linked by factor XIII to form the blood clot [41].

Many studies have confirmed FB as a pro-inflammatory effector. In addition to its ability to stimulate the proliferation of B-lymphocytes, T-lymphocytes and monocytes [42], immobilized FB and fibrin have high affinity to macrophage antigen 1 (MAC-1) and can activate neutrophils and monocytes [43-45]. In neutrophils, FB/MAC-1 interaction activates the NF-κB pathway, which is an anti-apoptotic and inflammatory cytokine-inducing pathway [43].

Role of Fibrinogen in MS

Deposition of fibrinogen (FB) in the CNS, after the disruption of the blood brain barriers, has been reported to induce immune reactions and increased recruitment of macrophages and lymphocytes into the CNS, which leads to demyelination. Fibrinogen has the ability to stimulate and recruit the $CD11b^+$ antigen-presenting cells, which enhances the recruitment and activation of myelin antigen-specific Th1 cells. In addition, the interaction between FB and αM βII- integrin receptor results in the activation of Talin-1, which interacts with Rap-1 and is essential for adhesion, migration and phagocytosis activity of antigen-presenting cells and lymphocytes [46].

Although there is a strong evidence that fibrinogen plays an important pro-inflammatory role in the development of MS, this might be limited to the deposited FB (insoluble FB), which is out of doubt incorporated in the recruitment of lymphocytes and macrophages. Thus, the sequence of events in this regards might be simplified as the disruption of the BBB → leakage of FB into CSF → deposition of FB perivascular → recruitment and chemotaxis of macrophages, T- and B- lymphocytes, which produce inflammatory cytokines and attack the myelin sheathes. This hypothesis might be supported by the finding that the fibrinogen leakage into CSF was not reported during the remission, but coincides with the fulminant MS [47].

On the contrary to the pro-inflammatory effects of immobilized FB/fibrin, sFB has the ability to inhibit lymphocytic antigen 1- dependent binding to the intracellular adhesion molecule 1 (ICAM-1) through a direct interaction with ICAM-1 [48], and to reduce IL8- activated neutrophils binding to ICAM-1- expressing cells, in addition to reducing the binding of neutrophils to TNFα- activated endothelium to 40%, under flow conditions [46].

Accordingly, blocking the pro-inflammatory functions of FB would ideally protect against its role in the recruitment of the inflammatory cells into CNS, which coincidence with disease progression and or exacerbation, without affecting its other vital functions, such as its role in the coagulation homeostasis. A monoclonal antibody, capable of binding the YC domain of FB, is able to inhibit the microglial adhesion and the Mac-1 binding to FB, which would significantly modify the symptoms of MS [49].

ROLE OF MESENCHYMAL STEM CELLS (MSCS)

MSCs are multipotent stromal cells that could be derived from adult bone marrow, as well as other tissues. These cells have the ability to differentiate into many cell lines, including neurons, which gives a great hope that MSCs could be a magic regenerative therapy to restore and correct any neurological deficit that results from nervous system injury or degenerative diseases *(see online supplementary video)*. MSCs can play an important role in neurogenesis, gliogenesis, remyelination and neural plasticity. The role of MSCs in MS and other autoimmune or inflammatory - based neurological diseases could be very significant and double faced. The disease manifestations and progression in MS are based on the neurological deficits that manifest with (and stay after) the inflammatory neurological injury. MSCs have the unique ability to restore the deficits through the regenerative differentiation ability, in addition to the control and or prevention of the autoimmune and inflammatory condition [50, 51].

MSCs are very tolerated by the immune system, where they can attenuate the immunity through modulating T cell activation and proliferation, either by a direct cell-cell interaction or via soluble factors. In addition, this action is independent on the major histocompatibility (MHC) matching because MSCs are characterized by low expression level of MHC I and II, and lack the T cell co-stimulating molecules CD80 and CD86. Moreover, the transplantation of allograft MSCs is used for the attenuation of the graft versus host reaction [52].

Recent studies have confirmed the principle that neurogenesis continues to take place during the adulthood as well. Neural stem cells (NSCs) takes responsibility of this function, however, they lack the ability of self-renewal and multipotency. Although the neuronal stem cells develope originally from the ectodermal niche, while MSCs belong to the mesoderm, both cells secrete many relevant growth factors, such as the nerve growth factor (NGF), the glial derived neurotrophic factor (GDNF), and the brain derived neurotrophic factor (BDNF). In addition, MSCs have shown great potential to differentiate into functional neurons. Moreover, MSCs secrete many immuno-modulatory factors that include interleukin 10 (IL10), transforming growth factor β (TGFβ), prostaglandin E2 (PGE2) and vascular endothelial growth factor (VEGF) [51, 52].

IL10 is one of the major anti-inflammatory cytokine that down regulates the expression of cytokines in the T helper-1 cells, as well as the major histocompatibility class-II antigens and costimulatory molecules on the surface of macrophages. Moreover, IL-10 antgonizes the activity of NF-κB and is capable of suppressing the production of IL8 in a dose dependent manner [53, 54]. While the adhesiveness of the BBB- forming cells increases significantly upon IL1β, TNFα, IFNγ, and Lipopolysaccharide stimulation, which increases the leukocytic (macrophages and lymphocytes) migration into the CNS, TGFβ, secreted by MSCs, has the ability to block those effects in a dose dependent manner [55]. PGE2 functions towards the inhibition of the proliferation, differentiation and function of the antigen presenting, as well as the cytotoxic cells, including the dentritic cells, macrophages and natural killer cells [56]. Meanwhile VEGF has neuroprotective effects, where it enhances the survival of Schwann cells and cerebellar granule neurons, and protects against the degeneration of the motor neurons [57]. Thus, the use of MSCs in the cases of MS, and other autoimmune based CNS diseases, could work towards the attenuation of the aggressive immune reactions, together with the potential regenerative recovery of the deficits.

Connick et al. from the University of Cambridge have conducted a clinical trial to assess the safety and efficiency of the intravenous administration of MSCs for the treatment of the secondary progressive MS. The use of 1.6 X 10(6) cells per kg body weight intravenous was able to improve the visual manifestations of the disease without any reported serious side effects [58].

The Therapeutic Recommendation

The intravenous and intrathecal adminstration of the monoclonal FB-YC domain antibodies (which blocks the binding of FB to Mac-1 and the microglial cells, blocking the role of FB in the enhancement of CNS lymphocytic migration), together with MSCs (priorly cultured to 90% confluence), could have the potential not only to attenuate the disease progression, but also to recover the deficits.

The major concern of this therapeutic strategy might be the risk of the incorporation of MSCs in cancer development, which can't be completely excluded without the application of long-term clinical studies. However, the behaviour of the bone marrow-derived MSCs (BMMSCs) in culture, in comparison to other multipotent cells, could provide some degree of tranquility.

In a previous report, the proliferative potential of human umbilical cord perivascular MSCs (HUCPVMSCs) was compared to that of the BMMSCs. The results showed similar proliferation rates at the begining, however, between days 7 and 14, the HUCPVMSCs had higher proliferation rate, and beyond 20 days, the BMMSCs responded to contact inhibition and stoped proliferation, while HUCPVMSCs continued to proliferate resulting in multilayering [59]. Similarly, HUCPVMSCs showed higher differentiation capacity.

MANAGEMENT OF THE ACUTE MS EXACERBATIONS

The current trend for the therapy of acute MS exacerbations relies on the use of intravenous cortisone infusion in a dose of 500-1000 mg/day for 3-5 days, followed by a 21 day tapering oral steroid therapy. Though this therapy is usually able to shorten the duration of the acute MS exacerbation, it can't be applied for long term because of the side effects of cortisone therapy. Nevertheless, the short term cortisone therapy during exacerbations is not free of side effects. The steroid therapy seems to decrease the oligodendrocyte-mediated repair of the demyelinated lesions, have a dramatic effects on the bone health, in addition to a group of systemic, metabolic and psychiatric complications [60]. In resistant cases, and based on the role of B-lymphocytes in the pathogenesis of MS, the application of plasma exchange (PE) can improve about 42.1% of the cases. However, serious side effects, such as myocardial infarction, arrhythmia and hemolysis, may occur in a rate of 2.8-6.5% [61, 62].

The use of the combined "MSCs - monoclonal FB-YC domain antibodies" during the acute MS exacerbations might have the potential to provide an effective therapeutic option, based on its rapid onset of action that is mediated by the monoclonal antibodies and the secreted mediators of MSCs. In the above recommended technique, the culture of the MSCs up to 90% confluence prior to adminstration will provide adequate mediators in the culture medium that will be potentiated by the further proliferation and the secretions of the cells *in vivo.* This innovative therapeutic strategy seems to worth further studying on the preclinical and clinical levels, in order to precisely assess its therapeutic value versus complications.

FUNDING

No funding was provided for the development of this work.

CONFLICT OF INTEREST

The intellectual properties included and described in this manuscript belong solely to the author. Reproduction or use of any of the included intellectual properties requires the written permission of the author. The author welcomes funding cooperation for experimental and clinical studies.

REFERENCES

[1] Torkildsen Ø, K-M Myhr, and L Bø. Disease-modifying treatments for multiple sclerosis – a review of approved medications. *Eur J Neurol.* 2016 Jan; 23(Suppl 1): 18–27.

[2] Politte Laura C., Jeff C. Huffman, and Theodore A. Stern. Neuropsychiatric Manifestations of Multiple Sclerosis. Prim Care Companion *J Clin Psychiatry.* 2008; 10(4): 318–324.

[3] Loma Ingrid, and Rock Heyman. Multiple Sclerosis: Pathogenesis and Treatment. *Curr Neuropharmacol.* 2011 Sep; 9(3): 409–416.

[4] Dhib-Jalbut S, Marks S. Interferon-b mechanisms of action in multiple sclerosis. *Neurology* 2010; 74(Suppl. 1): S17-24.

[5] The IFNB Multiple Sclerosis Study Group. Interferon beta-1b is effective in relapsing-remitting multiple sclerosis. I. Clinical results of a multicenter, randomized, double-blind, placebo-controlled trial. *Neurology* 1993; 43: 655–661.

[6] Jacobs LD, Cookfair DL, Rudick RA, et al. Intramuscular interferon beta1a for disease progression in relapsing multiple sclerosis. *Ann Neurol* 1996; 39: 285–294.

[7] Ebers GC, PRISMS (Prevention of Relapses Disability by Interferon b-1a Subcutaneously in Multiple Sclerosis) Study Group. Randomised double-blind placebo controlled study of interferon b-1a in relapsing/remitting multiple sclerosis. *Lancet* 1998; 352: 1498–1504.

[8] PRISMS Study Group and the University of British Columbia MS/MRI Analysis Group. PRISMS4: Longterm efficacy of interferon-b-1a in relapsing MS. *Neurology* 2001; 56: 1628–1636.

[9] Racke MK, LovettRacke AE, Karandikar NJ. The mechanism of action of glatiramer acetate treatment in multiple sclerosis. *Neurology* 2010; 74(Suppl. 1): S2530.

[10] Johnson KP, Brooks BR, Cohen JA, et al. Copolymer 1 reduces relapse rate and improves disability in relapsing remitting multiple sclerosis: results of a phase III multicenter, double-blind, placebo-controlled trial. *Neurology* 1995; 45: 1268–1276.

[11] Comi G, Filippi M, Wolinsky JS. European/Canadian multicenter, double-blind, randomized, placebo-controlled study of the effects of glatiramer acetate on magnetic resonance imaging – measured disease activity and burden in patients with relapsing multiple sclerosis. European/Canadian Glatiramer Acetate Study Group. *Ann Neurol* 2001; 49: 290–297.

[12] Comi G, Martinelli V, Rodegher M, et al. Effect of glatiramer acetate on conversion to clinically definite multiple sclerosis in patients with clinically isolated syndrome (PreCISe study): a randomised, double-blind, placebo-controlled trial. *Lancet* 2009; 374: 150311.

[13] Wolinsky JS, Narayana PA, O'Connor P, et al. Glatiramer acetate in primary progressive multiple sclerosis: results of a multinational, multicenter, double-blind, placebo-controlled trial. *Ann Neurol* 2007; 61: 14–24.

[14] Papadopoulou A, Kappos L, Sprenger T. Teriflunomide for oral therapy in multiple sclerosis. *Expert Rev Clin Pharmacol* 2012 Nov; 5: 617–628. doi:10.1586/ecp.12.56

[15] O'Connor P, Wolinsky JS, Confavreux C, et al. Randomized trial of oral teriflunomide for relapsing multiple sclerosis. *N Engl J Med* 2011; 365: 1293–1303. doi:10.1056/NEJMoa1014656.

[16] Confavreux C, O'Connor P, Comi G, et al. Oral teri- flunomide for patients with relapsing multiple sclerosis (TOWER): a randomised, double-blind, placebo-controlled, phase 3 trial. *Lancet Neurol* 2014; 13: 247–256. doi:10.1016/S1474-4422(13)70308-9.

[17] Linker RA, Gold R. Dimethyl fumarate for treatment of multiple sclerosis: mechanism of action, effectiveness, and side effects. *Curr Neurol Neurosci Rep* 2013; 13: 394.

[18] Gold R, Kappos L, Arnold DL, et al. Placebo-controlled phase 3 study of oral BG-12 for relapsing multiple sclerosis. *N Engl J Med* 2012; 367: 1098–1107.

[19] Fox RJ, Miller DH, Phillips JT, et al. Placebo-controlled phase 3 study of oral BG-12 or glatiramer in multiple sclerosis. *N Engl J Med* 2012; 367: 1087–1097.

[20] Rosenkranz T, Novas M, Terborg C. PML in a patient with lymphocytopenia treated with dimethyl fumarate. *N Engl J Med* 2015; 372: 1476–1478.

[21] Sweetser MT, Dawson KT, Bozic C. Manufacturer's response to case reports of PML. *N Engl J Med* 2013; 368: 1659–1661.

[22] Kappos L, Radue EW, O'Connor P, et al. A placebo controlled trial of oral fingolimod in relapsing multiple sclerosis. *N Engl J Med* 2010; 362: 387401.

[23] Calabresi PA, Radue EW, Goodin D, et al. Safety and efficacy of fingolimod in patients with relapsing remitting multiple sclerosis (FREEDOMS II): a double-blind, randomised, placebo-controlled, phase 3 trial. *Lancet Neurol* 2014; 13: 545–556.

[24] Cohen JA, Barkhof F, Comi G, et al. Oral fingolimod or intramuscular interferon for relapsing multiple sclerosis. *N Engl J Med* 2010; 362: 40215.

[25] Polman CH, O'Connor PW, Havrdova E, et al. A randomized, placebo-controlled trial of natalizumab for relapsing multiple sclerosis. *N Engl J Med* 2006; 354: 899–910.

[26] Rudick RA, Stuart WH, Calabresi PA, et al. Natalizumab plus interferon beta-1a for relapsing multiple sclerosis. *N Engl J Med* 2006; 354: 911–923.

[27] Clifford DB, De Luca A, Simpson DM, Arendt G, Giovannoni G, Nath A. Natalizumab-associated progressive multifocal leukoencephalopathy in patients with multiple sclerosis: lessons from 28 cases. *Lancet Neurol* 2010; 9: 438–446.

[28] Bloomgren G, Richman S, Hotermans C, et al. Risk of natalizumab-associated progressive multifocal leukoencephalopathy. *N Engl J Med* 2012; 366: 1870–1880.

[29] Brown JW, Coles AJ. Alemtuzumab: evidence for its potential in relapsing remitting multiple sclerosis. *Drug Des Devel Ther* 2013; 7: 131–138. doi: 10.2147/DDDT.S32687.

[30] Cohen JA, Coles AJ, Arnold DL, et al. Alemtuzumab versus interferon beta 1a as first-line treatment for patients with relapsing remitting multiple sclerosis: a randomised controlled phase 3 trial. *Lancet* 2012; 380: 1819–1828.

[31] Coles AJ, Twyman CL, Arnold DL, et al. Alemtuzumab for patients with relapsing multiple sclerosis after disease-modifying therapy: a randomised controlled phase 3 trial. *Lancet* 2012; 380: 1829–1839.

[32] Coles AJ, Wing MG, Molyneux P, et al. Monoclonal antibody treatment exposes three mechanisms underlying the clinical course of multiple sclerosis. *Ann Neurol* 1999; 46: 296–304.

[33] Paolillo A, Coles AJ, Molyneux PD, et al. Quantitative MRI in patients with secondary progressive MS treated with monoclonal antibody Campath 1H. *Neurology* 1999; 53: 751–757.

[34] Tuohy O, Costelloe L, Hill-Cawthorne G, et al. Alemtuzumab treatment of multiple sclerosis: long-term safety and efficacy. *J Neurol Neurosurg Psychiatry* 2015; 86: 208–215.

[35] Millefiorini E, Gasperini C, Pozzilli C, et al. Randomized placebo-controlled trial of mitoxantrone in relapsing remitting multiple sclerosis: 24 month clinical and MRI outcome. *J Neurol* 1997; 244: 153–159.

[36] Martinelli V, Radaelli M, Straffi L, et al. Mitoxantrone: benefits and risks in multiple sclerosis patients. *Neurol Sci* 2009; 30(Suppl. 2): S167–170.

[37] Hartung HP, Gonsette R, Konig N, et al. Mitoxantrone in progressive multiple sclerosis: a placebo controlled, double-blind, randomised, multicentre trial. *Lancet* 2002; 360: 2018–2025.

[38] Martinelli V, Cocco E, Capra R, et al. Acute myeloid leukemia in Italian patients with multiple sclerosis treated with mitoxantrone. *Neurology* 2011; 77: 1887–1895.

[39] Ellis R, Brown S, Boggild M. Therapy-related acute leukaemia with mitoxantrone: four years on, what is the risk and can it be limited? *Mult Scler* 2015; 21: 642–645.

[40] Ballabh P, Braun A, Nedergaard M. The blood-brain barrier: an overview: structure, regulation, and clinical implications. *Neurobiol Dis*. 2004 Jun;16(1):1-13.

[41] Hall Cecil E, Henry S. Slayter. The Fibrinogen Molecule: Its Size, Shape, and Mode of Polymerization. *The Journal of Biophysical and Biochemical Cytology*. 1958; 5 (1): 11–6.

[42] Jacques A. Hatzfeld, Antoinette Hatzfeld, and Joelle Maigne. Fibrinogen and its fragment D stimulate proliferation of human hemopoietic cells *in vitro*. *Cell Biology*. 1982; Vol. 79, pp. 6280-6284.

[43] Rubel C, Gomez S, Fernandez GC, Isturiz MA, Caamano J, Palermo MS. Fibrinogen–CD11b/CD18 interaction activates the NF-kappa B pathway and delays apoptosis in human neutrophils. *Eur J Immunol*. 2003; 33: 1429–38.

[44] Diamond MS, Springer TA. A subpopulation of Mac-1 (CD11b/CD18) molecules mediates neutrophil adhesion to ICAM-1 and fibrinogen. *J Cell Biol*. 1993; 120: 545–56.

[45] Ugarova TP, Yakubenko VP. Recognition of fibrinogen by leukocyte integrins. *Ann N Y Acad Sci*. 2001; 936: 368–85.

[46] Pillay J, Kamp VM, Pennings M, Oudijk E-J, Leenen LP, Ulfman LH, Koenderman L. Acute-phase concentrations of soluble fibrinogen inhibit neutrophil adhesion under flow conditions in vitro through interactions with ICAM-1 and MAC-1 (CD11b/CD18). *J Thromb Haemost*. 2013; 11: 1172–82.

[47] Ryu Jae Kyu, Mark A. Petersen, Sara G. Murray, Kim M. Baeten, Anke Meyer-Franke, Justin P. Chan, Eirini Vagena, Catherine Bedard, Michael R. Machado, Pamela E. Rios Coronado, Thomas Prod'homme, Israel F. Charo, Hans Lassmann, Jay L. Degen, Scott S. Zamvil & Katerina Akassoglou. Blood coagulation protein fibrinogen promotes autoimmunity and demyelination via chemokine release and antigen presentation. *Nature Communications* 6, Article number: 8164 (2015) doi: 10.1038/ncomms 9164.

[48] Corti Paola and Mark T. Gladwin. Is Nitrite the Circulating Endocrine Effector of Remote Ischemic Preconditioning? *Circ Res*. 2014; 114:1554-1557.

[49] Akassoglou Katerina. *Monoclonal antibodies*. Publication number WO2011041518 A1. Application number PCT/US2010/050873. Publication date 7 Apr 2011.

[50] Zeng R, Wang LW, Hu ZB, Guo WT, Wei JS, Lin H, Sun X, Chen LX, Yang LJ. Differentiation of human bone marrow mesenchymal stem cells into neuron-like cells *in vitro*. *Spine* (Phila Pa 1976). 2011 Jun; 36(13): 997-1005. doi: 10.1097/BRS.0b013e3181eab764.

[51] Salgado Antonio J., Joao C. Sousa, Bruno M. Costa, Ana O. Pires, António Mateus-Pinheiro, F. G. Teixeira, Luisa Pinto, and Nuno Sousa. Mesenchymal stem cells secretome as a modulator of the neurogenic niche: basic insights and therapeutic opportunities. *Front Cell Neurosci*. 2015; 9: 249.

[52] Eyal Ben-Ami, Sonia Berrih-Aknin and Ariel Miller. Mesenchymal stem cells as an immunomodulatory therapeutic strategy for autoimmune diseases. *Autoimmunity Reviews* 10 (2011) 410–415.

[53] Dobroslav Kyurkchiev, Ivan Bochev, Ekaterina Ivanova-Todorova, Milena Mourdjeva, Tsvetelina Oreshkova, Kalina Belemezova, and Stanimir Kyurkchiev. Secretion of immunoregulatory cytokines by mesenchymal stem cells. *World J Stem Cells*. 2014 Nov 26; 6(5): 552–570.

[54] Méndez-Samperio Patricia, Elizabeth García, Abraham Vázquez, and Janet Palma. Regulation of Interleukin-8 by Interleukin-10 and Transforming Growth Factor β in Human Monocytes Infected with *Mycobacterium bovis*. *Clin Diagn Lab Immunol.* 2002 Jul; 9(4): 802–807.

[55] Fabry Z, DJ Topham, D Fee, J Herlein, J A Carlino, M N Hart and S Sriram. TGF-beta 2 decreases migration of lymphocytes *in vitro* and homing of cells into the central nervous system in vivo. *The Journal of Immunology*. 1995. Vol. 155, Issue 1.

[56] Mallory Agard, Saja Asakrah, and Lisa A. Morici. PGE2 suppression of innate immunity during mucosal bacterial infection. *Front Cell Infect Microbiol*. 2013; 3: 45.

[57] Duffy AM, Bouchier-Hayes DJ, Harmey JH. Vascular Endothelial Growth Factor (VEGF) and Its Role in Non-Endothelial Cells: Autocrine Signalling by VEGF. In: Madame Curie Bioscience Database [Internet]. Austin (TX): *Landes Bioscience*; 2000-2013.

[58] Connick P, Kolappan M, Crawley C, Webber DJ, Patani R, Michell AW, Du MQ, Luan SL, Altmann DR, Thompson AJ, Compston A, Scott MA, Miller DH, Chandran S. Autologous mesenchymal stem cells for the treatment of secondary progressive multiple sclerosis: an open-label phase 2a proof-of-concept study. *Lancet Neurol.* 2012 Feb; 11(2):150-6.

[59] Baksh Dolores, Raphael Yao, Rocky S. Tuan. Comparison of Proliferative and Multilineage Differentiation Potential of Human Mesenchymal Stem Cells Derived from Umbilical Cord and Bone Marrow. *Stem Cells* 2007; 25:1384 –1392.

[60] Ontaneda Daniel and Alex D. Rae-Grant. Management of acute exacerbations in multiple sclerosis. *Ann Indian Acad Neurol.* 2009; 12(4): 264–272.

[61] Henze T, Prange HW, Talartschik J, Rumpf KW. Complications of plasma exchange in patients with neurological diseases. *Klin Wochenschr.* 1990; 68:1183–8.

[62] Couriel D, Weinstein R. Complications of therapeutic plasma exchange: A recent assessment. *J Clin Apher.* 1994; 9:1–5.

In: Stem Cells Origins of Cytokine Interactions
Editor: Prasad S. Koka
ISBN: 978-1-53615-302-6

Chapter 12

IMPACT OF THE FIBROBLAST MORPHOGENETIC FIELD INFORMATION ON HUMAN HEALTHY AND TUMOR CELL LINES *IN VITRO*

Dušan M. Marić[1,*], Zoran Milankov[2], Vesna V. Kojić[3], Dimitar S. Jakimov[3], Dzihan Abazovic[4], Dušica L. Marić[5] and Himanshu Bansal[6]

[1]Faculty of Medicine, University of Novi Sad, Serbia, Institute for Child and Mothers' Health, Novi Sad, Serbia
[2]Dental Clinic Tusk, Novi Sad, Serbia
[3]Oncology Institute of Vojvodina, University of Novi Sad, Serbia
[4]Emergency Medicine Center of Montehegro, Podgorica, Montenegro
[5]Faculty of Medicine, Department of Anatomy, University of Novi Sad, Serbia
[6]Consultant Orthopedic Surgeon, Anupam Hospital Rudrapur, Uttrakhand, India

ABSTRACT

Biological investigation was conducted to study *in vitro* changes in morphological properties and metabolic activity of human continuous cell lines after treatment with the morphogenetic field hologram information (the field). We used human tumor cell lines, including cervical (HeLa) and lung cancer (A549), as well as one normal fetal lung fibroblasts cell line (MRC-5) to investigate the effect of such treatment. Metabolic activity of investigated cell lines was evaluated by MTT assay. Results showed that MRC-5 cells exposed 3 h to the field had 50% increases in metabolic activity 48 h after treatment in comparison to untreated control cells. Similar but lower effect was observed in both tested tumor cell lines. Changes in cells depended also on the exposure time.

* Corresponding Author Email: ducamaric@gmail.com.

Morphology of the cells was examined by visual observations using light microscopy. Metabolic activity was evaluated by MTT assay.

Results of this study showed that application of the field has measurable influence on biological systems.

Keywords: stem cell, morphogenetic field, cell lines

INTRODUCTION

Morphogenetic field can be presented as a progenitor field which, during the embryonic development, influences the constitution of the morphological structural unit. These fields have clear spatial boundaries across which lie cells which will differentiate in various directions of the differentiational hierarchy. The specification of morphogenetic progenitor fields, organization of the boundaries in relation to neighboring fields and their subsequent regionalization or subdivision are mediated by intercellular signaling. This signaling is responsible for the genetic expression of tissue structure and is used in many processes of morphogenetic differentiation [1].

Storing informational structures in the physio-logical and biomechanical gradients emphasizes the importance of epigenetic transmission of informational structures that do not represent the DNA structure. Recent studies show the presence of alternative mechanisms that must be taken into consideration as parts of the morphogenetic field. The geometric form of the substrate on which the cells are, have an impact on the further cellular behavior [2].

An important part of the necessary information for the integration of cell activity in the work of cellular system is the positional information [3, 4]. It enables cells and tissues to form mutual cellular locations related to the formation of 3D structure. Traditional medium for the fields of positional information is a chemical gradient of some morphogenic molecule [5-7]. The expression of HOX gene in the adult human fibroblasts reveals that they encode their own position along three anatomical axes [8, 9]. Such 3D oriented fibroblasts form a body-wide informational network [10] through the body fascia, making a new organ that stores and processes spatial coordination via an interconnected cellular network. Positional information plays a major role in the pattern of behavior of cells and tissues as well as tissue remodeling throughout life.

It is now clear that bioelectrical properties of cells allow chemical as well as mutual cell communication. It is known for a long time that individual cells react to the effects of extra cellular fields of physiological strength. It is believed that bioelectrical properties form a field of positional information for migratory cellular morphogenesis [11].

Another type of information that is transmitted by morphogenetic field is the information about the scaffold that serves as a template for the formation and reconstruction of pattern. It is considered that tissue bioelectric properties (voltage gradients) can function as a pattern template [12].

I would add that there is a gradient force that defines the direction of the proliferation of cell structure and a way of polarization "electric pressure" as it is called by Nikola Tesla.

Target morphology is a form defined by scales of different size and intensity, as well as the levels of the organization, which require a biological system during development, which is being maintained against cellular aging by stress in life (remodeling and wound healing) and major injuries (regeneration). Models that include target morphology require detailed information, cellular and tissue processing, stressing systemic mechanisms and events during embryonic development and regeneration, as well as mechanisms that are not activated during neoplastic growth [13]. On the basis of cellular remodeling of bone fractures in newborns, it's possible that without special mechanical manipulation of the body during growth make remodeling of bone to the genetically defined form if no injury of the structure of the nerve (neural network stores the morphological information).

The idea that cancer is not new and can be considered a lack of control of morphogenetic field [14, 15] is based on that it may result from the loss of its natural order within the cellular microenvironment [16] and in the literature it appears as a contradiction in relation to the generally accepted theory of the disorder of DNA sequences or the gene expression leading to the formation of tumors [17]. The open debate is whether the development of tumor cells can be best understood through genetic analysis of cells/tissues, or the cellular microenvironment cells ie. Epigenetic mechanisms, play a critical role.

Cancer as a reversible physiological state has certain medical implications and effects that characterize the influence of cell environment on progression of neoplastic processes, and can influence the prevention processes and strategies of detection. Within the morphogenetic field mechanical pathways of cellular changes can lead to the reprogramming-rebooting of cells or to cancer normalization as opposed to approaches that lead to termination of tumor cells [18], although it can further fall behind malignantly intense proliferative response of altered cells [19].

Particularly interesting is the idea that the cancer is not a genetic disease specific loci but rather a kind of structure that attracts or induce or influence transcription of multidimensional spaces that describe the state of the cell [20] Huang et al. 2009 [21] points to an interesting paradox: while most studies tries to determine which gene is mutated during the malignant process, no one actually wondered whether normal cells such as. Adult neuron, peripheral blood cells or epithelial cells has the same genome. There is no clearly defined mutations that may explain the different celullar phenotypes during the formation of the cell line during development. Reversal of tumors (observed when cancer cells are placed in a normal embryonal environment), highlights the potential impact of tissue structures in the formation of a tumor changes [22-24]. Even though the dynamic physiological nature of cancer cells as a disorder of the regulatory process is the most important topic in molecular biology today, but the morphogenetic organization of tissue behind the tissue level has not been clearly understood.

In support of this we have an example of tumor regression and altered cancer tissue during radiation. In addition to physical activity, radioactive and gamma-radiation contain essential coherence of information, in terms of full outbreak and deletion of information morphic structure of irradiated tissue. The termination of radiation in this sense, the surrounding healthy tissue affect the environment, which receives a dose of blood stem cells, the progenitor blast through stages beginning to take on the form of the surrounding tissue in the foregoing mechanism.

We have a similar effect with apoptosis induced tissue poisons, ie. cytostatics. Cytostatics also erased morphic structure for such nonsinchronised (some would say Cacophonic) tissue. Only the surrounding healthy tissue Morphic introduce order into the environment during the period of reparation.

Here we have perceived and electromagnetic effect of "taking the field" who observe the techniques and the emission of telecommunication systems (radio, TV, mobile telephony).

It is obvious that the differentiation of cells depends on the expression of the gene, ie. DNA activity. Activity of cells is therefore phylogenetically determined. In each subsequent stage of cell division, the parent (zygote) and across progenitors, define specific cell receptors, which become sensitized to a specific activators, activation of the receptor is happening with the help of transmitters. Each transmitter activation happens in two ways: physically and electromagnectically. Each activation of the electromagnetic track leaves, can be done easily through biofeedback (bfb) system. In bfb activities emit electromagnetic waves and photons that have certain properties, which can be captured and saved certain way in the database [25].

The aim of this study was to investigate and measure *in vitro* the influence of the morphogenetic field hologram information on human continuous cell lines as a biological model system.

Materials and Methods

MRC-5 fibroblast cell line has been utilized in laboratory conditions in order to take informational structures of healthy fibroblasts.

Electromagnetic sample was taken above each "biological amplifier hologram" with which we recorded hologram, or morphic cell bfb information in the form of MP3 recording. The recording we used to activate the DS modulator.

Since the cellular receptors are in con- junction with bfb, they react to such a hologram, morphic (specific) electromagnetic excitation, and automatically trigger biofeedback in the cell. When it detects an activation signal that triggers biofeedback it leads to the activation of specific DNA sequences, which cause specific activities in the excited cells.

With the use of electromagnetic stimuli, specific cellular receptors of specific tissues are activated first. These specific receptors activate the DNA of cells by an electromagnetic stimulation (indirectly or probably directly) leading to further course of stimulation that goes by their natural course.

CELL LINES AND TREATMENT

Cell lines used in our study included: human cervical carcinoma *HeLa* (ATCC CCL2), the human lung cancer *A549* (ATCC CCL 185), as well as normal human fetal lung fibroblasts cell line *MRC-5* (ATCC CCL 171). Images of used continuous cell lines are shown in Figure 1.

Cells cultures were seeded in flasks (25 cm^2, 10 mL, Costar) in DMEM (Dulbecco's Modified Eagle's Medium, Sigma) with 4.5% of glucose and 2 mM L-glutamine, supplemented with 10% (v/v) fetal calf serum (FCS, Sigma), and antibiotics/antimicotics solution (Sigma), and incubated at 37 °C, with absolute humidity and 5% CO_2 atmosphere in total darkness (incubator HeraCell, Heraeus). Cells were sub-cultured twice a week and a single-cell suspension was obtained using 0.1% trypsin in EDTA (Serva). Cells were harvested in logarithmic phase of growth. Exponentially growing viable cells were used throughout the experiments. Cell viability, determined by trypan blue dye exclusion test (DET) was >90%.

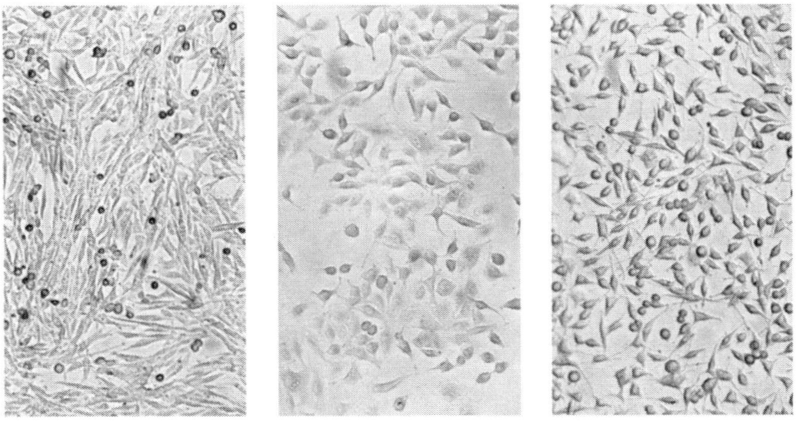

Figure 1. Morphology of used cell: cervical carcinoma HeLa (left), the lung cancer A549 (middle) and normal fetal lung fibroblasts cell line MRC-5 (right).

Cells were cultured in required amount and exposed to the morphogenetic field. Morphogenetic fields the electromagnetic carrier (384 KHz carrier from the DeltaSigma modulator) who carrying modulated hologramic information.

Hologramic information is one (in the nature of recorded signal) digitalized and modulated with previous recorded spectral picture of DNA absorption of in the process of recording) impacted pulsed (3 Hz) light, by emission trough a prepared cells in the flascon. Modulation is the process of varying one or more properties of a periodic waveform, called the carrier signal; we use a Delta Sigma modulation method, called "D-class." For patented device for generating this type of impulse see supplementary material.

The culture flasks with seeded cell lines HeLa, A549 and MRC-5 were fixed and stimulated in focused field with the properties of morphogenic hologramic information of living fibroblasts MRC 5 colony. Total stimulation time of control and treated cultures was 3 hours.

The control cultures of investigated cell lines were subjected to the same experimental conditions but without stimulation during 3h.

After the treatment, morphology of the cells was examined by visual observations using light microscopy and metabolic activity was evaluated by MTT assay.

One portion of control and stimulated cells was left in incubator to undergone another 10 subsequent passages and after that the MTT assay was performed again.

In Vitro MTT Assay

We evaluated metabolic activity of control and stimulated HeLa, A549 and MRC-5 cells by colorimetric tetrazolium MTT test [34], after 3 h exposure to morphogenetic field (electromagnetic wave with morphogenetic fibroblast information). Cell availability was determined in freshly isolated cells by quantitative colorimetric assay for metabolic viability which is based on reduction of tetrazolium salt MTT to a colored ad product, formazan by mitochondrial dehydrogenases in viable cells. Formason produced was dissolved with acidified sodium dodecyl sulphate (10% SDS-0.01Nhce) and optical density (OD) mixture with microplates spectropfotometer at 540nm. For *in vitro* investigation of metabolic activity cell lines were seeded in flat base 96 well microplates (Sarstedt) with a density of $5 \cdot 10^3$ cells in 90 μL volume of medium per well, in octuplicate for each sample. Seeded plates were left in incubator, allowing the cells to attach to the bottom. The percentage of metabolically active cells was calculated averaging results from two independent experiments conducted in quadruplicate for each experimental culture.

Morphology Analysis

For the morphological examination after treatment, cells were seeded directly on microscope slides. After reaching confluences, cells were stained with Giemsa and observed under light microscopy.

Immunohistochemistry Staining

Specimens of control and treated cells were stained with S100 immunohistochemical marker after treatment with electromagnetic wave with morphogenetic fibroblast information. Stained slides were analyzed under light microscope and images were taken with digital camera.

Data Analysis

All experiments in this study were performed at least twice. For in vitro MTT test two independent experiments were conducted in octuplicate for each culture of tested cells. Mean values, standard deviations (SD) and coefficients of variation (CV) were calculated for each concentration. All statistical processing was done in Microsoft Office Excel program.

RESULTS

Metabolic Activity

Results of metabolic activity of control and stimulated HeLa, A549 and MRC-5 cells, evaluated by MTT assay after 3 h exposure to electromagnetic wave with the fibroblast morphogenetic information, are given on Figure 2.

As it is shown, increase in MTT reduction by healthy MRC-5 cells exposed to field during 3 hours compared to untreated cells was observed) (Figure 2A).

By contrast, in tumor cells HeLa, and A549, there was a decrease of metabolic activity and cell proliferation by 25%.

The effect in MRC-5 cells, although reduced (about 20%), it was evident even after 10 passages of treatment (Figure 2–B).

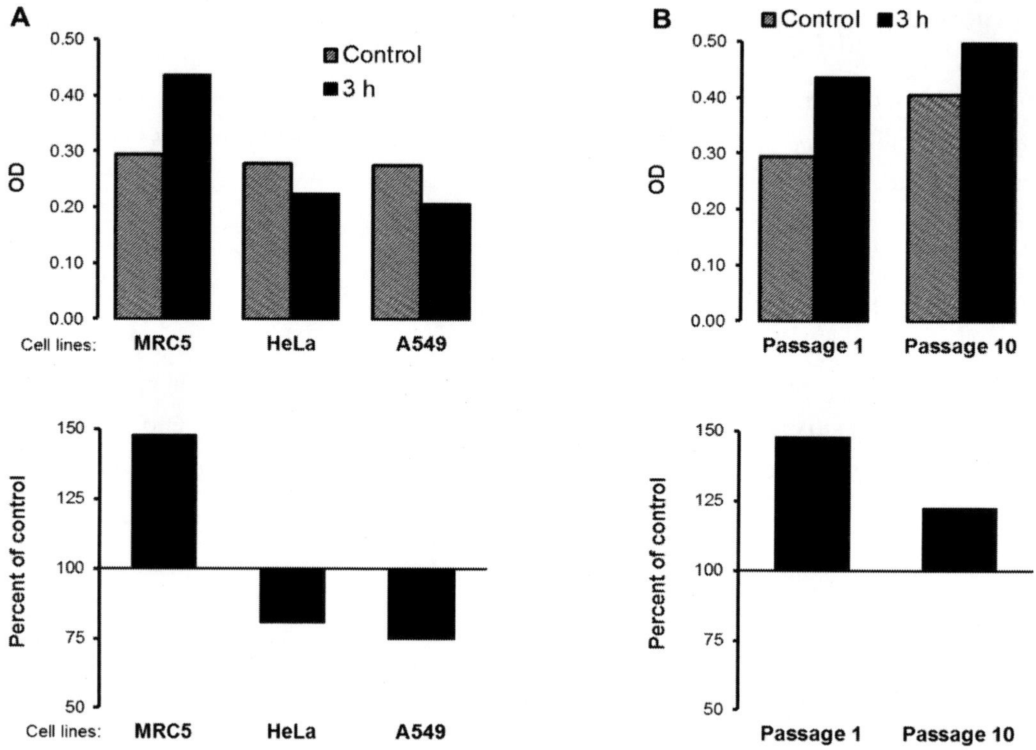

Figure 2. A – Metabolic activity of control and stimulated HeLa, A549 and MRC-5 cells evaluated by MTT assay, after 3 h exposure to morphogenetic fibroblast information B – Metabolic activity of stimulated MRC-5 cells immediately after treatment and after 10 passages.

Thereafter, we examine the effect on MRC-5 cells with treatment lasting 1 or 3 h and with readings after 24 h. Results are shown in Figure 3. The proliferation was increased after 1 h (17.9%) in comparison to 3 h (13.5%).

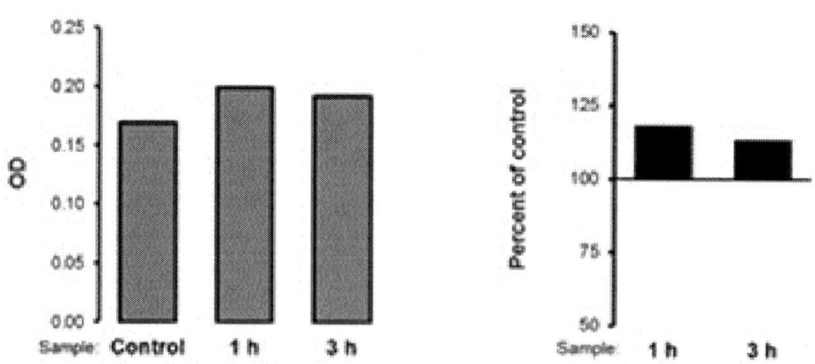

Figure 3. Effect of different duration (1 and 3 h) of applied electromagnetic field on MRC-5 cells.

Cell Morphology

The results of morphological analysis showed that the experimental group had no signs of change in the structure of cells compared to the control group. Images are presented in Figure 3.

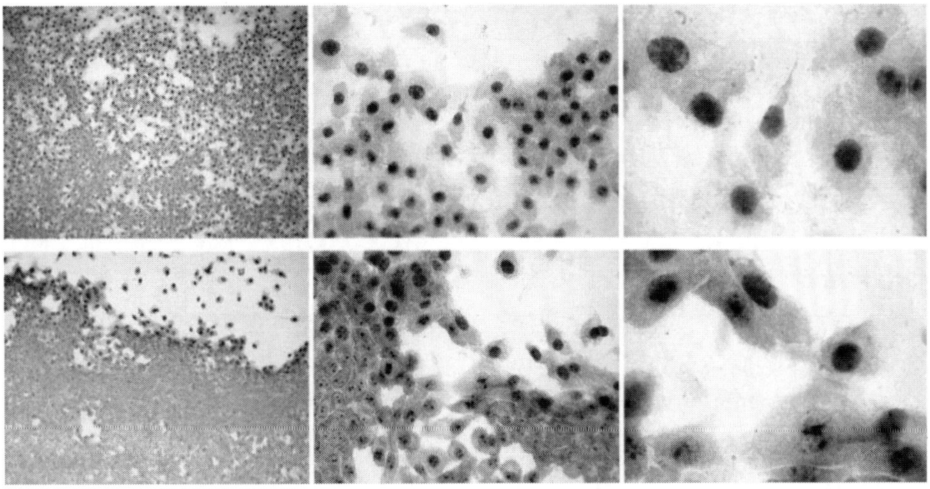

Figure 4. Specimens of MRC-5 cells stained with Giemsa. First row show control sample and the second cells after treatment with the morphogenetic information. Images were taken with Olympus Camedia 3040 on Olympus BX51 microscope. Upper row shows specimens of cells treated only with the field Left column are images of 10×15, column in the middle those of 40×15 and right of 100×15 magnification.

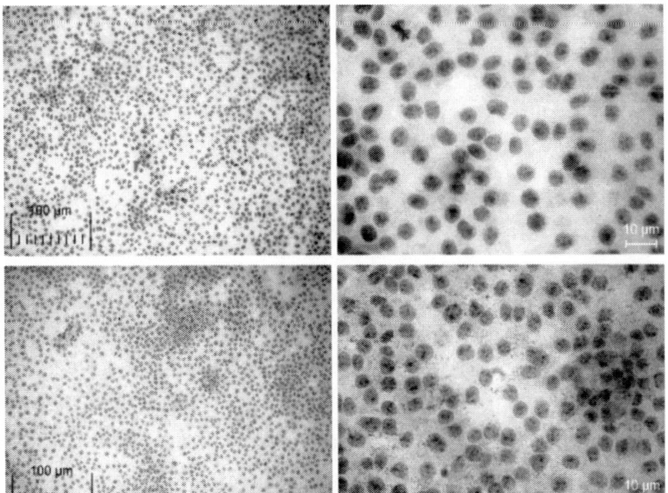

Figure 5. Specimens of MRC-5 cells stained with S100 immunohistochemical marker after treatment with the field Images were taken with Olympus Camedia 3040 on Olympus BX51 microscope. Upper row shows control specimens and bottom row specimen of treated cells. Left column are images of 10×15 and right of 40 × 15 magnification.

S100 Immunohistochemistry

Immunohistochemical staining of the specimens (Figure 4) showed that in treated specimen we spotted S100 positive cells intracytoplasmic granules.

Discussion

First of all, the concepts of morphogenitic field should be confined, which are essentially information has very personalised caracter. Such information must have its physical carrier. Carriers of information are of physical nature of electromagnetic character (see electromagnetic spectrum) and every electromagnetic wave, including gamma-radiation, may be a carrier of such information.

This information represents a software, which on a specific way interacts with genetic material (and various biochemical molecules of lower level), causing the expression of genes in parallel with biochemical mechanisms [26].

Therefore just activated structures which contain both biochemical and electromagnetic stimuli show "cellular signaling" to their full extent. Such signaling is an integral part of the cell biofeedback. Proof of the existence of that kind of structure we find in the removal of pericellular fluid, which is recovered but non-activated, does not allow the activation of cellular receptors (in addition to its own contended biochemical and general physical factors), till the moment while the entire cellular mechanism does not activate with the restoration of the information of electromagnetic signals:

1) contact with already activated cells of the same type.
2) recorded electromagnetic informational structure.

The hypothesis that the carcinoma is basically a phenomenon at the level of multicellular organization is supported by data from a series of experiments. As the cells are more scattered and disorganized within the monolayer cultures, they are more prone to carcinogenesis and changes than the cells in the context of symmetric healthy tissue [27].

So one could interpret the regeneration and recovery of certain cellular structures in humans: UV irradiation, laser beam and certain substances that can have a therapeutic effect but also to potentiate the induction of cancer in tissues with a high degree of three-dimensional tissue disorganization [28, 29].

In accordance with that, some data show that tumor genesis is being induced when cells are surrounded by their neighboring or other cells or tissues, and can be defined as a system of spatial inhibition or physical barriers [30].

In contrast, the re-establishment of proper interaction of cancer cells in humans can reverse the behavior of malignant cells, even in the presence of significant genetic damage [31].

Numerous authors emphasize the suppressive nature of signals from adjacent tissue [32]. One of the potential mechanisms of the cellular protections in relation to the tumor genesis is the planar cells polarity [33]. Planar cell polarity represents a collection of protein components designed to function as a tumor suppressor by coordinating the function of cells at a great distance [34, 35].

It is clear that the loss of polarity can launch the formation of tumors in Drosophila [36].

The planar cell polarity allows cells to align in an orthogonal axis relative to their apical-basal mutual polarity. It is also interesting to discuss the impacts that function at a higher level of organization than the local tissue, as well as specific inhibitory signals such as morphostatic signal which is related to the position and orientation of objects in 3 dimensions, but in the context of a larger morphology.

It seems that the situation within the host is an important factor in the genesis of tumors, which is not a genetic mutation, but may mean the divergence in power or positioning of the morphogenetic field. Disruption of normal topographical relations of tissues, tends to cause cancer, indicating the possibility that the morphogenetic field can change encoded anatomy or cause may be difficulties in forming structure of cells and primitive tissues on geometric borders of neighboring fields, and normally the rat ovarian tissue placed in a normal rat spleen results in malignant neoplasms. Also, the implantation of mouse embryo in adult individuals causes teratocarcinoma, perhaps due to disruption of the host and the embedded morphogenetic field of structure, such as transformation [37, 38].

Morphogenetic field should be the most active and "accessible" during embryogenesis. It is therefore not surprising that, despite the high malignancy and euploidy, tumor cells integrated in the host embryo cells become integrated as a normal tissue [39, 40].

Neuroblastoma in childhood has a high rate of spontaneous regression [41]. Human metastatic melanoma cells, injected into zebra embryos do not acquire neoplastic phenotype, but form tumors when injected in the aftermath of organogenesis [42]. Similarly, the implanted sarcoma progresses in 80% of adult rats, but only in 6.4% of embryos of rats. Similar data has been recently revealed for fish and other types of embryos which were capable to tame aggressive cancer cells when they were implemented in the host [43, 44].

These data are consistent with the concept of morphogenetic fields as they point to the potency of active patterns for normalization of cancer (over-ride genetic defects); They are less compatible with signaling at the cell level; and embryos have a high level of a large growth factors which otherwise would be able to potentiate the growth of tumors. Tumors can be described as wounds that do not heal - territories of disturbance and cell growth without appropriate patterns and programs that reach a definite status [45, 46].

Morphogenetic information delivering to a particular cell or tissue means "informing" them about polarity, voltage membranes as type I receptor that need to be activated. In our experiment, we can see intense metabolic activity of a primitive group of fibroblasts, whose morphogenetic hologram matrix has been carried out to carcinoma cells. In contrast to "healthy" fibroblasts, after reception of cellular information, carcinoma loses its metabolic activity of more than 30% of the population, and change its phenotype form. This modality of transmission of the information is not new, but it is not routinely used in the daily practice. Even progressive passage of treated carcinoma cell colonies did not regain its original shape which would indicate a permanent morpho-functional cellular change that could play a role in everyday practice, which in principle is consistent with the works of Beloussov and Levin [47, 48, 49]. Analyzing endogenous voltage gradients during intercellular communication is surely needed and is considered to be equally important [50].

Hence, stimulation with holographic matrix of morphogenetic field structure, may represent the future of the treatment of such conditions in healthy individuals [51] also so is the treatment of the stem cells of the tumor itself in the sense of reduction or neutralization of the tumor [52], with the stem cells is promisine.

There are several hypotheses for tumor genesis and the question is whether the tumor change is essentially a cellular defect or an effect of the community. Why cells (or small group of cells) in normal tissue initiate occurrence of tumor? Morphogenetic field can be at the heart of this dilemma. Tumor can arise from failure of hosts to impose or transpose necessary information in a particular area. On the other hand, it is possible that tumor cells are those that have ceased to be oriented towards the morphogenetic field of signals (e.g., microgravity disorder) [53, 54].

An important analysis of the function of tumor cells in the organism is the bioelectric cellular signaling. It is considered that disturbance in morphogenetic factors as well as molecular factors is involved in the genesis of the tumor. There is more evidences that the disorder of bioelectric signaling can lead to the lack of normal pattern behavior and can lead to cancerous growth. Ion channels, ion pumps, and ion compounds can be identified as oncogenes [55, 56, 57].

S 100 protein exists normally in cells of neural crest origin (Schwann cells and melanocytes), chondrocytes, adipocytes, macrophages myoepithelial cells. It's considered that they have both intracellular and extracellular role and that they are able to regulate the process of phosphorylation, transcription factors, Ca^{2+} homeostasis dynamics of the cytoskeleton, enzymatic activity. Reinforcement of intensity can be interpreted by intense amplification of metabolic activity of fibroblasts, or a change in function of the cells of cancer. S 100 is not normally found in these tumors portals [58].

Interphase is 23 hours, and mitosis is one hour. Three hours is period needed [59] that biofeedback system forms at least two phases of receptor differentiation and to form differentiation in desired cellular lineage.

Receptors are very important category in differentiation and are formed during biofeedback process.

Biofeedback is forming through the intracellular and extracellular mechanism that have electro-magnetic nature (caused by cellular biopotencial which is periodical by its nature; that "waving" causes electromagnetic effect; the most powerful effects are in microtubule, because they have the structure of electromagnetic coil [6]. Microtubules are like coils with isolators: alpha and beta tubules are arranged like that; there is continuously present phenomenon: 24.12 nm diameter exactly octave higher; there are 13 reels; each reel 1 tone and 1 frequent shift: means that tubules will start to work as a pump if there are in resonancy. Depending the resonant state of tubules develops different cellular pattern (cellular morphology. [61]. In that manner, established bio-feedback, forms morphological characteristics of cells in 3 D.

Conclusion

The effect of field was investigated on a continuous human cell lines HeLa, A549 and MRC-5. The following conclusions can be made:

- MRC-5 cells exposed to field during 3 h had an increase in metabolic activity in comparison to untreated cells, which was observed 48 h after treatment.
- The results of morphological analysis, conducted after cell treatment, showed no change in cell morphology.
- Immunohistochemical results showed that increasing metabolic activity on cellular level in fibroblast group, and changing and impairment of metabolic function of treated carcinoma cells.
- According to the results of this study, the greatest effect was observed 48 h after the 3-hour treatment, suggesting that the effect of the event takes time.
- Morphogenetic hologram information given through electromagnetic wave can influence changes in stem cells and their environment.

Supplementary Material

Details about used experimental methods and complete data sets of the results obtained in this study can be found in the supplement available in the online version of this article, at the journal web site.

In the first phase, the recording of nerve impulses and storing them in a database.

Recorded Signals

All nerve signals are in the nature digital. The device that is used is therefore specific digital electromagnetic stimulator cell receptor.

The benefits of the Delta Sigma (DS) modulation principle of activation.

Modulator produced electromagnetic oscillation at a frequency of about 400 KHz. So this becomes a carrier wave frequency of the recorded digital information bfb. In the modulator is sent to the recorded signal is applied to the trigger circuits 555, leading to initiatives stimulating output circuits 555. This bike is connected to an antenna coil 0.39 mH. Such a coil acting as an antenna sends electromagnetic signal into space.

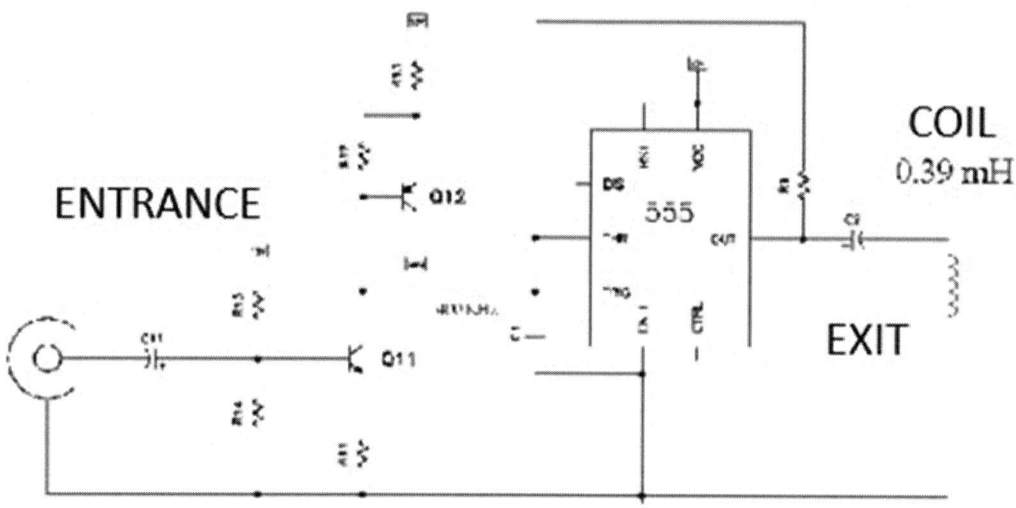

Figure 6. At the entrance DS modulator is supplied excitation signal morphic hologram.

The device consists of:

1. Emission coil,
2. Stimulus coil,
3. Delta-Sigma modulator,
4. Source excitation signal,
5. Emission coil inductance is 0.49 mH,
6. Coil stimulus is done with wheel 555,
7. Delta-Sigma modulation with two NPN transistors whose signal goes to the pin 555,
8. Source emission coil in the measurement was performed with tone generator.

Morphic information we sampled from healthy fibroblasts from standard cell cultures MRC 5.

Conflict of Interest

The authors declare that they have no conflict of interest.

References

[1] Davidson EH. Later embryogenesis: Regulatory circuitry in morphogenetic fields. *Development.* 1993 Jul 1; 118(3):665-90.

[2] Chen CS, Mrksich M, Huang S, Whitesides GM, Ingber DE. Geometric control of cell life and death. *Science.* 1997 May 30;276(5317):1425-8.

[3] Mathews J, Levin M. Gap Junctional Signaling in Pattern Regulation: Physiological network connectivity instructs growth and form. *Developmental neuro-biology*. 2016 Jun 1.

[4] Law R, Levin M. Bioelectric memory: modeling resting potential bistability in amphibian embryos and mammalian cells. *Theoretical Biology and Medical Modelling*. 2015 Oct 15;12(1):1.

[5] Ashe HL, Briscoe J. The interpretation of morphogen gradients. *Development.* 2006 Feb 1;133(3):385-94.

[6] Gurdon J. Developmental biology and the redirection or replacement of cells. Philosophical Transactions of the Royal Society of London B: *Biological Sciences.* 1999 Dec 29;354(1392):1967-76.

[7] Teleman AA, Strigini M, Cohen SM. Shaping morphogen gradients. *Cell.* 2001 Jun 1;105(5):559-62.

[8] Chang HY, Chi JT, Dudoit S, Bondre C, van de Rijn M, Botstein D, Brown PO. Diversity, topographic differentiation, and positional memory in human fibroblasts. *Proceedings of the National Academy of Sciences*. 2002 Oct 1;99(20):12877-82.

[9] Rinn JL, Bondre C, Gladstone HB, Brown PO, Chang HY. Anatomic demarcation by positional variation in fibroblast gene expression programs. *PLoS Genet.* 2006 Jul 28;2(7):119.

[10] Langevin HM, Cornbrooks CJ, Taatjes DJ. Fibroblasts form a body-wide cellular network. *Histochemistry andcell biology*. 2004 Jul 1;122(1):7-15.

[11] Yao L, Pandit A, Yao S, McCaig CD. Electric field-guided neuron migration: a novel approach in neurogenesis. *Tissue Engineering Part B: Reviews.* 2011 Mar 2;17(3):143-53.

[12] Adams DS, Levin M. Endogenous voltage gradients as mediators of cell-cell communication: strategies for investigating bioelectrical signals during pattern formation. *Cell and tissue research*. 2013 Apr 1; 352(1):95-122.

[13] Teufel T. Wholes that cause their parts: organic self-reproduction and the reality of biological teleology. Studies in History and Philosophy of Science Part C: *Studies in History and Philosophy of Biological and Biomedical Sciences*. 2011 Jun 30;42(2):252-60.

[14] Baker SG, Cappuccio A, Potter JD. Research on early-stage carcinogenesis: Are we approaching paradigm instability? *Journal of Clinical Oncology*. 2010 Jul 10;28(20):3215-8.

[15] Soto AM, Sonnenschein C. The somatic mutation theory of cancer: growing problems with the paradigm? *Bioessays*. 2004 Oct 1;26(10):1097-107.

[16] Ingber DE. Can cancer be reversed by engineering the tumor microenvironment? In *Seminars in cancer biology* 2008 Oct 31 (Vol. 18, No. 5, pp. 356-364). Academic Press.

[17] Fan Y, Bergmann A. Apoptosis-induced compensatory proliferation. The Cell is dead. Long live the Cell! *Trends in cell biology*. 2008 Oct 31;18(10):467-73.

[18] Dinicola S, D'Anselmi F, Pasqualato A, Proietti S, Lisi E, Cucina A, Bizzarri M. A systems biology approach to cancer: fractals, attractors, and nonlinear dynamics. *Omics: a journal of integrative biology*. 2011 Mar 1;15(3):93-104.

[19] Huang S, Ingber DE. Shape-dependent control of cell growth, differentiation, and apoptosis: switching between attractors in cell regulatory networks. *Experimental cell research*. 2000 Nov 25;261(1):91-103.

[20] Bissell MJ, Radisky D. Putting tumours in context. *Nature Reviews Cancer*. 2001 Oct 1;1(1):46-54.

[21] Bizzarri M, Cucina A, Conti F, D'Anselmi F. Beyond the oncogene paradigm: understanding complexity in cancerogenesis. *Acta biotheoretica*. 2008 Sep 1;56(3): 173-96.

[22] Bizzarri M, Cucina A, M Biava P, Proietti S, D'Anselmi F, Dinicola S, Pasqualato A, Lisi E. Embryonic morphogenetic field induces phenotypic reversion in cancer cells. *Review article*. *Current pharmaceutical biotechnology*. 2011 Feb 1;12(2):243-53.

[23] Weaver VM, Gilbert P. Watch thy neighbor: cancer is a communal affair. *Journal of cell science*. 2004 Mar 15;117(8):1287-90.

[24] Aaron RK, Boyan BD, Ciombor DM, Schwartz Z, Simon BJ. Stimulation of growth factor synthesis by electric and electromagnetic fields. *Clinical orthopaedics and related research*. 2004 Feb 1;419:30-7.

[25] Milankov Z. "*The effect of the aura*" ISBN 978¬86¬84863¬29-6) Word press 2012, Novi Sad Serbia

[26] Levin M. Endogenous bioelectrical networks store non-genetic patterning information during development and regeneration. *The Journal of physiology*. 2014 Jun 1;592(11):2295-305.

[27] Parodi S, Brambilla G. Relationships between mutation and transformation frequencies in mammalian cells treated *"in vitro"* with chemical carcinogens. *Mutation Research/Reviews in Genetic Toxicology.* 1977 Dec 31; 47(1):53-74.

[28] Sample A, He YY. Autophagy in UV Damage Response. *Photochemistry and Photobiology.* 2016 Dec 1.

[29] Eguchi G, Watanabe K. Elicitation of lens formation from the 'ventral iris' epithelium of the newt by a carcinogen, N-methyl-N′-nitro-N-nitrosoguanidine. *Development*. 1973 Aug 1;30(1):63-71.

[30] Mesnil M, Crespin S, Avanzo JL, Zaidan-Dagli ML. Defective gap junctional intercellular communication in the carcinogenic process. *Biochimica et Biophysica Acta (BBA)-Biomembranes.* 2005 Dec 20;1719(1):125-45.

[31] Barcellos-Hoff MH. It takes a tissue to make a tumor: epigenetics, cancer and the microenvironment. *Journal of mammary gland biology and neoplasia.* 2001 Apr 1;6(2):213-21.

[32] Soto AM, Sonnenschein C. The somatic mutation theory of cancer: growing problems with the paradigm? *Bioessays.* 2004 Oct 1;26(10):1097-107.

[33] Sun SD, Purdy AM, Walsh GS. Planar cell polarity genes Frizzled3a, Vangl2, and Scribble are required for spinal commissural axon guidance. *BMC neuroscience.* 2016 Dec 12;17(1):83.

[34] Klezovitch O, Vasioukhin V. *Cadherin signaling: keeping cells in touch. F1000 Research*. 2015;4(F1000 Faculty Rev).

[35] Lee M, Vasioukhin V. Cell polarity and cancer–cell and tissue polarity as a non-canonical tumor suppressor. *Journal of cell science*. 2008 Apr 15;121(8):1141-50.

[36] Wodarz A, Näthke I. Cell polarity in development and cancer. *Nature cell biology.* 2007 Sep 1;9(9):1016-24.

[37] Blackiston D, Adams DS, Lemire JM, Lobikin M, Levin M. Transmembrane potential of GlyCl-expressing instructor cells induces a neoplastic-like conversion of melanocytes via a serotonergic pathway. *Disease Models and Mechanisms.* 2011 Jan 1;4(1):67-85.

[38] Morokuma J, Blackiston D, Adams DS, Seebohm G, Trimmer B, Levin M. Modulation of potassium channel function confers a hyperproliferative invasive phenotype on embryonic stem cells. *Proceedings of the National Academy of Sciences.* 2008 Oct 28;105(43): 16608-13.

[39] Astigiano S, Damonte P, Fossati S, Boni L, Barbieri O. Fate of embryonal carcinoma cells injected into postimplantation mouse embryos. *Differentiation.* 2005 Dec 1;73(9-10):484-90.

[40] Illmensee K, Mintz B. Totipotency and normal differentiation of single teratocarcinoma cells cloned by injection into blastocysts. *Proceedings of the National Academy of Sciences*. 1976 Feb 1;73(2):549-53.

[41] Nakagawara A. Molecular basis of spontaneous regression of neuroblastoma: role of neurotrophic signals and genetic abnormalities. *Human cell.* 1998 Sep;11(3):115-24.

[42] Haldi M, Ton C, Seng WL, McGrath P. Human melanoma cells transplanted into zebrafish proliferate, migrate, produce melanin, form masses and stimulate angiogenesis in zebrafish. *Angiogenesis*. 2006 Sep 1;9(3):139-51.

[43] Hendrix MJ, Seftor EA, Seftor RE, Kasemeier-Kulesa J, Kulesa PM, Postovit LM. Reprogramming metastatic tumour cells with embryonic microenvironments. *Nature Reviews Cancer*. 2007 Apr 1;7(4):246-55.

[44] Kasemeier-Kulesa JC, Teddy JM, Postovit LM, Seftor EA, Seftor RE, Hendrix MJ, Kulesa PM. Reprogramming multipotent tumor cells with the embryonic neural crest microenvironment. *Developmental Dynamics*. 2008 Oct 1;237(10):2657-66.

[45] Pierce GB, Speers WC. Tumors as caricatures of the process of tissue renewal: prospects for therapy by directing differentiation. *Cancer research*. 1988 Apr 15;48(8):1996-2004.

[46] Riss J, Khanna C, Koo S, Chandramouli GV, Yang HH, Hu Y, Kleiner DE, Rosenwald A, Schaefer CF, Ben-Sasson SA, Yang L. Cancers as wounds that do not heal: differences and similarities between renal regeneration/repair and renal cell carcinoma. *Cancer research*. 2006 Jul 15;66(14):7216-24.

[47] Beloussov LV, Grabovsky VI. Information about a form (on the dynamic laws of morphogenesis). *BioSystems.* 2007 Feb 28;87(2):204-14.

[48] Beloussov LV, Lakirev AV. Generative rules for the morphogenesis of epithelial tubes. *Journal of theoretical biology*. 1991 Oct 21;152(4):455-68.

[49] Levin M. The wisdom of the body: future techniques and approaches to morphogenetic fields in regenerative medicine, developmental biology and cancer. *Regenerative medicine*. 2011 Nov;6(6):667-73.

[50] Adams DS, Levin M. Endogenous voltage gradients as mediators of cell-cell communication: strategies for investigating bioelectrical signals during pattern formation. *Cell and tissue research.* 2013 Apr 1;352(1): 95-122.

[51] Kwon HJ, Lee GS, Chun H. Electrical stimulation drives chondrogenesis of mesenchymal stem cells in the absence of exogenous growth factors. *Scientific Reports.* 2016;6.

[52] Ledda M, Megiorni F, Pozzi D, Giuliani L, D'Emilia E, Piccirillo S, Mattei C, Grimaldi S, Lisi A. Non ionising radiation as a nonchemical strategy in regenerative medicine: Ca 2+-ICR "*In Vitro*" effect on neuronal differentiation and tumorigenicity modulation in NT2 cells. *PloS one.* 2013 Apr 9;8(4):61535.

[53] Höckel M. Morphogenetic fields of embryonic development in locoregional cancer spread. *The Lancet Oncology.* 2015 Mar 31;16(3):e148-51.

[54] Pisanu ME, Noto A, De Vitis C, Masiello MG, Coluccia P, Proietti S, Giovagnoli MR, Ricci A, Giarnieri E, Cucina A, Ciliberto G. Lung cancer stem cell lose their stemness default state after exposure to microgravity. *BioMed research international.* 2014 Sep 7;2014.

[55] Becchetti A. Ion channels and transporters in cancer. 1. Ion channels and cell proliferation in cancer. *AmericanJournal of Physiology-Cell Physiology.* 2011 Aug 1;301(2):C255-65.

[56] Prevarskaya N, Skryma R, Shuba Y. Ion channels and the hallmarks of cancer. *Trends in molecular medicine.* 2010 Mar 31;16(3):107-21.

[57] Arcangeli A, Crociani O, Lastraioli E, Masi A, Pillozzi S, Becchetti A. Targeting ion channels in cancer: a novel frontier in antineoplastic therapy. *Current medicinal chemistry.* 2009 Jan 1;16(1):66-93.

[58] Marenholz I, Heizmann CW, Fritz G. S100 proteins in mouse and man: from evolution to function and pathology (including an update of the nomenclature). *Biochemical and biophysical research communications.* 2004 Oct 1;322(4):1111-22.

[59] http://www.bio.bg.ac.rs/materijali_katedri/Nastanak%20celija-%20deobe_611_1382004339.pdf.

[60] http://vle.du.ac.in/mod/book/print.php?id=13685&chapterid=30331.

[61] https://www.youtube.com/watch?v=wvJAgrUBF4w.

In: Stem Cells Origins of Cytokine Interactions
Editor: Prasad S. Koka ISBN: 978-1-53615-302-6

Chapter 13

THE LNCRNA HOTAIR WAS REGULATED BY POLYACRYLIC ACID RESINS IN HUMAN GINGIVAL FIBROBLASTS

J. Syed[1], G. Spagnuolo[2] and A. Procino[2,*]
[1]Advanced Technology Dental Research Laboratory,
Department of Oral Basic and Clinical Science,
Faculty of Dentistry, King Abdulaziz University, Saudi Arabia
[2]Department of Neurosciences, Odontostomatologic and Reproductive Sciences,
Medical School "Federico II" of Naples, Italy

ABSTRACT

Although, the studies about the genotoxic and cytotoxic effects of methacrylate resins have increased considerably, the understanding of molecular regulatory basis is still poorly understood. HOTAIR is a non-coding RNA related to several diseases; originally, it was found in colorectal, gastric cancer and hepatoma. Moreover, HOTAIR is one of the most important regulator of the Epithelial-Mesenchymal Transition process. The aim of this paper was to investigate the ability of three commercial acid resins (Unifast, JetKit and Duralay) to control the HOTAIR gene expression. The cytotoxic activity in HGFs, measured by MTT assay, was always conditioned by high concentrations of the resins. On the other hand, the proliferation rate does not change in HGFs treated with Unifast and Duralay, while Jet Kit is a negative regulator of the cell proliferation. The HOTAIR gene expression was regulated in HGFs treated with Unifast and Duralay, while JetKit did not induce change in charge of HOTAIR expression.

In this study, the ability of different commercial polyacrylic acid resins to regulate lncHOTAIR gene has been demonstrated for the first time. Moreover, no data have previously been reported about the HOTAIR gene expression in HGFs. Several

* Corresponding Author Email: alfredo.procino@unina.it.

experimental evidences identify HOTAIR as a crucial regulator of EMT; pathologically, any alteration of EMT process leads to several diseases. Particularly, overgrowth of gingival epithelium is due to the deregulation of EMT specific markers. In conclusion, our results could lead to better understanding the role of HOTAIR in different gingival pathologies related to the EMT deregulation.

Keywords: ncRNAs, HOTAIR, resins, HGFs

Introduction

The development of dental materials used in clinical practice, is a research area of considerable interest [1]. Methacrylates, and similar acrylic resins, are the most used polymeric materials in dental prosthesis; their chemical composition varies considering the different acrylic resins available from different chemical industrial products; the mostly incorporated particles include monomer (methyl methacrylate [MMA]), matrix (polymethyl methacrylate [PMMA]). The polymethylmetha-crylates (PMMA), have wide applications like the individual impression trays, orthodontic devices, implants and artificial crowns [2]. The molecular mechanisms that are behind the genotoxic and cytotoxic effect due to the resins is not clearly established yet, but is deeply investigated, considering their role in the biocompatibility of new dental materials. Moreover, residuals released by methacrylate (MMA, PMMA, etc.) are able to induce apoptosis, necrosis and enhance death in human U937 monoblastoid cells [3]. Chemical constituents and products released after the polymeric conversion, are considered responsible of several adverse responses: (i) localized and systemic inflammation, (ii) toxicological reactions, (iii) carcinogenic/mutagenic effects. However, gingival inflammation and periodontal bone resorption can occur, due to aging and degradation of acrylic resins (porosity, marginal integrity) [3, 4]. The polyacrylic resins regulate the synthesis and release of cytokines, chemokines, pro-inflammatory proteins in several cell phenotype [5]. It has already been demonstrated, the ability of acid resins to regulate the Interleukin-6 (IL-6), C-Reactive Protein (C-RP), TNF-alpha gene expression, in human gingival affected by several diseases [6, 7, 8, 9].

The analysis of data, produced by the human genome sequencing, has established that only 2% of the genes encode proteins while over 90% of the genome produces non-coding RNAs (ncRNAs) [10, 11]. Based on nc-RNAs size, it has been possible to classify these nucleic acids into two classes: (i) Small RNAs (ncRNAs), including microRNAs (miRNAs); (ii) long non-coding RNAs (lincRNAs) [12, 13]. The miRNAs are involved in the control of normal cell activity, through the transcription of specific genes. Conversely, miRNAs deregulation is related to morphological and physiological disorders (diabetes, cancer, etc.) [14, 15].

The lncRNAs are molecules of 200 nucleotides in length. Abnormal behavior of lncRNAs is related to several morphological and physiological diseases like neoplastic transformation, synpolydactyly, mainly by means of smart regulation of genome activity [16]. Moreover, the lncRNAs regulate several processes inside the cell like chromatin structure, through molecular interactions among RNA, DNA and proteins. Therefore, the lncRNAs represent a network of information that determines the normal cellular functions [16, 17]. It has been identified that lncRNA titled HOX antisense intergenic RNA (HOTAIR) transcribed by HOXC locus are located on chr 12q13.13, exactly between HOXC11 and HOXC12 [18-20]. HOTAIR is the first lncRNA that controls the gene transcription mainly by means of trans- mechanism. In particular, HOTAIR is able to block the transcription of the HOXD locus on chromosome 2, exactly from HOXD9 to HOXD13. Moreover, HOTAIR is involved in the regulation of Epithelial-Mesenchymal Transition (EMT), crucial event related to several biological processes [21, 22]. It has already been demonstrated, the up-regulation of HOTAIR, in several diseases like Celiac Disease (CD) and neoplastic transformation, firstly in colon and gastric cancer mainly by means of the interaction with miRNA196a, a microRNA located on chromosome 2 inside the HOXC locus between HOXC9 and HOXC10 [23-26]; afterwards, HOTAIR was identified in the Oral Squamous Cell Carcinoma (OSCC) [27]

The aim of this study is to demonstrate the role of different commercial polyacrylic acid resins in the control of lncRNA HOTAIR gene expression in Human Gingival Fibroblasts (HGFs).

METHODS

Human Gingival Fibroblasts (HGFs) Culture

Gingival tissue was obtained from healthy 20-30 year-old patients, under informed consent approved by an Institutional Review Board (University of Napoli "Federico II"). The connective layer was dissected from the gingival samples by means of a surgical blade. Tissue fragments were washed twice with phosphate buffered saline (PBS), transferred into tissue culture dishes in Dulbecco's Modified Eagle Medium (DMEM) supplemented with: 10% fetal bovine serum (FBS), 2mM glutamine, 100U/ml of penicillin, 100 mg/ml of streptomycin, at 37°C in a humidified atmosphere of 5% CO_2 in air. After 10 days, fragments were removed and released, fibroblast started proliferating. Once the confluence was reached, cells were washed with PBS, detached from the culture dishes using a brief treatment with trypsin/EDTA for 5 minutes and re-cultured until confluent monolayer was again obtained [9].

Polyacrylic Resins

The resins were prepared by mixing the materials within the well-plates; it has been performed the cytotoxic assay (MTT) in order to identify the exact concentration of the resins eligible for the experiments. The cytotoxic experiments were performed using resin concentration from 0,001 mM to 1mm; moreover, the assay was performed, choosing two different time ranges: 3 and 30 minutes after mixing the materials; subsequently, the well-plates were covered with HGF specific medium for 24h, in order to demonstrate the release of the resin residuals. Subsequently, the HGFs were stimulated with the conditioned medium for 24h; finally, the cytotoxic assay (MTT) was used for analyzing the cell viability [9].

MTT Assay

MTT, measuring the mitochondrial dehydro-genase activity. HGF cells were seeded into 96-well flat-bottomed tissue culture plates respectively at 10^4 cells/well. After 24 hours of incubation, the culture medium was replaced with 200μL/well of dilutions. Elapsed further 24 hours, the medium was substituted with 100μL/well of MTT solution (1 mg/ml) in PBS, and the cells were incubated for an additional hour at 37°C in a 5% CO2 atmosphere. After the solution was removed, 100 μL/well of DMSO were added and the plates were swirled gently for 10 min. The optical density of each well was immediately measured in a spectrophotometer (Sunrise™, Tecan, Mannederf/Zurich, Switzerland) at 590 nm.a The optical density of cells cultured in the medium along with saline solution without the miniscrew extracts was used as control for 100% of cell viability similarly as a reference for determining ~~of~~ the cytotoxicity (%) in the assay. At least four independent experiments were performed in quadrup-licate. Data by four independent experiments were statistically analyzed (Mann-Whitney U-test, $p < 0.05$) [28].

Alamar Blue Assay

HGFs (5Å ~ 104 cells/well) were seeded in 96-multiwell plates. After 24 h at 37°C, 0.2 mL of pure extract of each medium conditioned by the resins, was added to the wells. Cells incubated in complete medium formed the untreated control group. Cell proliferation of four replicate samples was assessed by AlamarBlue assay (AB) at 24h, 7, and 14 days. AlamarBlue assay uses a visible blue fluorogen probe resazurin, which is reduced to a red fluorescent compound (resorufin) by cellular redox enzymes of the mitochondrial respiratory chain. Viable cells continuously convert resazurin to resorufin, thereby generating a quantitative measure of viability and proliferation [28]. The AB assay was performed according to the manufacturer's protocol (BioSource International, Camarillo, CA, USA). At predetermined culture intervals (1, 3, 7, 14 days), 200 μL of AlamarBlue dye were added directly into culture media at a final concentration of 10% and the plates were returned to the incubator for 4 h. As a negative control, AB was added to the medium without cells. The percentage AB reduction was calculated from the values of optical density at 540 and 590 nm using the manufacturer's formula. Each experiment was performed three times (n = 3) in quadruplicate. Statistical analysis was performed by the non-parametric one-way ANOVA test (with a p value < 0.05 considered significant).

RNA Extraction and Analysis

Total RNA was isolated from HGFs culture cells, using RNeasy Micro-Kit (Qiagen GmbH, Hilden, Germany) following the manufacturer's instructions. Samples were treated with RNase-free DNase (Qiagen GmbH) to prevent amplification of genomic DNA. One microgram RNA was subjected to cDNA synthesis for 1 h at 37°C using the Ready to go You-Primer First-Strand Beads kit (Amersham Biosciences cod. 27-9264-01) in a reaction mixture containing 0,5 mg random hexamers (GeneAmp RNA PCR Random Hexamers Set N808-0127 Applied Biosystems, Foster City, CA) [29].

Real Time-PCR (QRT-PCR)

QRT-PCR will be performed using Taq-Man or technology (QRT-PCR StepOne). This assay uses a specific oligonucleotide probe, annealing between the two primer sites, which is labelled with a reporter fluorophore and a quencher. Cleavage of the probe by the exonuclease activity of Taq polymerase during strand elongation releases the reporter from the probe resulting in an increase in reporter emission intensity owing to its separation from the quencher. This increment in net fluorescence is monitored in real-time during each PCR amplification. The cDNA, will be used for real-time PCR performed in 48-well optical reaction plates with cDNA equivalent to 100ng RNA in a volume of 25 μl reaction containing Taqman Universal Master Mix (Applied Biosystem 4304437). Optimized concentrations of FAM-labelled probe and specific forward and reverse primer for lncRNA HOTAIR (Applied Biosystem) from Assay on Demand. The results will be analyzed using a comparative method, and the values will be normalized to the β-actin and GAPDH expression as endogenous controls [29].

RESULTS

MTT Assay

Figures 1-2 represents the outcomes of HGFs treated with JetKit, Unifast and Duralay at two different time points (3 and 30 minutes). The HGFs viability was related to the resins concentration; high concentrations of the JetKit, Unifast and Duralay were always coupled with the low viability of HGFs. Therefore, considering the results of MTT, we decide to use 0,01mM as an appropriate concentration for our experiments, because the viability of the cells was over 80% after treatment with the acid resins. Moreover, 0,01mM represent the concentration certainly similar to that used in the dental practice.

Alamar Blue Assay

The cell proliferation has been studied through Alamar blue assay. The HGFs were treated with polyacrylic resins and the proliferation was followed for 24h, 7 and 14 days. Although the viability of the cells, did not change during the period of the experiments; while we observed some differences about the rate of viability related to the different resins used. In details, HGFs viability was limited by JetKit resin, compared with control. Conversely Unifast and Duralay did not affect the proliferation of HGFs but improved the viability, compared with the control (Figure 3).

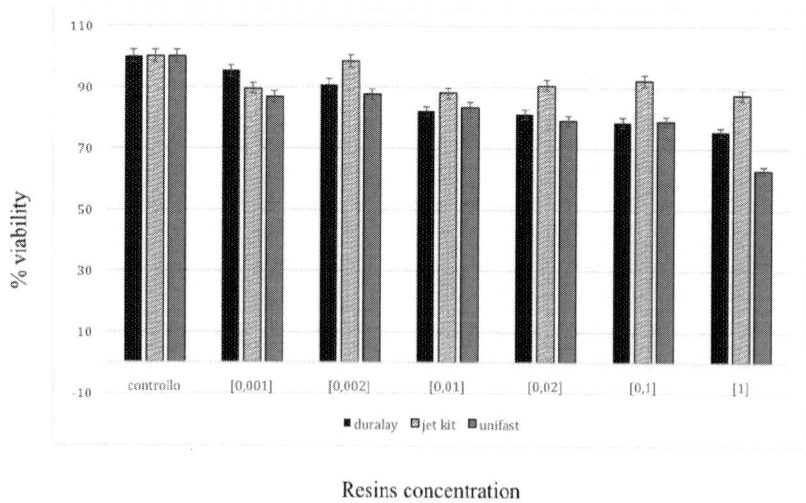

Figure 1. MTT assay in HGFs treated for 3 min with Duralay JetKit and Unifast. (0.01 mM was associated with a low rate of cell death).

QRT-PCR

The lncRNA HOTAIR gene expression has been analyzed through RealTime-PCR assay (QRT-PCR), in Human Gingival Fibroblasts treated with the specific medium conditioned by the resins Jet Kit, Duralay and Unifast prepared with described in the methods (indirect method: to 3 and 30 minutes after mixing). The first experiment was performed, stimulating the HGFs for 3 minutes and analyzing the HOTAIR gene expression. The outcomes did not show changes in the HOTAIR gene behavior after treatment with the resins. Conversely, we observed a different behavior in HGFs treated with Unifast, Duralay and JetKit (Figure 4). The HOTAIR mRNA was always hyper-expressed in HGFs treated with Unifast and Duralay resins compared with control; while JetKit was not able to regulates the HOTAIR gene expression in HGFs.

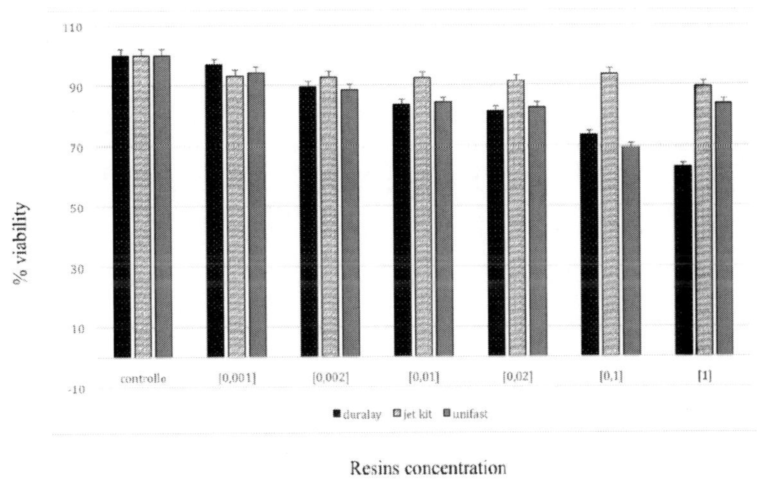

Figure 2. MTT assay in HGFs treated for 30 min with Duralay JetKit and Unifast. (0.01 mM was associated with a low rate of cell death).

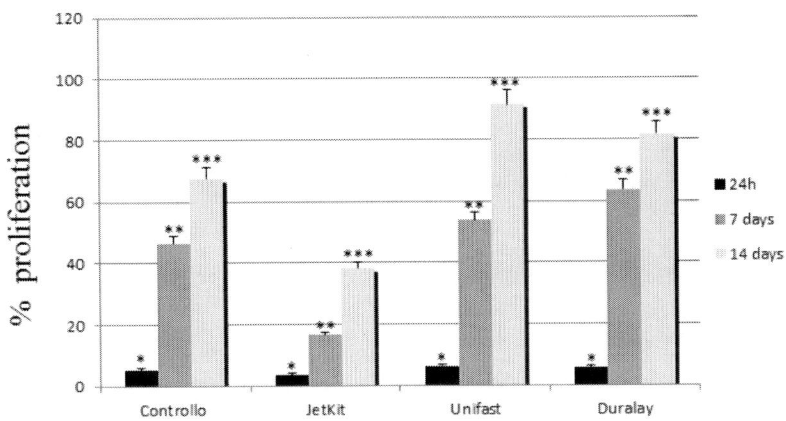

Figure 3. Alamar blue assay in HGFs treated with acid resins for 24h, 7 and 14 days.

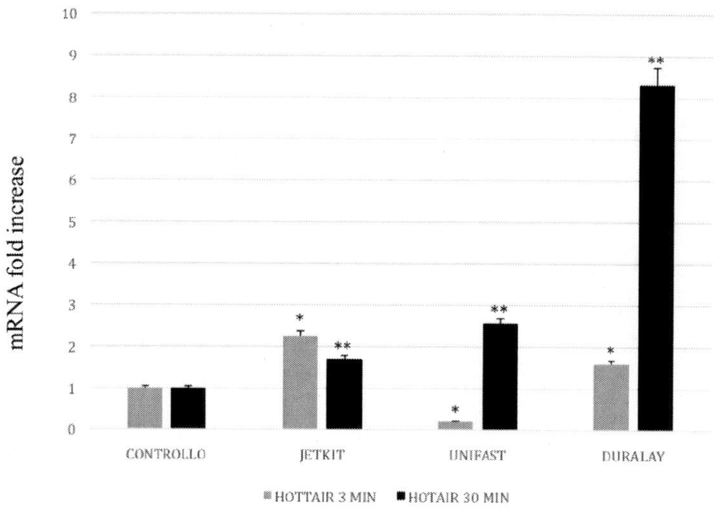

Figure 4. lncRNA HOTAIR gene expression in HGFs treated with resins at 0,01 mM (3 min. VS 30 min).

DISCUSSION

Several studies suggest that HOTAIR is able to control the epigenetic regulation of DNA through a deep relationship with chromatin, involving the recruitment of complexes which are able to identify specific sequences repressing and/or silencing several target genes [30, 31]. Moreover, HOTAIR is a powerful predictor of poor prognosis in different neoplastic disease, mainly by means of the EMT regulation process [32]. The lncRNA HOTAIR epigenetically promotes EMT silencing miRNA34a and interacting with Polycomb-Responsive-Element-2 (PRC2) [33]. Recently, it has been demonstrated that deregulation of EMT process, caused by abnormal activity of specific markers, leads to different disorders against human gingival epithelial cells. In particular, the overgrowth of gingival epithelium, *in vivo*, is due to the deregulation of several molecular targets that usually control the EMT process [34]. Therefore, the plasticity of epithelial tissue and uncontrolled alteration of the Epithelium-Mesenchyme transition, provides support to the hypothesis of the excessive gingival growth; this phenomenon determines the pathogenesis of fibrotic disease due to an increase of interaction between epithelium and connective tissue layers induced by a loss of the basement membrane [35]. In conclusion, based on the results obtained, we hypothesized that HOTAIR might induce the alteration of the cell memory program in HGFs due to deregulation of the EMT process, caused by direct contact between the gingival tissue with polyacrylic acid resins and this event is time dependent. Despite the results obtained, further studies and experiments, will be useful to better understand the role of ncRNAs (miRNA and lncRNAs) for the control of pathogenesis of gingival tissue.

REFERENCES

[1] Tammaro, L; Vittoria, V; Calarco, A; Petillo, O; Riccitiello, F; Peluso, G. Effect of layered double hydroxide intercalated with fluoride ions on the physical, biological and relase proprieties of a dental composite resin. *J Dent*, 2014 Jan, 42(1), 60-7.

[2] Rahul, B; Shaily, MB; Hongjun, L; Brajendra, M. *Biocompatible Denture Polymers – A Review*. Department of Metallurgical & Materials Engineering, Colorado School of Mines, Golden, CO 8040 2009.

[3] Cimpan, MR; Matre, R; Cressey, LI; Tysnes, B; Lie, SA; Gjertsen, BT; Skaug, N. The effect of heat-and auto-polymerized denture base polymers on clonogenicity, apoptosis, and necrosis in fibroblasts: denture base polymers induce apoptosis and necrosis. *Acta Odontol Scand.*, 2000 Oct, 58(5), 217-28.

[4] Oppenheimer, BS; Oppenheimer, ET; Danishefsky, AP; Stout, AP; Eirich, FR. Further Studies of Polymers As Carcinogenic Agents In Animals. *Cancer Res*, 1995, 15, 334-340.

[5] Kalburgi, V; Sravya, L; Warad, S; Vijayalaxmi, K; Sejal, P; Hazeil, Dj. Role of systemic markers in periodontal diseases: a possible inflammatory burden and risk factor for cardiovascular diseases. *Ann Med Health Sci Res.*, 2014 May, 4(3), 388-92.

[6] Khalaf, H; Lönn, J; Bengtsson, T. Cytokines and chemokines are differentially expressed in patients with periodontitis: possible role for TGF-β1 as a marker for disease progression. *Times new May*, 67(1), 29-35.

[7] Xin, H; Wang, YH; Liu, H. Zhonghua Kou Qiang Yi Xue Za Zhi. *Expression of cytokines of IL6 family in gingival fibroblasts by Toll-like receptor-2.*, 2012 Sep, 47(9), 523-7.

[8] Xin, H; Wang, YH; Liu, H; Zhonghua Kou Qiang Yi Xue Za Zhi. *Expression of cytokines of IL6 family in gingival fibroblasts by Toll-like receptor-2.*, 2012 Sep, 47(9), 523-7.

[9] Borelli, B; Zarone, Z; Rivieccio, V; Riccitiello, V; Simeone, M; Sorrentino, R; Rengo, S; Spagnuolo, G; Procino, A. Polyacrylic resins regulate transcriptional control of, interleukin-6, gp80, and gp130 genes in human gingival fibroblasts. *J Oral Science*, IN PRESS.

[10] Stein, LD. Human genome: end of the beginning. *Nature*, 431, 915-916, 2004.

[11] Ponting, CP; Belgard, TG. Transcribed dark matter: meaning or myth? *Hum Mol Genet*, 2010, 19, R162- R168.

[12] Brosnan, CA; Voinnet, O. The long and the short of noncoding RNAs. *Curr Opin Cell Biol*, 2009, 21, 416-425.

[13] Costa, FF. Non-coding RNAs: new players in eukaryotic biology. *Gene*, 2005, 357, 83-94.

[14] Lu, J; Getz, G; Miska, EA; et al. MicroRNA expression profiles classify human cancers. *Nature*, 2005, 435, 834-838.

[15] Farazi, TA; Spitzer, JI; Morozov, P; Tuschl, T. miRNAs in human cancer. *J Pathol*, 2011, 223, 102-115.

[16] Huarte, M; Rinn, JL. Large non-coding RNAs: missing links in cancer? *Hum Mol Genet*, 2010, 19, R152-R161.

[17] Spitale, RC; Tsai, MC; Chang, HY. RNA templating the epigenome: long noncoding RNAs as molecular scaffolds. *Epigenetics*, 2011, 6, 539-543.

[18] He, S; Liu, S; Zhu, HT. The sequence, structure and evolutionary features of hotair in mammals. *BMC Evolut Biol*, 2011, 11, 102.

[19] Kogo, R; Shimamura, T; Mimori, K; et al. Long noncoding rna hotair regulates polycomb-dependent chromatin modification and is associated with poor prognosis in colorectal cancers. *Cancer Res*, 2011, 71, 6320-6.

[20] Rinn, JL; Kertesz, M; Wang, JK; Squazzo, SL; Xu, X; Brugmann, SA. Functional demarcation of active and silent chromatin domains in human hox loci by noncoding RNAs. *Cell*, 2007, 129, 1311-23.

[21] Sollid, LM. Molecular basis of celiac disease. *Annu Rev Immunol*, 2000, 18, 53-81.

[22] Wang, J; Chen, D; He, X; et al. Downregulated lincRNA HOTAIR expression in ovarian cancer stem cells decreases its tumorgeniesisand metastasis by inhibiting epithelial-mesenchymal transition. *Cancer Cell Int*, 2015, 15, 24.

[23] Capuano, M; Iaffaldano, L; Tinto, N; et al. MicroRNA-449° overexpression, reduced NOTCH1 signals and scarce goblet cells characterize the small intestine of celiac patients. *PLoS One*, 2011, 6, e29094.

[24] Center, MM; Jemal, A; Ward, E. International trends in colorectal cancer incidence rates. *Cancer Epidem Biomar*, 2009, 18(6), 1688-94.

[25] Sollid, LM. Molecular basis of celiac disease. *Annu Rev Immunol*, 2000, 18, 53-81.

[26] Procino, A. lncRNA HOTAIR, HOXC11 and HOXC12 Gene Expressions are Upregulated in CaCo2 Cells Treated with P31-43 Toxic Peptide. *Current Signal Transduction Therapy*, 2016, 11, 33-36.

[27] Wu, J; Xie, H. Expression of long noncoding RNA-HOX transcript antisense intergenic RNA in oral squamous cell carcinoma and effect on cell growth. *Tumour Biol.*, 2015 Nov, 36(11), 8573-8.

[28] Procino, A; Simeone, M; Spagnuolo, G. Role of IL6 and Its Receptors (gp80, gp130) In TEGDMA and HEMA Cytotoxicity. *Int. J. of Res. Stud. in Sci., Eng. and Tech.*, 2011, Volume 2, Issue 7, July PP 74-76.

[29] Procino, A. The Paralogous Group HOX 13 Discriminates between Normal Colon Tissue and Colon Cancer. *Molecular and Genetic Medicine*, 2014, 8, 3.

[30] Padua Alves, C; Fonseca, AS; Muys, BR; et al. The lincRNA Hotair is required for epithelial-to-mesenchymal transition and stemness maintenance of cancer cell lines. *Stem Cells*, 2013, 31(12), 2827-32.

[31] Tano, K; Akimitsu, N. Long non-coding RNAs in cancer progression. *Front Genet*, 2012, 3, 219.

[32] Wang, J; Chen, D; He, X; et al. Downregulated lincRNA HOTAIR expression in ovarian cancer stem cells decreases its tumorgeniesis and metastasis by inhibiting epithelial-mesenchymal transition. *Cancer Cell Int*, 2015, 15, 24.

[33] Liu, YW; Sun, M; Xia, R; et al. LincHOTAIR epigenetically silences miR34a by binding to PRC2 to promote the epithelial-tomesenchymal transition in human gastric cancer. *Cell Death Dis*, 2015, 6, e1802.

[34] Sume, SS; Kantarci, A; Lee, A; Hasturk, H; Trackman, PC. *Epithelial to mesenchymal transition in gingival overgrowth.*, 2010 Jul, 177 (1), 208-18.

[35] Kantarci, A; Nseir, Z; Kim, YS; Sume, SS; Trackman, PC. Loss of basement membrane integrity in human gingival overgrowth. *J ent Res.*, 2011 Jul, 90(7), 887-93.

In: Stem Cells Origins of Cytokine Interactions
Editor: Prasad S. Koka
ISBN: 978-1-53615-302-6

Chapter 14

A CONCEPT OF INDUCING INNATE REGENERATION THROUGH CELL-FREE REGENERATIVE MEDICINE: A POSSIBLE THERAPEUTIC APPROACH FOR THE ISCHEMIC CEREBROVASCULAR INSULT

Jyothi Navada Suryanarayana **and** ***Prasad S. Koka***
Haffkine Institute, Mumbai, India

ABSTRACT

Cerebro vascular insult is commonly known as stroke which is caused by interruption of blood supply and rupture of blood vessels or due to the abnormal vascular structure. This leads to focal tissue loss and death of neural cell population. Even though stem cell transplantation is considered as likely treatment choice, it will be associated with certain unfavorable outcomes such as tumor formation, immune mediated graft rejections. Studies have shown that regenerative effects promoted by stem cells in larger part through their secreted bio-factors and exosomes. Neural stem cells are present in some parts of the adult brain they participate in regeneration events. As the result ischemic shocks, capability of synthesizing trophic factors and signaling molecules are severely compromised in injured areas of Central Nervous System. Indeed, several experimental studies shed lights on practical applicability of neurotrophic factors (NTFs), MSCs secretome and exosomes in inducing regenerative repair in the nervous system. In the case of stroke, the major concern is to protect the penumbra by limiting consequential cell death, further enhancing the regeneration and repair. Therefore, providing neuroprotective and regenerative support through exogenous NTFs, stem cell secretome and exosomes or by stimulating their endogenous secretion can help to compensate tissue loss, regain neural plasticity and functional recovery after stroke. Intranasal

* Corresponding Author Email: jyothi.navada@gmail.com.

administration of NTFs and stem cells derived neuro therapeutics ensures efficient, safe and direct delivery to central nervous system which bypasses blood brain barrier. This cell-free therapeutic strategy appears to be safe and effective also eliminates the invasive transplantations and its associated adverse side effects.

Keywords: neurotrophic factors, stem cell secretome, exosomes, stroke, neural regeneration

INTRODUCTION

Cerebral ischemic insult results from blood vessel occlusion or damage leads to loss of neurons and glial cells in the brain or spinal cord. Among neural cells, neurons are particularly sensitive to various injuries such as ischemia, hypoxia, trauma and infections. Penumbra is the area surrounding an ischemic event where the cells potentially destined for the infraction but not get irreversibly injured. The neurons that are deprived of oxygen and metabolic substrates lead to structural damage to neuronal networks and functional impairments of sensation, movement and cognition. Due to the failure of energy dependent process in neurons distorts the transmembrane ionic gradients which lead to ion and water imbalance. This is in turn responsible for the apoptotic and necrotic cascades. However, the penumbra may viable for certain hours after the onset of ischemic shock due to minimal blood supply through collateral arteries. As the time elapse, the size of the penumbra tends to decline. Therefore, a major concern is to rescue the penumbra by inhibiting cell death. Further, enhancing the activity-dependent rewiring and synapse strengthening could open a time-limited therapeutic window immediately after ischemia, during which the greatest gain in recovery occurs.

It has long been believed that lineages of neurons and microglia become progressively segregated during development and will not differentiate again once they have matured. However, now it is revealed, through many studies, that neural stem cells (NSC) are present even in adult mammalian brain. Neurogenesis and oligodendrogenesis takes place continuously throughout the lifespan. Multipotent stem cells can be isolated from the adult brain which resides in the hippocampal dentate gyrus (DG) and subventricular zone (SVZ) of striatum. Some studies have shown that new neurons are also generating from other region of adult brain such as amygdala [11, 14 and 15], hypothalamus [12], cortex, striatum [16] and substantia nigra [17]. However, it appears that neurogenesis in these regions takes place under known physiological conditions at very low levels [65].

The germinal precursors of the subventricular region provide new neurons to the olfactory bulb, which comprises migrating neuroblasts, immature precursors, astrocytes and ependymal cells. The SVZ astrocytes act as neural stem cells in both the normal and regenerating brain [69]. Hilus, subgranular layer and granular cell layer are the three layers of DG which generates stem cells for the hippocampus. In response to ischemic injury the subgranular zone of DG enhances the proliferation of neuronal and glial progenies [1, 70, 71]. Then these derived progenies from the stem cells selectively migrate into lesioned sites of the brain and differentiate into new neurons/glial cells there by improving the functional deficits in rat stroke models of focal ischemia and cerebral hemorrhage [3, 72]. This compensatory adaptive response to ischemic injury could restore the brain function after ischemia.

During post-ischemic events, the sparse ongoing proliferation in the brain may be attributable to a limitation in the availability of certain trophic factors and regulatory cues even when the progenitor cells are present. Neuronal progenitor cells (NPCs) located just beneath the SVZ undergoes differentiation and produce new cells after ischemia in neonatal brains. Newly generated cells subsequently migrate to the infract region and differentiate into neuronal precursor cells in order to compensate the lost neural cells as the neonatal brain possess much greater plasticity [117]. Regenerating axons respond selectively to trophic cues to rejuvenate and grow towards the target organs which they originally innervated [93]. So, recreating the plasticity by providing exogenous supporting factors can be a potential therapy in treating perinatal brain disorders.

Growth factors, cytokines, chemokines and neuro regulatory factors play an essential role in neurogenesis. The ischemia/hypoxia induced progenitor cell proliferation and plasticity can be modulated by providing external factors such as stem cells secretome (SCS), NTFs, stem cell derived exosomes. Therefore, these cell-free neuro-therapeutics can induce the innate regeneration in a much more efficient, simpler and noninvasive manner.

Neurotrophic Factors (NTFs)

Neurotrophic factors are a family of biomolecules, nearly all of which are peptides or small proteins. NTFs specifically influence the neuronal functions. It induces neuronal differentiation, survival and growth during embryonic development. Once after maturation it helps by continued maintenance of neurons and promotes regeneration after injury. They also involved in the maintenance of synaptic plasticity and formation of long lasting memories. NTFs are classified into families according to the type of binding receptors. (Refer Table 1 for classification and functions of NTFs).

NTFs control intracellular signaling pathways that sculpted neuronal circuits during brain development and fundamentally regulate plasticity as well as cell survival in the adult brain. Normal differentiation of neural stem cells to form neurons and glial cells are also under the influence of NTFs. NTFs play significant role in proliferation, migration, differentiation and production of the extracellular matrix of the nervous system through influencing the target cells by autocrine, paracrine or juxtracrine mechanisms. NTFs such as BDNF, FGF, NGF, GDNF and TGF-β1 secretions are up regulated in brain cells under certain pathological conditions including ischemia. The different temporal and cellular distribution profiles of NTF families which are induced in various neural cell types (neuronal, glial, pial) at particular stage after ischemia suggests the different roles of NTFs for each neural cell type. However, the NTFs levels secreted at the site of injury will not be sufficient to generate compensatory effects of NTFs.

Several studies have shown that administration of exogenous NTFs after nerve injury mimics the effect of target organ derived trophic factors. The Neurotrophins BDNF, NGF and NT3 show a well-defined and selective beneficial effect on the survival and phenotypic expression of primary sensory neurons in dorsal root ganglia and of motor neurons in spinal cord after axotomy [4]. *In vitro* assays have shown that NTFs enhances both axonal and dendritic growth. Evidence from the experiment using conditional knockouts are by crossing Neurotrophin (NT) knockouts with mouse mutants lacking pro-apoptotic genes. The results suggested that long distance peripheral sensory axon growth *in vivo* is NTF dependent.

There are many neuronal growth factors and it would be impossible to give a comprehensive review on the role of each NTF in neuronal regeneration. Therefore, this article will only focus on some relevant facts of NTFs which supports the theme of utilizing NTFs in regenerative therapy for cerebral ischemia.

Nerve Growth Factors (NGFs)

NGF is the first discovered NTF. NGF is specific for a subset of primary sensory neurons and sympathetic neurons [95]. Increased expression of NGF and P75 promotes the Schwan cell proliferation. Proliferated Schwan cells and recruited macrophages at the site of injury together secrete mitogen and growth factors which play a role in axonal regeneration and remyelination [94]. NGF and other factors released by Schwan cells may potentiate axonal regeneration by up regulating the expression of neural cams [90]. In neurons, NGF helps to inactivate the 'bad' protein by phosphorylating it via PI-3 Kinase/PKB signaling pathway. Under these conditions Bcl-2 can inhibit the activity of Bax. This prevents cytochrome-C release from mitochondria and blocks the activation of 'procaspase 9' by Apaf -1[73]. NGF had shown to enhance retinal ganglion cell survival and induce the recovery of retinal neurons after ischemic insults in rat models. Intra ocular NGF treatment in induced transient complete ischemia of cat's retina was effective in enhancing the survival and function recovery of retinal neurons [2]. NGF is also responsible for increased BDNF synthesis in a majority (~90%) of trkA expressing DRG cells [4, 89].

Table 1. Classification and roles of NTFs

Type of Receptors	**Neurotrophic Factor Family**	**Neurotrophic Factors as Ligands**	**Roles of NTF**
Tyrosin Kinase Receptors	EGF Receptor Family	EGF	Induce cell proliferation and differentiation, generates CNS progenitor cells [120, 121]
		HB-EGF	Regulates survival of mid brain Dopaminergic neurons [122].
		TGFα	Stimulate Neuron cell proliferation in the adult injured brain [123].
	Eph Family	Ephrin (LERK)	Negatively regulates adult neurogenesis [124].
	FGF Receptor Family	FGF 1-9	Induce angiogenesis, Inhibit apoptosis [126].
	HGF Receptor Family	HGF	Promote cell progenitor cell mobilization, proliferation, induce myelin repair, angiogenesis, inhibit autoreactive T-cell responses
		MSP	Suppresses NO synthesis in the injured hypoglossal nuclei, play a role in brain plasticity and regeneration [113].
	Insulin Receptor Family	Insulin	NSC self-renewal, neurogenesis, cognition and sensory functions, stem cell homeostasis in CNS. [108]
		IGF-1	Regulates cell proliferation, division, migration and development, inhibit apoptosis and NSC maintenance in adult CNS [127].
	NGF Receptor Family	NGF	Protect neural cells, axonal regeneration and remyelination [138].
		BDNF	Promote neuronal cell differentiation, survival, GLIAL scar formation and prevention of cell death [91 and 92, 103].
		NT3	Axonal regeneration, stimulates neurite out growth, restores sensory motor conduction velocity [4]
		NT4/5	Promotes neurite extension and survival of cerebellar granule cells [111]
	PDGF Receptor Family	PDGF A & B	Induces cell proliferation, Oligodendrogenesis, Immune regulation. [1, 9]
		CSF-1	Induce stem/progenitor cell proliferation and promote neuronal differentiation [62]
		Steel Factor	Supports survival of melanogenic subpopulation [128].
	Ret Receptor Family	GDNF	Facilitate regeneration of the target cells, promote neuronal cell survival, axonal growth, prevents cell death [129].

Table 1. (Continued)

Type of Receptors	Neurotrophic Factor Family	Neurotrophic Factors as Ligands	Roles of NTF
		Neuturin	Development and maintenance of neurons [130].
		Artemin	Generation, survival and growth of sympathetic neurons at different stages of development [131].
		Persephin	Promote the survival of motor neurons and dopaminergic neurons [132].
	Tie Family	Angiopoitein-1 & 2	Induces angiogenesis, promote cell survival. [62]
	VEGF Receptor Family	VEGF-A, B & C	Induce angiogenesis, Inhibit apoptosis and immune regulatory [133].
		PIGF	Postnatal neurogenesis.
Cytokine Receptors		CNTF	Promote neuronal cell survival and antioxidant effects [34].
		LIF	Stimulates the progenitor proliferation [9] Neural stem/progenitor cell expansion [100]
		IL-2	Influence neurodevelopment in brain and auto immunity [107]
		IL-6	Stimulate stem or progenitor cell proliferation and Induce angiogenesis [62]
Serine/Threonine Kinase Receptor (TGF-β Super Family)	BMP Family	BMP2-8	Promote neurogenesis, angiogenesis and regulate tissue homeostasis [134].
		OP1 & 2	Increases dendritic growth and branching in dissociated embryonic cortical neurons [112]
		GDF 5-7	Regeneration, neuronal differentiation and maintain cell survival [135].
	Activin Family	Activin βA & βB	Regulates neuronal regeneration by suppressing inflammatory response in adult CNS [106]
	TGF-β Family	TGF-β1~3	Induces stem cell differentiation and suppress inflammation [136].

In vivo administration of NGF enhances the expression of Calcitonin Gene Related Protein (CGRP) and Substance P (SP) in motor neurons, small and medium sized sensory neurons which are previously deprived due to nerve injury [118, 119]. CGRP when synthesized in ventral horn of this spinal cord contribute to the regeneration of nervous tissue after injury. On the other hand, SP and its induced cytokines promote multiplication of cells required for repair, replacement and growth of new blood vessels.

Brain Derived Neurotrophic Factor (BDNF)

Brain Derived Neurotrophic Factor (BDNF) is one of the prominent and most widely distributed factors within the CNS. BDNF acts through trkB. BDNF is a potent stimulator of neurogenesis in adult mammalian CNS. This displays trophic effects on a wide range of nerve cells including hippocampal, dopaminergic, cerebellar and cortical neurons *in vitro* [7]. BDNF efficiently prevents neuronal cell death after traumatic and toxic nerve injuries and crucially promotes synaptic as well as axonal plasticity associated with learning memory and sensory motor recovery [5]. BDNF stimulates neuro plasticity at several levels including axonal growth, dendritic branching and synaptic transmission [91, 92].

Intracerebroventricular administration of BDNF) to the adult rat fore brain in conjunction with the cell proliferation marker bromodeoxyuridine (Brdu), resulted in numerous $Brdu^+$ cells not only in SVZ lining the infused lateral vertical but also in the striatal and septal parenchyma. A higher number of $Brdu^+$ cells were found in a discrete region of the thalamus and hypothalamus bordering the third ventricle. This experimental outcome demonstrates that adult brain parenchyma may regenerate new neurons which could replace those cells lost as a result of injury or disease [7].

Intravenous treatment of BDNF after focal cerebral ischemia in the model of temporary Middle Cerebral Artery Occlusion (MACO) rats resulted in 50% reduction in infract volume. This BDNF treatment significantly downregulates the pro-apoptotic 'Bax' expression in cortical neurons of the penumbra and upregulates the expression of anti-apoptotic protein 'Bcl-2' [6].

Daily intravenous bolus application of BDNF during the first five days after cortical stroke induces hippocampal neurogenesis and significantly improved sensory motor recovery during a 6-week period. BDNF substantially increased neurogenesis in the DG and enhanced the migration of SVZ progenitor cells to the nearby striatum of the ischemic hemisphere. BDNF acts as a modulator of neuro-genesis in the brain and improves a long-term functional neurological outcome after cerebral ischemia [5].

Glutamate triggered peroxide accumulation leads to excitotoxicity by interfering with the Ca^{2+} homeostasis, which is thought to be one of the major causes of cellular death after ischemia [101]. BDNF protects the neurons against neurotoxicity by suppressing the glutamate triggered peroxide accumulation and also by reducing cytotoxic actions of Nitric Oxide (NO) [102, 103].

Ciliary Neurotrophic Factor (CNTF)

Ciliary Neurotrophic Factor (CNTF) is a member of cytokine family. It has trophic effects on ciliary, motor, sensory, sympathetic, and hippocampal and retinal neurons.

CNTFs expressed almost exclusively in the nervous system and it regulates the patterns of normal neurogenesis.

CNTF consistently promotes moto-neuron survival in neonatal animal models following axotomy and *in vitro*. CNTF has also shown a direct effect in adult moto-neuron survival *in vivo* [96, 97]. Systemic CNTF administration enhances muscle fiber innervation and intra muscular nerve branching as well as increases the myelination of regenerative axons [98, 99].

Bicistronic $rAAV_2$ encoding CNTF and green fluorescent fusion protein ($rAAV_2$ – CNTF – GFP) was injected in to rats injured eyes. Fluorogold mediated retrograde labeling method revealed that there was a significant increase in the retinal ganglionic cell (RGC) survival and axonal regeneration [26].

When CNTF is injected into the fore brain of adult C57BL/6 mice, it increased the number of cells ($Brdu^+$) in the both neurogenic regions. In DG, CNTF also enhanced the differentiation of precursors into neurons. In SVZ, CNTF induces the differentiation of precursors into astrocytes ($GFPA^+$) in proportion of ($GFAP^+/Brdu^+$) cells from 1% to 2%. CNTF also enhanced the distance of migration of new neurons [29].

A study on post ischemic regulatory changes in the expression of CNTF and its receptors CNTRFα had shown differential regulation of CNTF and CNTRFα in activated astrocytes of the post ischemic hippocampus. This suggests that they are involved in the regulation of astrocyte responses and neuronal reorganization occurring after ischemic insults [33].

A recent study has shown that in MCAO adult male C57BL/6 mice, CNTF expression paralleled changes in the number of proliferated $Brdu^+$ SVZ cells. However, stroke induced proliferation was absent in CNTF -/- mice [34]. This indicated the fact that endogenous and exogenous CNTF mediates the stroke induced neurogenesis in the adult CNS.

Glial Derived Neurotrophic Factor (GDNF)

Glial derived neurotrophic factor (GDNF) belongs to the transforming growth factor beta (TGF-β) superfamily. GDNF plays a very potent role in promoting differentiation and survival of dopaminergic and motor neurons. Now it has been shown to have prominent trophic activity on many neural cell types of CNS. GDNF has got neuro protection activity against neurotoxin mediated brain damages caused due to elevated intracellular free radicals and mitochondrial respiratory enzymes [36].

In injury induced adult motor-neuron de-generation mouse models, injuring motor-neuron axons at the exit point of the nerve from the spinal cord resulted in a completed loss of motor-neurons by six weeks following surgery. *In vitro* and *in vivo* GDNF treatment prevented the half of the loss and enhanced the population of surviving motor-neurons [27].

In an experiment, TAT protein of 11 amino acid length encoded by Human Immunodeficiency Virus (HIV) with a capacity to cross the blood brain barrier was coupled with GDNF to create TAT-GDNF fusion protein. TAT-GDNF (0.6 nmol) or vehicle was intravenously applied to induced MCAO rat models. This resulted in the significant reduction of the caspase-3 in immune reactive and DNA fragmented cells and viable neurons were increased in the striatum, also the size of the infraction was significantly reduced [28].

Intracerebro ventricular infusion of GDNF and insulin like growth factor-1 (IGF-1) in to ischemic rat models had significantly increased ischemia induced progenitor cell proliferation by 4-fold in DG and by 2-fold in the SVZ [38]. Acute GDNF infusion in ischemia/hypoxia induced neonatal rats shown to protect neurons against oxidative stress through prominently increased levels of glutathione peroxidase, superoxide dismutase and catalyze activity [37, 39]. GDNF regulates endogenous levels of X-linked inhibitor of apoptotic protein and neuronal apoptosis inhibitory protein in motor neurons after sciatic nerve axotomy [41]. Kitagawa et al. (1998) experimentally shown that the topical application of GDNF causes a major reduction in the infract size by 48% accompanied by the mark able reduction in TUNEL and caspase positive cells in the affected areas of MACO rats.

Intrathecal administration after axotomy enhances the IB4 labeling in DRG. The IB4 Isolectin specifically recognizes microglial cells in the activated amoeboid or phagocytic states during cerebral ischemia. Isolectin IB4 recognizes the functional ret for GDNF, 'Ret' under the normal and ischemic condition of the brain [4, 44]. GDNF treatment simulates the activation of Ret – dependent signaling and its downstream effector ERK in dose and time dependent manner in a population of IB4 positive cells of the CA_3 region in hippocampal organotypic cultures [43]. In the hippocampus, the transient fore brain ischemia induced the expression of Ret mRNA with a peak effect at 12 hours whereas administration of GDNF through intra hippocampal microinjection (1.0 μg) induced the Ret mRNA expression with a peak at 6 hours – 12 hours in sham operated rats [40, 42]. Ret activation in microglia might provide a trophic signaling after brain damage in the post ischemic brain [44]. N-methyl-D-aspartate (NMDA) binds to a type of glutamate receptor that participates in excitatory neurotransmission which may particularly involve in cell damage and cognitive defects when stimulated excessively. Upon NMDA exposure, an increase of Ret phosphorylation occurred in microglial cells in CA_1 and CA_3 following GDNF exposure. Such neuroprotection upon excitotoxicity involves a GDNF mediated modulation of microglial cell activity [43].

Neurotrophin-3 (NT-3)

Neurotrophin-3 (NT-3) plays a key role in neuronal differentiation during development and brain injury. Continuous infusion of NT-3 following axotomy had shown to be effective in restoring sensory motor conduction velocity in dose dependent manner [4, 88].

Intrathecal administration of NT-3 enhanced the expression of P75 and recovery of receptor trkC in DRG and induced neurite outgrowth [4]. NT-3 also appears to promote the regeneration of various sub population of DRG neurons [86, 87]. *In vivo* targeted administration of NT-3 to the cut sciatic nerve using fibronectin mats increased the rate and amount of axonal regeneration [4, 86].

Leukemia Inhibitory Factor (LIF)

Leukemia inhibitory factor (LIF) participates in the expansion of neural stem/progenitors after perinatal hypoxia/ischemia. LIF receptor (LIFR)/9p130 receptor heterodimer induces Notch1 to maintain NSP population during development. LIF induces stat3 phosphorylation in neuro sphere which enhances the Notch1 expression and increases notch activation *in vivo*. Notch signaling plays a major role in the expansion of neural tripotent stem cells NSP within the SVZ subsequent to ischemia or perinatal hypoxia [100].

Vascular Endothelial Growth Factors (VEGFs)

Ischemia mediated hypoxic condition induces VEGF gene expression in brain cells. But, the extent of sensitivity to induce VEGF gene may vary across the neural cell type. Within 2-5 days after ischemic insults the neovascularization begins and continues up to months.

Therefore, rapid and sustained induction of VEGF in neurons may help to reconstruct the vascular system in affected area [1]. VEGFs play a significant role in angiogenesis which is essential in the treatment of ischemia. Human Mesenchymal Stem Cells (hMSC) were transfected with angiopoietin-1 gene and VEGF gene (Ang-vegf-hMSC) and intravenously infused to the induced MCAO. Rats receiving Ang-vegf-hMSC had shown significant structural and functional recovery which is revealed by MRI and behavioral analysis [32].

MESENCHYMAL STEM CELLS, THEIR SECRETOME AND EXOSOMES IN NEURO THERAPEUTICS

At present, stem cell therapy has been a promising treatment option for most of the neurological disorders including stroke, Alzheimer's disease, Huntington's disease, Parkinson's disease, Amyotrophic lateral Sclerosis. Among many qualifying stem cell sources, mesenchymal stem cells and neural precursor cells are extensively investigated for their ability to communicate with the host cells upon transplantation in experimental CNS diseases.

MSCs are self-renewing clonal precursors of non-hematopoietic cells, generally isolated from the bone marrow (BM-MSC). MSCs can also be obtained from the alternative tissue sources including adipose tissues, placenta, dental pulp, umbilical cord blood (HVCPVCS), the Wharton's Jelly (WJSCS), olfactory mucosa, lung, spleen, deciduous teeth and also from the brain [9]. Therefore, MSCs can be taken from the autologous and allogenic sources when compared to neural progenitors. MSCs both from autologous and allogenic sources had shown beneficial clinical outcomes in a wide range of therapeutics due to their multi-lineage differentiation potential and by providing trophic support to neighboring cells [26, 61].

Several studies have shown that after transplanting MSCs or intravenous administration of BM-MSCs, these cells were able to promote neurogenesis, survival of neuroblasts, behavioral improvement and reduce the volume of lesion in stroke models [20, 21] by modulating immune responses, angiogenesis, apoptosis, oxidation levels, migration and differentiation through stimulating surrounding cells [25]. Like BM-MSCs, adipose stromal cells (ASCs) therapy also enhances angiogenic and neurogenic process [10]. But there are certain serious risk factors may have also associated with stem cells transplantations, including Immune mediated rejection, senescence induced genetic instability or loss of function, limited cell survival. Another major problem is the possibility of malignant transformations [64]. Moreover, highly purified populations of neural cell types derived from MSCs are NSCs may contain other neuronal or glial cell types that may not produce predictable interaction with the host neurons [3].

The disproportion between low rates transplanted/administered MSCs and tremendous functional recovery and persistence denotes an indirect primary mechanism other than the structural integration into ischemic region [24]. Indeed, in recent years it is becoming largely accepted that regenerative effects included by MSCs are mainly because of their secreted factors [18]. MSCs secretome is composed of a variety of proteic soluble fractions such as growth factors, cytokines, chemokines, metabolites and bioactive lipids [9, 10] that regulate physiology of host cells in an autocrine are paracrine manner which orchestrating in multiple interactions with surrounding microenvironment. In addition to secretomes, a vesicular fraction composed of micro-vesicles and exosomes which are involved in delivering proteins cargos and genetic contents (RNA and miRNA) to the target cells. Now it is clear that engrafting MSCs at the sight of nerve injury is not necessary [64]. Therefore, by harnessing the therapeutic properties of secretome factors and exosomes in cell-free regenerative protocols can promote innate regeneration through regulated differentiation.

Stem Cell Derived Exosomes as Cell-Free Neurotherapeutics

Exosomes are the minute, membrane bound nano vesicle (30-150 nm). They are endocytically originated from various cell types either constitutively or upon induction. They participate in the complex intercellular communication process by shuttling cargoes between different cells. Exosomes are derived from specialized intracellular compartments such as multi-vesicular bodies (MVBs) or late endosomes. The bilayer encapsulation of exosomes allowing cargo to travel long distances by providing balanced micro-environment to ensure the protection against degradation [8]. Once secreted, they can interact with the cells localized in the proximity of their origin or they can deliver the contents to the target cells which are situated remotely.

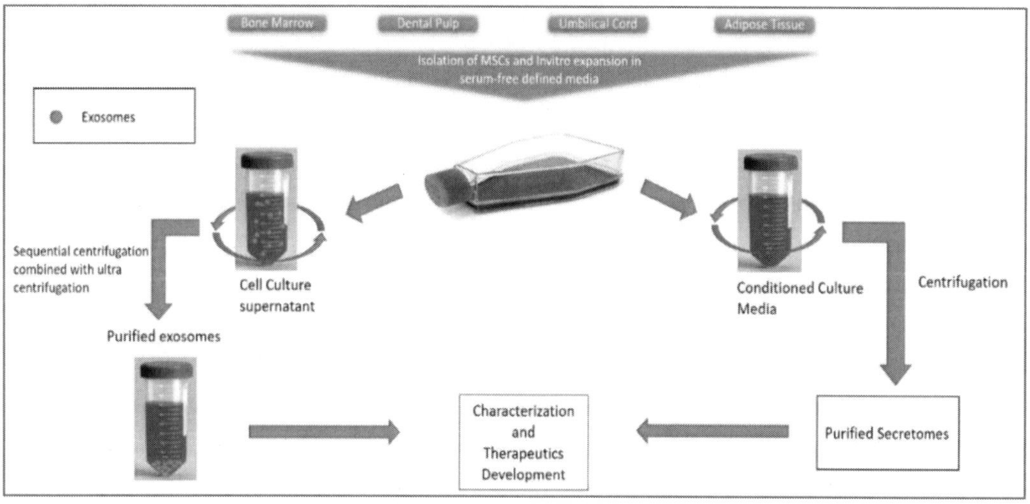

Figure 1. MSCs are isolated from bone marrow, dental pulp, umbilical cord and adipose tissues are subjected to *in vitro* expansion. Exosomes and secretomes of MSC are purified from the culture medium for the therapeutics development.

Characteristic Features of Exosomes

The potential clinical values of exosomes in regenerative therapies can be attributed to its features such as

- They potentially mediate the regenerative repair *in vivo*. They are safer since they are non-viable and will not form tumors.
- Less immunogenic and negligible allo immunogenicity.
- These nano vesicles are better tolerated in the body since they are naturally produced, ubiquitously present in the biological fluids and have an intrinsic homing ability.
- Long half-life in circulations and their ability to cross the blood brain barriers, more importantly, they potentially mediate the regenerative repair *in vivo*. [8].
- Exosomes can modify the physiology of the recipient cells via horizontal transfer of bioactive contents such as cytokines, growth factors, lipids, mRNA and miRNA or pre-miRNA.
- Exosomes can be stored without using toxic cryo preservatives at -20°C for 6 months with their biochemical activity intact.
- Encapsulation and protection of their content from degradation *in vivo*.

The unique characteristic compositions of the membrane proteins and lipids exosomes play a vital role in membrane trafficking, secretion/uptake and fusion with target cells [61 and 62]. Because of the reason the exosomes are originated from MVBs they contain transport and fusion proteins (annexes, GTPases and flotillin), tetraspanins (CD9, CD63, CD81 and CD82), heat shock proteins (HSP70, HSP90), proteins involved in MVB biogenesis (TSG101 and Alix), lipid related proteins and phospholipases [84, 85]. The tightly packed areas of sphingolipids and cholesterol generally known as lipid rafts are present in the exosomal membrane which serves as platforms to execute certain key activities such as membrane trafficking, cell adhesion and endocytosis. [81, 82 and 83].

Of the cell types known to produce exosomes the clinically tested human MSC is the most prolific exosomes producer which are readily available, proliferative and immunosuppressive. MSC exosomes exhibited potential therapeutic values in animal disease models [53]. Among the stem cell derived exosomes MSC derived exosomes are widely used in neurological disease researches [47]. When MSC exosomes administered locally or systemically, they retain the homing properties and mimics the therapeutic effects of the MSCs. Like their cell source, MSC exosomes regulates optimal tissue functioning by influencing housekeeping biological processes which are critical for maintaining tissue homeostasis which enables to regain vital cellular functions and promotes repair and regeneration.

miRNA Mediated Therapeutic Activities of Exosomes

The number of administered MSCs recruited in the ischemic brain after treatment is far too small to be in contact with all the parenchymal cells. MSCs release miRNA contained in their secreted exosomes which alter gene expression in the recipient neural cells [46, 51]. Restoration of tissue function in multiple disease/injury models and induce beneficial *in vitro* effects of MSC derived exosomes are mainly mediated by miRNA. miRNA is a class of non-coding RNA molecule consisting of 19-24 nucleosides, involved in homeostasis, past translational modification of gene expression [104]. Exosomes ensure the protected internal micro environment and deliver the cargoes to the target cells. miRNA packaging occurs non-randomly and specific miRNA populations are preferentially sorted and shuttled into MVBs [105].

An experiment conducted by Yu et al. [137] in adult zebra fish model of spinal cord injury, this study has revealed the role miR-133b in functional recovery. They inhibited miR-133b expression by using morpholino antisense oligonucleotides. This resulted in a significantly impaired locomotor recovery and regeneration of axons due to decreased miR-133b expression. MCAO rats were treated with MSCs, there was a considerable increase in miR-133b level in the ipsilateral hemisphere and when MSCs were exposed to ipsilateral ischemic extracts *in vitro* from rats subjected to MCAO, there was a significant increase in miR-133b levels in MSCs and their exosomes.

When neurons and astrocytes are treated with MSC derived exosomes enriched factors exposed to 72 hrs. post MCAO brain extracts in primary culture shown a significant increase in miR-133b levels and neurite branch number. This study demonstrates the fact that transfer of miR-133b to brain parenchymal neural cells via exosomes in turn regulates the neurite outgrowth [46]. Lee et al. (2014) reported that BM-MSCs delivered miR-124 and miR-145 to neural progenitor cells (NPCs) and astrocytes via contact independent and exosomes mediated mechanism [51].

MSC exosomes also can downregulate target gene expression. For example, RhoA protein expression was downregulated through miR-133b in neurons. Inhibition of RhoA stimulates neurite outgrowth and there by improves functional recovery after stroke [46, 48]. At the site of injury Connective tissue growth factor (CTG-F) is considered as a major inhibitor of axonal growth. MSC exosomes mediate the transfer of miR-133b to astrocytes may down regulate CTG-F expression which subsequently leads to increase in neurite outgrowth and diminishes the glial scar [46, 47].

Under certain circumstances brain produces a higher number of neuronal stem cells that release exosomes packed with a cocktail of miRNA and protein. The NSC neuron lineages are exposed to a complex mix of exosomes within the neurogenic niche. Exosomes secreted locally by different niche cells can modify the biochemistry of other niche cells and regulates the progression of different neurogenic stages.

Therapeutic Applicability of Exosomes Supported by Animal Studies

Some studies have shown that in animal models after traumatic brain injury exosomes treatment greatly enhances the number of newly generated endothelial cells in the lesion boundary zone and newly formed immature and mature neurons in the dentate gyrus. There were considerable improvements in the functional recovery at least in part by promoting neurovascular remodeling (neurogenesis and angiogenesis) and by reducing neuro inflammations [49].

Recently, Zahang et al. (2015 & 2016) conducted experiments on rat models of Traumatic Brain Injury (TBI). They have demonstrated that systemic administration of cell-free exosomes generated by human MSCs (hMSCs) (with treatment initiated 24-hour post injury) leads to the activation of GFAP+ astrocytes and suppression of CD68+ microglia/macrophages. Thus, a significant reduction in neuro inflammation was observed in the treated TBI rat models.

In another study, adult male wistar rats were subjected to two hours of MCAO. Intravenous administration of cell-free MSCs derived exosomes significantly enhanced the number of newly formed double cortin (a maker of neuroblasts) and Von Williberand factor (a marker of endothelial cells). There was a great increase in the axonal density and synaptophysin-positive areas along the ischemic boundary zone of the cortex and striatum. This resulted in neurogenesis, neurite remodeling and improved functional recovery after induced stroke [57].

Maraia Micci et al. (2016) conducted experiments to study the role of NSC exosomes in protecting synapses. She treated mouse hippocampal slice cultures with cultured rat NSC's exosomes and performed oligomers of recombinant Aβ42 (Amyloid β 42 promotes depolarization of synaptic membrane) were added. Pre-treatment with NSC exosomes spared hippocampal synapses from accumulating Aβ42, prevented synaptic excitability and preserved the long-term potentiation [59].

In addition to these, in cortical neurons menstrual stem cell exosomal enriched fractions induced the growth of the longest neurite. When compared to effects of BM-MSC exosomes enriched fractions on neurite outgrowth of dorsal root ganglia neurons, exosomes from both the cell types had similar impacts [58].

MSC Secretome Mediated Neuronal Regeneration

A broad spectrum of factors secreted from MSCs generally referred as MSC secretome. MSCs secrete a panel of biologically active cytokines (CNTF, GDNF, TGF-β and VEGF), interleukins (IL-1β, IL-6 and IL-8), C-C ligands (CCL-2, CCL-5 and CCL-23), classical growth factors, extracellular matrix metallo proteinases (MMP), tissue metallo proteinase inhibitor (TIMP) and tissue remodeling enzymes. These have therapeutic relevance for the neurogenic, neuro protective, angiogenic, anti-inflammatory and immune regulatory activities [24, 70].

By using LC-MS/MS, antibody array, cDNA microarrays, quantitative RT-PCR techniques 201 unique proteins are identified in the cultured medium [80]. MSCs secrete an array of functionally important factors that play a vital role in the healing of injured tissue micro environment. It secretes chemo attractants one such example is the PEDF which stimulates the migration of fibroblasts that eventually participate in scaffold building [79]. MSCs can secrete a wide range of neuro regulatory factors they promote neurogenesis by triggering endogenous neural progenitor's proliferation and differentiation. They also rescue the cells in the affected area by inhibiting apoptosis, modulating immune conditions, promoting glial scar formation, neurite outgrowth, neovascularization, and by providing relevant neuro protection [18, 65]. MSC secretomes were shown to promote the functional recovery by modulating the actions of neurogenic niches and neural progenitors in animal stroke models [10].

The MSC secretomes containing cysteine rich protein 61 (Cyr61) stimulated an angiogenic response *in vitro* and neovascularization *in vivo*. Cyr61 depleted secretome completely abrogates angiogenic inducing capabilities. Hence secretome contributes to promoting neovascularization, a key activity needed for regeneration and repair of injured tissues [60].

Therapeutic Roles of MSCs Secretome Supported by Experimental Studies

MSCs secreted factors such as hepatocyte growth factor (HGF), fibroblast growth factor (FGF) -II, BDNF and platelet derived growth factor (PDGF) – AB was shown to inhibit the autoreactive T-cell responses and stimulate the endogenous oligodendrogenesis in rodent MS models. The *in vivo* studies have shown that MSC conditioned media and HGF enhanced the myelin repair and immune modulation, which is mediated through tyrosine kinase receptor C-met [74, 75, 76]. In an experiment, mouse bone marrow stromal cell derived neuro progenitor cells (mMSC-NPC) secreted a key factor prosaposin (PSAP), which rescued the matured neurons from apoptotic cell death [68].

Marrow derived stem cells expresses genes encoding a variety of arteriogenic cytokines. Factors secreted by human MSCs during culture when delivered through peripheral blood stream following injury in TBI animal models, promoted arteriogenesis through the paracrine mechanism and were sufficient to protect against TBI. The effects can be enhanced by hypoxic preconditioning [23].

In another study, concentrated conditioned medium from human MSCs cultured under normoxic and hypoxic conditions were infused through tail vein of rats subjected to TBI. Rats treated with hypoxic preconditioned secretome performed considerably better in both motor and cognitive functional tests subsequent postmortem evaluation at the 4-day time point revealed that both normoxic and hypoxic preconditioned secretome treated rats had a greater number of newly forming neurons and a significant reduction in apoptosis as well as infract volumes.

The administration of human umbilical cord perivascular cells (HUCPVCS) secretome derived from a mesenchymal progenitor population residing within the Wharton' jelly of the umbilical cord was able to potentiate the neural differentiation survival and viability of different neuronal and glial cell populations *in vitro* and *in vivo* [19].

In a study, neurotrophic activity and therapeutic potential for neuro degenerative conditions of human dental pulp stem cells (hDPSCs) were investigated as well as compared with human BM-MSCs and human adipose derived MSC (hAMSCs) on axotomized adult rat retinal ganglion cells *in vitro*. hDPSCs secreted greater levels of various growth factors including NGF, BDNF, VEGF than BMMSC or hAMSC and promoted significant multifactorial paracrine mediated RGC survival and neurite outgrowth [63].

Enhanced NPCs secretion of BDNF and VEGF assessed on the cell culture supernatants were shown to stimulate a robust SOD-2 mediated anti-apoptotic and antioxidant effects on neurons [78]. NPCs secreted PDGF-AA and FGF-II activates the proliferations of host oligodendrocyte progenitor cells (OPCs) and promotes their maturation into re-myelinating oligodendrocytes [77].

Spinal cord injury (SCI) rat models were treated with the secretome of apoptotic human peripheral blood mono nuclear cells enhanced recruitment of $CD68^{+}$ cells, activated the ERK1/2 pathway in the spinal cord with concomitant reduction of oxidative stress by downregulation of NO synthase expression and upregulation of CXCL-1. This resulted in substantial functional recovery, angiogenesis, reduced cavity formation and eventually attenuated secondary damage [73].

Intranasal Route for Administration of Cell-Free Neurotherapeutics

Most of the drugs are not being effectively delivered using conventional drug delivery approach to CNS because of its complexity. CNS has limited accessibility to blood compartment due to a number of barriers [115]. The BBB is the stringent barrier for delivery of neurotherapeutics *in vivo*. But, BBB is not present at the interface between nasal epithelium and CNS. Intranasally delivered therapeutic molecules can pass freely through intracellular junctions of the olfactory epithelium and extracellularly reach the olfactory bulbs of the CNS. The possible mechanisms through which molecules can directly enter cerebrospinal fluid or to the brain are by rapid perineuronal or slow transneuronal routes. Both the olfactory and trigeminal nerves innervate the nasal cavity providing direct connections with the CNS. The direct transport of neurotherapeutics including macro molecules and low molecular weight drugs from intranasal route to CNS is possible because of the unique olfactory and trigeminal nerve pathways [116]. Intranasal drug delivery system is a practical, non-invasive, safe and convenient method of bypassing BBB to deliver therapeutic agents to the brain and spinal cord which eliminates the need for invasive or systemic administration procedures and its potential side effects [114].

Experimental Evidence of Intranasal Route for Drug Delivery

Intranasal administration of either FGF-II or Heparin Binding Epidermal Growth Factor (HB-EGF) increases neurogenesis in the SVZ of mouse brain [139]. Intranasal administration of IGF-I gives protection against focal ischemic brain damage in rats following MCAO [140]. Intranasally administered NGF reversed the degeneration and memory loss in mouse Alzheimer's disease model [141]. Intranasal administration of nerve growth factor (NGF) rescues recognition memory deficits in AD11 anti-NGF transgenic mice [145].

Intranasally administered FGF-II and EGF have shown to stimulate neurogenesis in adult CNS of animals. In human patients with Alzheimer's and mild cognitive impairments, intranasal drug delivery of insulin enhanced the memory, attention and functioning [114].

Deferoxamine (DFO) is a high affinity iron chelator, has been shown to be neuroprotective in rat models of TBI, hemorrhage, Parkinson's and Alzheimer's disease. Intranasal delivery of DFO bypassed the BBB and delivered extra cellularly to the CNS along the olfactory and trigeminal nerve pathways within minutes. Intranasal DFO provides increased brain exposure and significant protection in rat ischemic stroke model [142].

Intranasal administration of native horse radish peroxidase in rats and primates provided anatomical evidence that the larger tracer molecules reach the olfactory bulbs of the CNS via intracellular junctions of the olfactory epithelium [143].

Intranasal administration of 20IU arginine-vasopressin (AVP) potentially enhanced the amplitude of Event Related Potentials (ERP). In contrast, intravenous administration of AVP leading to higher plasma AVP levels but remained ineffective [144].

Intranasally administered FGF-II and EGF have shown to stimulate neurogenesis in adult CNS of animals. In human patients with Alzheimer's and mild cognitive impairments, intranasal drug delivery of insulin enhanced the memory, attention and functioning [114].

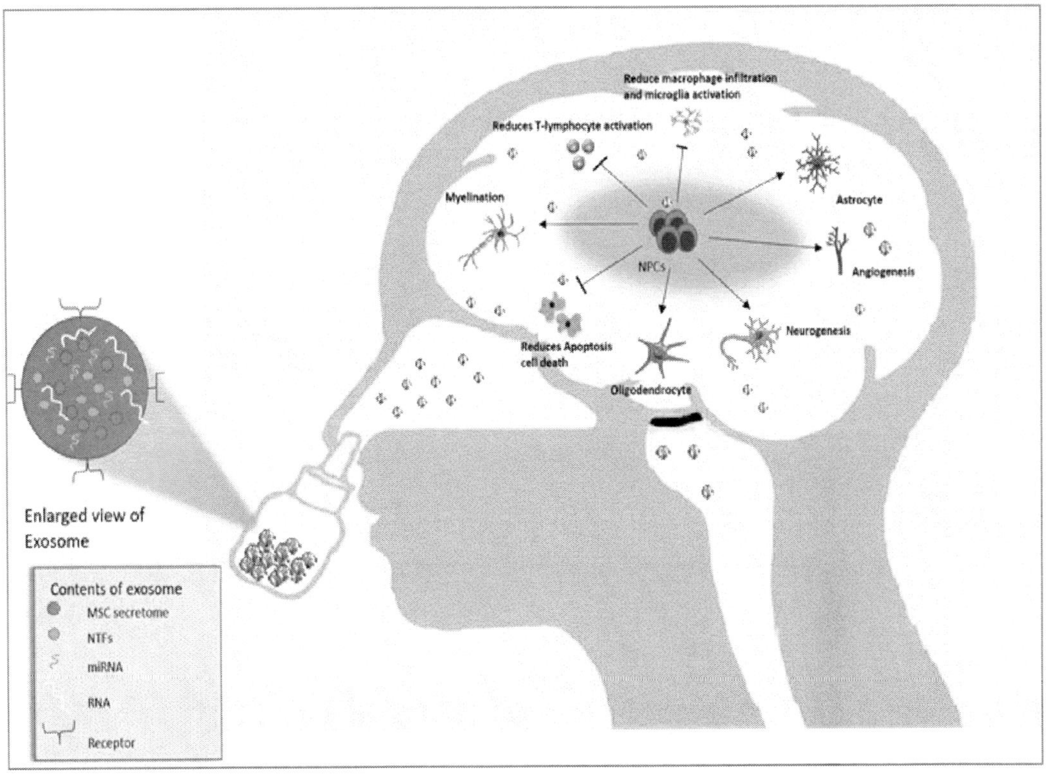

Figure 2. A schematic diagram representing intranasal delivery of cell-free neuro therapeutic bypassed the blood brain barrier (BBB) and transported throughout the brain via olfactory and trigeminal nerve pathways. NTFs, stem cell derived secretome encapsulated in MSC exosomes or in combination, delivered via intranasal route helps an ischemic CNS to regain regenerative capability and its function.

Therefore, an efficient strategy of delivering exosomes – encapsulated neurotrophic factors, stem cell secretomes in combination or individually to the CNS via intranasal route could potentially induce regenerative repair in a cell-free and non-invasive manner (Refer Figure 2).

In summary, the cell-free regenerative hypothesis can be an alternative outlook on the use of stem cells in regenerative neurology. Magnificent sensory locomotor functional recovery after cerebral ischemia can be achieved by utilizing therapeutic potentials of stem cell secreted bio-factors rather than implanting stem cells themselves.

In conclusion, by harnessing therapeutic potentials of the NTFs, stem cell secretome and exosomes we can restore the neural plasticity and brain functioning of ischemic brain by supplying adequate trophic cues for regeneration and repair. Extensive research on dosage optimization and clinical trials in this direction could provide effective, safe yet simple and convenient cell-free regenerative therapeutic strategy which drastically minimizes the severity of the stroke.

REFERENCES

[1] Abe, Koji, Therapeutic Potential of Neurotrophic Factors and Neural Stem Cells against Ischemic Brain Injury. *Journal of Cerebral Blood Flow and metabolism.* 20:1393-1408 © 2000.

[2] Siliprandi, Renata, Roberto Canella, and Giorgio Carmignoto. Nerve Growth Factor Promotes Functional Recovery of Retinal Ganglion Cells After Ischemia. *Investigative Ophthalmology & Visual Science*, November 1993, Vol. 34, No. 12.

[3] Kim Seung U., and Jean de Vellis. Review Stem Cell-Based Cell Therapy in Neurological Diseases. *Journal of Neuroscience Research* 87:2183–2200 (2009).

[4] Terenghi, Giorgio, Peripheral nerve regeneration and neurotrophic factors. *J. Anat.* (1999) 194, pp. 1–14.

[5] Schabitz, Wolf-Rudiger, Tobias Steigleder, Christiana M. Cooper-Kuhn, PhD; Stefan Schwab, Clemens Sommer, Armin Schneider, H. Georg Kuhn, PhD. Intravenous *Brain-Derived Neurotrophic Factor Enhances Poststroke Sensorimotor Recovery and Stimulates Neurogenesis.* DOI: 10.1161/ STROKEAHA.106.477331.

[6] Schabitz, Wolf-R., Clemens Sommer, Werner Zoder, Marika Kiessling, Markus Schwaninger, Stefan Schwab, Intravenous Brain-Derived Neurotrophic Factor Reduces Infarct Size and Counterregulates Bax and Bcl-2 Expression after Temporary Focal Cerebral Ischemia. *Stroke*. 2000; 31:2212-2217.

[7] Pencea, Viorica, Kimberly D. Bingaman, Stanley J. Wiegand, and Marla B. Luskin. Infusion of Brain-Derived Neurotrophic Factor into the Lateral Ventricle of the Adult Rat Leads to New Neurons in the Parenchyma of the Striatum, Septum, Thalamus, and Hypothalamus. *The Journal of Neuroscience*, September 1, 2001, 21(17):6706–6717.

[8] Vishnubhatla, Indira, Randolph Corteling, Lara Stevanato, Caroline Hicks and John Sinden. The Development of Stem Cell-derived Exosomes as a Cell-free Regenerative Medicine. *Journal of Circulating Biomarkers*. DOI: 10.5772/58597.

[9] Dragoa, Denise, Chiara Cossettic, d, Nunzio Iracic, d, Edoardo Gaudee, Giovanna Muscoe, Angela Bachib, Stefano Pluchinoa, c, d. The stem cell secretome and its role in brain repair. *Biochimie.* Volume 95, Issue 12, December 2013, Pages 2271–2285.

[10] Salgado, A.J., Sousa, J.C., Costa, B.M., Pires, A.O., Mateus-Pinheiro, A., Teixeira, F.G., Pinto, L. and Sousa, N. (2015), Mesenchymal stem cells secretome as a modulator of the neurogenic niche: basic insights and therapeutic opportunities. *Front. Cell. Neurosci*. 9:249.

[11] Bernier, P.J., Bedard, A., Vinet, J., Levesque M., and Parent A. (2002), Newly generated neurons in the amygdala and adjoining cortex of adult primates. *Proc. Natl. Acad. Sci. USA* 99, 11464–11469. doi: 10.1073/pnas.172403999.

[12] Fowler, C.D., Johnson, F., and Wang, Z. (2005), Estrogen regulation of cell proliferation and distribution of estrogen receptor-alpha in the brains of adult female prairie and meadow voles. *J. Comp. Neurol.* 489, 166–179. doi: 10. 1002/cne.20638.

[13] Fowler, C.D., Liu, Y., Ouimet, C., and Wang, Z. (2002), The effects of social environment on adult neurogenesis in the female prairie vole. *J. Neurobiol.* 51, 115–128.doi:10.1002/neu.10042.

[14] Gonçalves, L., Silva, R., Pinto-Ribeiro, F., Pêgo, J.M., Bessa, J.M., Pertovaara, A. et al. (2008), Neuropathic pain is associated with depressive behavior and. induces neuroplasticity in the amygdala of the rat. *Exp. Neurol.* 213, 48–56. doi: 10.1016/j.expneurol.2008.04.043.

[15] Kokoeva, M.V., Yin, H., and Flier, J.S. (2005), Neuro genesis in the hypothalamus of adult mice: potential role in energy balance. *Science* 310, 679–683. doi: 10. 1126/science.1115360.

[16] Bédard, A., Gravel, C., and Parent, A. (2006), Chemical characterization of newly generated neurons in the striatum of adult primates. *Exp. Brain Res.* 170, 501–512.doi:10.1007/s00221-005-0233-5.

[17] Yoshimi, K., Ren, Y.R., Seki, T., Yamada, M., Ooizumi, H., Onodera, M., et al. (2005), Possibility for neurogenesis in substantia nigra of parkinsonian brain. *Ann. Neurol.* 58, 31–40.doi:10.1002/ana.20506.

[18] Teixeira, F.G., Carvalho, M.M., Sousa, N., and Salgado, A.J. (2013), Mesenchymal stem cells secretome: a new paradigm for central nervous system regeneration? *Cell Mol. Life Sci.* 70, 3871–3882.doi:10.1007/s00018-013-1290-8.

[19] Teixeira, F.G., Carvalho, M.M., Neves-Carvalho, A., Panchalingam, K.M., Behie, L.A., Pinto, L. et al. (2015), Secretome of Mesenchymal progenitors from the Umbilical cord acts as modulator of neural/glial proliferation and differentiation. *Stem Cell Rev.* 11, 288–897. doi: 10.1007/s12015014-9576-2.

[20] Komatsu, K., Honmou, O., Suzuki, J., Houkin, K., Hamada, H., and Kocsis, J.D. (2010), Therapeutic time window of mesenchymal stem cells derived frombone marrow after cerebral ischemia. *Brain Res.* 1334, 84–92.doi: 10.1016/j.brainres. 2010.04.006.

[21] Zheng, W., Honmou, O., Miyata, K., Harada, K., Suzuki, J., Liu, H., et al. (2010), Therapeutic benefits of human mesenchymal stem cells derived from bone marrow after global cerebral ischemia. *Brain Res.* 1310, 8–16. doi: 10.1016/j. brainres.2009.11.012.

[22] Chang, Ching-Ping, Chung-Ching Chio, Chong-Un Cheong, Chien-Ming Chao, Bor-Chieh Cheng and Mao-Tsun Lin, Hypoxic preconditioning enhances the therapeutic potential of the secretome from cultured human mesenchymal stem cells in experimental traumatic brain injury. *Clinical Science* (2013) 124, 165–176.

[23] Kinnaird, T., Stabile, E., Burnett, M.S., Lee, C.W., Barr, S, Fuchs, S. and Epstein, S.E. (2004) Marrow-derived stromal cells express genes encoding a broad spectrum of arteriogenic cytokines and promote *in vitro* and *in vivo* arteriogenesis through paracrine mechanisms. *Circ. Res.* 94, 678–685.

[24] Heidi R. Hofer and Rocky S. Tuan, Secreted trophic factors of mesenchymal stem cells support neurovascular and musculoskeletal therapies. *Stem Cell Research & Therapy* (2016) 7:131.

[25] Liang, X., Ding, Y., Zhang, Y., Tse, H., Lian, Q. Paracrine mechanisms of mesenchymal stem cell-based therapy: current status and perspectives. *Cell Transplant*. 2014; 23:1045–59.

[26] Hellström, Mats, Margaret A. Pollett, and Alan R. Harvey. Post-Injury Delivery of rAAV2-CNTF Combined with Short-Term Pharmacotherapy Is Neuroprotective and Promotes Extensive Axonal Regeneration after Optic Nerve Trauma. *Journal of Neurotrauma*. December 2011, 28(12): 2475-2483.

[27] Li, L., W. Wu, L.F. Lin, M. Lei, R.W. Oppenheim, and L.J. Houenou, Rescue of adult mouse motoneurons from injury-induced cell death by glial cell line-derived neurotrophic factor-*Proc. Natl. Acad. Sci. USA* Vol. 92, pp. 9771-9775, October 1995 Neurobiology).

[28] Kilic, Ülkan, Ertugrul Kilic, Gunnar P.H. Dietz, Mathias Bähr Intravenous TAT-GDNF Is Protective After Focal Cerebral Ischemia in Mice. *Stroke*. 2003; 34:1304 -1310.

[29] Emsleya, J.G., T. Hagga, Endogenous and exogenous ciliary neurotrophic factor enhances forebrain neurogenesis in adult mice *Experimental Neurology* Volume 183, Issue 2, October 2003, Pages 298–310.

[30] Takeshi Hayashia, Masanori Iwaia, Tomoaki Ikedab, Guang Jina, Kentaro Deguchia, Shoko Nagotania, Hanzhe Zhanga, Yoshihide Seharaa, Isao Naganoa, Mikio Shojia Tsuyomu Ikenoueb, Koji Abea Neural precursor cells division and migration in neonatal rat brain after ischemic/hypoxic injury *Brain Research* Volume 1038, Issue 1, 15 March 2005, Pages 41–49.

[31] Covey, M.V., S.W. Levison, Leukemia inhibitory factor participates in the expansion of neural stem/progenitors after perinatal hypoxia/ischemi *Neuroscience* Volume 148, Issue 2, 24 August 2007, Pages 501–509.

[32] Kentaro Toyamaa, Osamu Honmoua, Therapeutic benefits of angiogenetic gene-modified human mesenchymal stem cells after cerebral ischemia *Experimental Neurology* Volume 216, Issue 1, March 2009, Pages 47–55.

[33] Chun Kun Parkb, Won-Kyu Jua, Hans-Dieter Hofmannc, Matthias Kirschc, Joon Ki Kangb, Myung-Hoon Chuna, Mun-Yong Leea, Differential regulation of ciliary neurotrophic factor and its receptor in the rat hippocampus following transient global ischemia. *Brain Research*. Volume 861, Issue 2, 14 April 2000, Pages 345–353.

[34] Seong Su Kanga,, Matthew P. Keaseya, b, Sheila A. Arnolda, Rollie Reida, Justin Geraldsa, Theo Hagga, Endogenous CNTF mediates stroke-induced adult CNS neurogenesis in mice. *Neurobiology of Disease*. Volume 49, January 2013, Pages 68–78.

[35] Yun Wang, Shinn-Zong Lin, Ai-Lin Chiou, Lawrence R. Williams and Barry J. Hoffer Glial Cell Line-Derived Neurotrophic Factor Protects against Ischemia-Induced Injury in the Cerebral Cortex. *Journal of Neuroscience* 1 June 1997, 17 (11) 4341-4348.

[36] Bowenkamp, K.E., Hoffman, A.F., Gerhardt, G.A., Henry, M.A., Biddle, P.T., Hoffer, B.J., Granholm, A.E. (1995) Glial cell line-derived neurotrophic factor supports survival of injured midbrain dopaminergic neurons. *J Comp Neurol* 355:479–489.

[37] Chao, C.C., Lee, E.H. Neuroprotective mechanism of glial cell line-derived neurotrophic factor on dopamine neurons: role of antioxidation. *Neuropharmacology*. 1999; 38:913–916.

[38] Dempsey, Robert J., Kurt A. Sailor, Kellie K. Bowen, Kudret Tureyen and Raghu Vemuganti. Stroke-induced progenitor cell proliferation in adult spontaneously hypertensive rat brain: effect of exogenous IGF-1 and GDNF. *Journal of Neurochemistry,* 2003, 87, 586–597.

[39] Iwata-Ichikawa, E., Kondo, Y., Miyazaki, I., Asanuma, M., Ogowa, N. Glial cells protect neurons against oxidative stress via transcriptional up-regulation of the glutathione synthesis. *J Neurochem*. 1999;72: 161–167. 33.

[40] Miyazaki, H., Nagashima, K., Okuma, Y., Nomura, Y. Expression of Ret receptor tyrosine kinase after transient forebrain ischemia is modulated by glial cell line derived neurotrophic factor in rat hippocampus. *Neurosci Lett*. 2002; 318:1–4.

[41] Perrelet, D., Ferri, A., Liston, P., Muzzin, P., Korneluk, R.G. Kato, A.C. IAPs are essential for GDNF-mediated neuroprotective effects in injured motor neurons *in vivo*. *Nat Cell Biol*. 2002; 4:175–179.

[42] Hiroyuki Miyazakia, Kazuo Nagashimab, Yasunobu Okumaa, Yasuyuki Nomuraa, Expression of Ret receptor tyrosine kinase after transient forebrain ischemia is modulated by glial cell line-derived neurotrophic factor in rat hippocampus, *Neuroscience Letters*. Volume 318, Issue 1, 18 January 2002, Pages 1–4.

[43] Boscia, F., Esposito, C.L., Di Crisci, A., de Franciscis, V., Annunziato, L. and Cerchia, L. (2009a) GDNF selectively induces microglial activation and neuronal survival in CA1/CA3 hippocampal regions exposed to NMDA insult through Ret/ERK signalling. *PLoS ONE* 4, e6486.

[44] Boscia F., Esposito C.L., Casamassa A., de Franciscis V., Annunziato L. and Cerchia L. The isolectin IB4 binds to RET tyrosine kinase receptor in micrglia. *Journal of neurochemistry*, volume 126, 2013.

[45] Perrelet, D., A. Ferri, P. Liston, P. Muzzin, R.G. Korneluk & A.C. Kato. IAPs are essential for GDNF-mediated neuroprotective effects in injured motor neurons *in vivo*. *Nature Cell Biology* 4, 175 - 179 (2002)

[46] Hongqi Xin, Yi Li, Ben Buller, Mark Katakowsk I, Yi Zhang, Xinli Wang, Xia Shang, Zheng Gang Zhang, Michael Choppa, Exosome-Mediated Transfer of miR-133b from Multipotent Mesenchymal Stromal Cells to Neural Cells Contributes to Neurite Outgrowth. *Stem Cells* 2012; 30:1556 –1564.

[47] Duisters, R.F., Tijsen A.J., Schroen B. et al. miR-133 and miR-30 regulate connective tissue growth factor: Implications for a role of microRNAs in myocardial matrix remodeling. *Circ Res* 2009; 104: 170–178; 176p following 178.

[48] Dergham, P., Ellezam B., Essagian C. et al. Rho signaling pathway targeted to promote spinal cord repair. *J Neurosci* 2002; 22: 6570–6577.

[49] Zhang, Y., Chopp, M., Meng, Y., Katakowski, M., Xin, H., Mahmood, A., et al. (2015). Effect of exosomes derived from multipluripotent mesenchymal stromal cells on functional recovery and neurovascular plasticity in rats after traumatic brain injury. *J. Neurosurg*. 122, 856–867. doi: 10.3171/2014.1. JNS14770.

[50] Zhang, Y., Chopp, M., Zhang, Z.G., Katakowski, M., Xin, H., Qu, C., et al. (2016). Systemic administration of cell-free exosomes generated by human bone marrow derived mesenchymal stem cells cultured under 2D and 3D conditions improves functional recovery in rats after traumatic brain injury. *Neurochem. Int.* doi: 10.1016/j.neuint.2016.08.003 [Epub ahead of print.

[51] Zhang, Y., Chopp, M., Liu, X.S., Katakowski, M., Wang, X., Tian, X., et al. (2016a). Exosomes derived from mesenchymal stromal cells promote axonal growth of cortical neurons. *Mol. Neurobiol.* doi: 10.1007/s12035-016-9851-0[Epubahead ofprint].

[52] Yongxiang Ya, Yuqin Ye, Xinhong Su, Jun He, Wei Bai and Xiaosheng He. MSCs-Derived Exosomes and Neuroinflammation, Neurogenesis and Therapy of Traumatic Brain Injury. *Front. Cell. Neurosci*. 11:55.

[53] Yeo, R.W., Lai, R.C., Zhang, B., Tan, S.S., Yin, Y., Teh, B.J., et al. (2013) Mesenchymal stem cell: an efficient mass producer of exosomes for drug delivery. *Adv. Drug Deliv. Rev*. 65, 336–341. doi: 10.1016/j.addr.2012.07.001.

[54] Zhuang, X., Xiang, X., Grizzle, W., Sun, D., Zhang, S., Axtell, R.C., et al. (2011) Treatment of brain inflammatory diseases by delivering exosome encapsulated anti-inflammatory drugs from the nasal region to the brain. *Mol. Ther*. 19, 1769–1779. doi: 10.1038/mt.2011.164.

[55] Marote, Ana, Fábio G. Teixeira, Bárbara Mendes-Pinheiro and António J. Salgado (2016) MSCs-Derived Exosomes: Cell-Secreted Nanovesicles with Re-generative potential. *Front. Pharmacol*. 7:231.

[56] Merino, C., Zuñiga, F.A., Escudero, C.A., Ormazabanl, V., Reyes, C., Salomon, C., et al. (2016). Mesenchymal stem cell-derived extracellular vesicles promote

angiogenesis: new alternatives for application to human health. *Front. Physiol.* 7:24. Doi: 10.3389/fphys.2016.00024.

[57] Xin, H., Li, Y., Cui, Y., Yang, J.J., Zhang, Z.G., and Chopp, M. (2013a). Systemic administration of exosomes released from mesenchymal stromal cells promote functional recovery and neurovascular plasticity after stroke in rats. *J. Cereb. Blood Flow Metab.* 33, 1711–1715.doi:10.1038/jcbfm.2013.152.

[58] Lopez-Verrilli, M.A., Caviedes, A., Cabrera, A., Sandoval, S., Wyneken, U., and Khoury, M. Mesenchymal stem cell-derived exosomes from different sources selectively promote neuritic outgrowth. *Neuroscience* 320, 129–139.doi: 10.1016/ j.neuroscience.2016.01.061.

[59] http://www.alzforum.org/news/conference-coverage/stem-cell-exosomes-restoring-zzzs-new-ideas-protect-brain.

[60] Estrada, R., Li N., Sarojini H., An J., Lee M.J., Wang E. Secretome from mesenchymal stem cells induces angiogenesis via Cyr61. *J. Cell. Physiol.*, 219: 563–571. doi:10.1002/jcp.21701.

[61] Lavoie, Jessie R., Michael, Rosu-Myles, Uncovering the secrets of mesenchymal stem cells. *Biochimc,* volume 95, Issue 12, Dec 2013, pages 2212-2221.

[62] Hyun Ok Kim, Seong-Mi Choi, and Han-Soo Kim. Mesenchymal Stem Cell-Derived Secretome and Microvesicles as a Cell-Free Therapeutics for Neurodegenerative Disorders. *Tissue Engineering and Regenerative Medicine*, Vol. 10, No. 3, pp. 93-101 (2013)

[63] Mead, B., Logan, A., Berry, M., Leadbeater, W., Scheven B.A. Paracrine-mediated neuroprotection and neuritogenesis of axotomised retinal ganglion cells by human dental pulp stem cells: comparison with human bone marrow and adipose-derived mesenchymal stem cells. *Journal.* pone.0109305.

[64] Helena Kupcova Skalnikova Proteomic techniques for characterisation of mesenchymal stem cell secretome. *Biochimie.* Volume 95, Issue 12, December 2013, Pages 2196–2211.

[65] Salgado, A.J., Sousa J.C., Costa B.M., Pires A.O., Mateus-Pinheiro A., Teixeira F.G., Pinto L. and Sousa N. (2015) Mesenchymal stem cells secretome as a modulator of the neurogenic niche: basic insights and therapeutic opportunities. *Front. Cell. Neurosci.* 9:249.

[66] Carvalho, M., Miguel, G. Teixeira, Fabio, L. Reis, Rui, Sousa, Nuno, J. Salgado, Antonio. Mesenchymal Stem Cells in the Umbilical Cord: Phenotypic Characterization, Secretome and Applications in Central Nervous System Regenerative Medicine, Current Stem. *Cell Research & Therapy,* Volume 6, Number 3, September 2011, pp. 221-228(8).

[67] Teixeira, Fábio G., Miguel M. Carvalho, Nuno Sousa, António J. Salgado. Mesenchymal stem cells secretome: A new paradigm for central nervous system

regeneration? *Cellular and Molecular Life Sciences*. October 2013, Volume 70, Issue 20, pp 3871–3882.

[68] Li, N., Sarojini H., An, J., Wang E. Prosaposin in the secretome of marrow stroma-derived neural progenitor cells protects neural cells from apoptotic death. *J. Neurochem*. (2010) 112, 1527–1538.

[69] Doetsch, Fiona, Isabelle, Caillé, Daniel A. Lim, Jose Manuel García-Verdugo, Arturo Alvarez- Buylla. *Subventricular Zone Astrocytes are Neural Stem Cells in the Adult Mammalian Brain*. Volume 97, Issue 6, p703–716, 11 June 1999.

[70] Takagi, Y, Nozaki K. Takahashi J., Yodoi J., Ishikawa M. Hashimoto N. (1999) Proliferation of neuronal precursor cells in the dentate gyrus is accelerated after transient forebrain ischemia in mice. *Brain Res* 831:283-287.

[71] Liu, J., Solway, K., Messing, R.O., Sharp, F.R. (1998) Increased neurogenesis in the dentate gyrus after transient global ischemia in gerbils. *J Nel/rosci* 18:7768-7778.

[72] Kim, Seung U., and Jean de Vellis. ReviewStem Cell-Based Cell Therapy in Neurological Diseases. *Journal of Neuroscience Research* 87:2183–2200 (2009).

[73] Kitagawa, H., Hayashi T., Mitsumoto Y., Koga N., Itoyama Y., Abe K (l998b) Reduction of ischemic brain injury by topical application of GDNF after permanent MCAO in rats. *Stroke* 29: 1417-1422.

[74] Thomas, Haider, Romana Höftberger, Beate Rüger, Michael Mildner, Roland Blumer Andreas Mitterbauer, Tanja Buchacher, Camillo Sherif, Patrick Altmann, Heinz Redli, Christian Gabriel,. *The secretome of apoptotic human peripheral blood mononuclear cells attenuates secondary damage following spinal cord injury in rats*. Volume 267, May 2015, Pages 230–242.

[75] Gerdoni, E., B. Gallo, S. Casazza, S. Musio, I. Bonanni, E. Pedemonte, R. Mantegazza, F. Frassoni, G. Mancardi, R. Pedotti, A. Uccelli. Mesenchymal stem cells effectively modulate pathogenic immune response in experimental autoimmune encephalomyelitis, *Ann. Neurol.* 61 (2007) 219e227.

[76] J. Zhang, Y. Li, M. Lu, Y. Cui, J. Chen, L. Noffsinger, S.B. Elias, M. Chopp, Bone marrow stromal cells reduce axonal loss in experimental autoimmune encephalomyelitis mice, *J. Neurosci. Res.* 84 (2006) 587e595.

[77] Bai, L., D.P. Lennon, A.I. Caplan, A. DeChant, J. Hecker, J. Kranso, A. Zaremba, R.H. Miller, Hepatocyte growth factor mediates mesenchymal stem cellinduced recovery in multiple sclerosis models, Nat. Neurosci. 15 (2012) 862e870. Voulgari-Kokota, R. Fairless, M. Karamita, V. Kyrargyri, V. Tseveleki, M. Evangelidou, B. Delorme, P. Charbord, R. Diem, L. Probert, Mesenchymal stem cells protect CNS neurons against glutamate excitotoxicity by inhibiting glutamate receptor expression and function, *Exp. Neurol.* 236 (2012) 161e170.

[78] Madhavan, L., V. Ourednik, J. Ourednik, Neural stem/progenitor cells initiate the formation of cellular networks that provide neuroprotection by growth factor-modulated antioxidant expression, *Stem Cells* 26 (2008) 254e265.

[79] Estrada, R., N. Li, H. Sarojini, J. An, M.J. Lee, E. Wang, Secretome from mesenchymal stem cells induces angiogenesis via Cyr61, *J. Cell Physiol.* 219 (2009) 563e571.

[80] Liu, C.H., S.M. Hwang, Cytokine interactions in mesenchymal stem cells from cord blood, *Cytokines* 32 (2005) 270e279.

[81] Subra, C., Laulagnier, K., Perret, B. and Record, M. (2007) Exosome lipidomics unravels lipid sorting at the level of multi vesicular bodies. *Biochimie* 89: 205-212.

[82] Trajkovic, K., Hsu, C., Chiantia, S., Rajendran, L., Wenzel, D., et al. (2008) Ceramide triggers budding of exosome vesicles into multi vesicular endosomes. *Science* 319: 1244-1247.

[83] Tan, S.S., Yin Y., Lee, T., Lai, R.C., Yeo, R.W., et al. (2013) Therapeutic MSC exosomes are derived from lipid raft microdomains in the plasma membrane. *Journal of Extracellular Vesicles* 2.

[84] Lai, R.C., Arslan, F., Tan, S.S., Tan, B., Choo, A., et al. (2010) Derivation and characterization of human fetal MSCs: an alternative cell source for large-scale production of cardioprotective microparticles. *Journal of Molecular and Cellular Cardiology* 48: 1215-1224.

[85] Kim, H.S., Choi, D.Y., Yun, S.J., Choi, S.M., Kang, J.W., et al. (2012) Proteomic analysis of micro-vesicles derived from human mesenchymal stem cells. *Journal of Proteome Research* 11: 839-849.

[86] Sterne, G.D., Brown, R.A., Green, C.J., Terenghi, G. (1997a) Neurotrophin 3 delivered locally via fibronectin mats enhances peripheral nerve regeneration. *European Journal of Neuroscience 9*, 1388–1396.

[87] Verge, V.M.K., Richardson, P.M., Wiesenfeld-Hallinz, Ho=KFELT T. (1995) Differential influence of nerve growth factor on neuropeptide expression *in vivo*. A novel role in peptide suppression in adult sensory neurons. *Journal of Neuroscience* 15, 2081–2096.

[88] Munson, J.B., Mcmahon, S.B. (1997) Effects of GDNF onaxotomised sensory and motor neurons in adult rats. *European Journal of Neuroscience* 9, 1126–1129.

[89] Cho, H.J., Kim, SY, Park, M.J., Kim, D.S., Kim, J.K., Chu, M.Y. (1997) Expression of mRNA for brain-derived neurotrophic factor in the dorsal root ganglion following peripheral inflam-mation. *Brain Research* 749, 358–362.

[90] Friedlander, D.R., Grumet M., Edelman G.M. (1986) Nerve growth factor enhances expression of neuron-glial cell adhesion molecule in PC12 cells. *Journal of Cell Biology* 102, 413–419.

[91] Horch, H.W. Local effects of BDNF on dendritic growth. *Rev Neurosci.* 2004; 15:117–129.

[92] Lu, B. BDNF and activity-dependent synaptic modulation. *Learn Mem.* 2003; 10:86–98.

[93] Gu, X., Thomas P.K., King, R.H.M. (1995) Chemotropism in nerve regeneration studied in tissue culture. *Journal of Anatomy* 186, 153–163.

[94] Reynolds, M.L., Woolf, C.J. (1993) Reciprocal Schwann cell-axon interaction. *Current Opinion in Neurobiology* 3, 683–693.

[95] Levi-Montalcini, R. (1987). The nerve growth factor: thirty-five years later. *EMBO Journal* 6, 1145–1154.

[96] Arakawa Y., Sendtner M., Thoenen H. (1990) Survival effect of ciliary neurotrophic factor (CNTF) on chick embryonic motoneurons in culture: comparison with other neurotrophic factors and cytokines. *Journal of Neuroscience* 10, 3507–3515.

[97] Sendtner, M., Gotz, R., Holtmann, B., Thoenen, H. (1997) Endogenous ciliary neurotrophic factor is a lesion factor for axotomised motoneurons in adult mice. *Journal of Neuro-science* 17, 6999–7006.

[98] Sahenk Z., Seharaseyon J., Mendell, J.R. (1994) CNTF potentiates peripheral nerve regeneration. *Brain Research* 655, 246–250.

[99] Ulenkate H.J.L.M., Kaal, E.C.A., Gispen, W.H., Jennekensfgi (1994) Ciliary neurotrophic factor improves muscle fibrere innervation after facial nerve crush in young rats. *Acta Neuropathologica* (Berlin) 88, 558–564.

[100] Covey, M.V., S.W. Levison. Leukemia inhibitory factor participates in the expansion of neural stem/progenitors after perinatal hypoxia/ischemia. *Neuroscience*. Volume 148, Issue 2, 24 August 2007, Pages 501–509.

[101] Choi, D.W., Hartley, D.M. Calcium and glutamate-induced cortical neuronal death. In: Waxman SG, ed. Molecular and Cellular Approaches to the Treatment of *Neurological Disease*. New York, NY: Raven Press; 1993: 23–24.

[102] Mattson, M.P., Lovell, M.A., Furukawa, K., Markesberry, W.R. Neurotrophic factors attenuate glutamate-induced accumulation of peroxides, elevation of intracellular Ca21 concentration, and neurotoxicity and increase antioxidant activities in hippocampal neurons. *J Neurochem*. 1995; 65:1740–1751.

[103] Kume, T., Kouchiyama, H., Kaneko, S., Maeda, T., Kaneko, S., Akaike, A, Shimohama, S, Kihara, T., Kimura, J., Wada, K., Koizumu, S. BDNF prevents NO mediated glutamate cytotoxicity in cultured cortical neurons. *Brain. Res.* 1997; 756:200–204.

[104] Redis, R.S., Calin, S., Yang, Y., You, M.J. and Calin, G.A. (2012) Cell-to-cell miRNA transfer: from body homeostasis to therapy. *Pharmacology and Therapeutics* 136: 169-174.

[105] Gibbings, D.J., Ciaudo, C., Erhardt, M. and Voinnet, O. (2009) Multivesicular bodies associate with components of miRNA effector complexes and modulate miRNA activity. *Nature Cell Biology* 11: 1143-1149.

[106] Abdipranoto-Cowley, A., Park, J.S., Croucher, D., Daniel, J., Henshall, S., Galbraith, S., Mervin, K., Vissel, B. Activin A is essential for neurogenesis following neuro-degeneration. *Stem Cells*. 2009 Jun; 27(6):1330-46.

[107] Petitto, John M., Zhi Huang, Grace K. Ha, Daniel Dauer. Interleukin-2 and Septo hippocampal Neurons: Neurodevelopment and Autoimmunity. *The Neuro Immunological Basis of Behavior and Mental Disorders*. pp 131-143.

[108] Ziegler, Amber N., Steven W. Levison & Teresa L. Wood. Insulin and IGF receptor signalling in neural-stem-cell homeostasis. *Nature Reviews Endocrinology* 11, 161–170 (2015).

[109] Jiao, Jian-wei, David A. Feldheim, and Dong Feng Chen. *Ephrins as negative regulators of adult neurogenesis in diverse regions of the central nervous system.* Current Issue > vol. 105 no. 25 > Jian-wei Jiao, 8778–8783.

[110] Andres, Rosa, Alison Forgie, Sean Wyatt, Qi Chen, Frederic J. de Sauvage, Alun M. Davie. Multiple effects of artemin on sympathetic neurone generation, survival and growth. *Development* 2001 128: 3685-3695;

[111] Gao, Wei-Qiang, J. Lisa Zheng, and Mona Karihaloo. Neurotrophin-4/5 (NT-4/5) and Brain-Derived Neurotrophic Factor (BDNF) Act at Later Stages of Cerebellar Granule Cell Differentiation. *The Journal of Neuroscience,* April 1995, E (4): 2656-2667.

[112] Cline, Hollis, Anirvan Ghosh, Yuh-Nung Jan, Josh Sanes, Dendritic Development. *Fundamental Neuro-science* (Fourth Edition), 2013.

[113] Stella, Maria Cristina, Alessandro Vercelli, Mariaelena Repici, Antonia Follenzi, and Paolo M. Comoglio. Macrophage Stimulating Protein Is a Novel Neurotrophic Factor. *Mol Biol Cell.* 2001 May; 12(5): 1341–1352.

[114] Hanson Leah R., and William H Frey-II, Intranasal delivery bypasses the blood-brain barrier to target therapeutic agents to the central nervous system and treat neurodegenerative disease. *BMC Neuroscience* 2008.

[115] Vyas, Tushar K., Shahiwala, Aliasgar, Marathe, Sudhanva, Misra, Ambikanandan. Intranasal Drug Delivery for Brain Targeting. *Current Drug Delivery,* Volume 2, Number 2, April 2005, pp. 165-175(11).

[116] Chandrakantsing Vijaysing Pardeshi & Veena Shailendra Belgamwar. Direct nose to brain drug delivery via integrated nerve pathways bypassing the blood–brain barrier: an excellent platform for brain targeting. *Expert Opinion on Drug Delivery.* 16 Apr 2013.

[117] Nadareishvili, Zurab, and John Hallenbeck, Neuronal Regeneration after Stroke. *The New England journal of medicine.* 2013.

[118] Michael, G.J., S. Averill, A. Nitkunan, M. Rattray, D.L.H. Bennett, Q. Yan, and J.V. Priestley. Nerve Growth Factor Treatment Increases Brain-Derived Neurotrophic Factor Selectively in TrkA-Expressing Dorsal Root Ganglion Cells

and in Their Central Terminations within the Spinal Cord. *The Journal of Neuroscience,* November 1, 1997, 17(21):8476–8490.

[119] Lindsay, R.M., C. Lockett, J. Sternberg, J. Winter. Neuropeptide expression in cultures of adult sensory neurons: Modulation of substance P and calcitonin gene-related peptide levels by nerve growth factor. *Neuroscience*. Volume 33, Issue 1, 1989, Pages 53–65.

[120] Vescovi, Angelo L., Brent A. Reynolds, Douglas D. Fraser, Samuel Weiss. bFGF regulates the proliferative fate of unipotent (neuronal) and bipotent (neuronal/ astroglial) EGF-generated CNS progenitor cells. *Neuron.* Volume 11, Issue 5, November 1993, Pages 951-966.

[121] Reynolds, Brent A., Wolfram Tetzlaff, and Samuel Weiss. A Multipotent EGF-Responsive Striatal Embryonic Progenitor Cell Produces Neurons and Astrocytes. *The Journal of Neuroscience*, November 1992, 12(11): 4565-4574.

[122] Farkas, L. & Krieglstein, K. Heparin-binding epidermal growth factor-like growth factor (HB-EGF) regulates survival of midbrain dopaminergic neurons. *Journal of Neural Transmission*. March 2002, Volume 109, Issue 3, pp 267–277.

[123] Anchan, Raymond M., Thomas A. Reh, John Angello, Arthur Balliet Macie Walker. EGF and TGF-α stimulate retinal neuroepithelial cell proliferation *in vitro. Neuron*. Volume 6, Issue 6, June 1991, Pages 923-936.

[124] Holmberg, Johan, Annika Armulik, Kirsten-André Senti, Karin Edoff, Kirsty Spalding, Stefan Momma, Rob Cassidy, John G. Flanagan, and Jonas Frisén. *Ephrin-A2 reverse signaling negatively regulates neural progenitor proliferation and neurogenesis*. Cold Spring Harbor Laboratory Press ISSN 0890-9369/05.

[125] Friesel, R.E. and T. Maciag. Molecular mechanisms of angiogenesis: fibroblast growth factor signal transduction. *The FASEB Journal*. vol. 9 no. 10 919-925.

[126] Bai, Lianhua, Donald P Lennon, Arnold I Caplan, Anne DeChant, Jordan Hecker, Janet Kranso, Anita Zaremba & Robert H Miller. Hepatocyte growth factor mediates mesenchymal stem cell–induced recovery in multiple sclerosis models. *Nature Neuroscience* 15, 862–870 (2012) doi:10.1038/nn.3109.

[127] Tee Jong Huat, Amir Ali Khan, Soumya Pati, Zulkifli Mustafa, Jafri Malin Abdullah and Hasnan Jaafar. IGF-1 enhances cell proliferation and survival during early differentiation of mesenchymal stem cells to neural progenitor-like cells. *BMC Neuroscience* 201415:91. DOI: 10.1186/1471-2202-15-91.

[128] Wehrle-Haller, B., J.A. Weston. Soluble and cell-bound forms of steel factor activity play distinct roles in melanocyte precursor dispersal and survival on the lateral neural crest migration pathway. *Development* 1995 121: 731-742.

[129] Blesch, Armin, Mark H. Tuszynski. GDNF gene delivery to injured adult CNS motor neurons promotes axonal growth, expression of the trophic neuropeptide CGRP, and cellular protection. *J. Comp. Neurol.*, 436: 399–410. doi:10.1002/ cne.107.

[130] Heuckerotha, Robert O., Hideki Enomoto, John R Grider, Judith P. Goldenb, Julie A. Hankea, Alana Jackman, Derek C. Molliver, Mark E. Bardgett, William D. Snider, Eugene M. Johnson Jr. Gene Targeting Reveals a Critical Role for Neurturin in the Development and Maintenance of Enteric, Sensory, and Parasympathetic *Neurons. Neuron.* Volume 22, Issue 2, February 1999, Pages 253–263.

[131] Andres, Rosa, Alison Forgie, Sean Wyatt, Qi Chen, Frederic J. de Sauvage, Alun M. Davies. Multiple effects of artemin on sympathetic neuron generation, survival and growth. *Development* 2001 128: 3685-3695.

[132] Milbrand, Jeffrey, J. de Sauvage, Timothy J. Fahrner, Robert H. Baloh, Melanie L. Leitner, Malú G. Tansey, Patricia A. Lampe, Robert O. Heuckeroth, Paul T., Kelli S. Persephin, a Novel Neurotrophic Factor Related to GDNF and Neurturin. *Neuron*. Volume 20, Issue 2, February 1998, Pages 245-253.

[133] Shweiki, Dorit, et al. "Vascular endothelial growth factor induced by hypoxia may mediate hypoxia-initiated angiogenesis." *Nature* 359.6398 (1992): 843-845.

[134] Kan, I., Barhum, Y., Melamed, E. et al. Mesenchymal Stem Cells Stimulate Endogenous Neurogenesis in the Subventricular Zone of Adult Mice Stem Cell. *Rev and Rep* (2011) 7: 40.

[135] Hegarty, Shane V., Aideen M. Sullivan, Gerard W. O'Keeffe. BMP2 and GDF5 induce neuronal differentiation through a Smad dependant pathway in a model of human midbrain dopaminergic neurons. *Molecular and Cellular Neuroscience.* Volume 56, September 2013, Pages 263-271.

[136] Watabe, Tetsuro and Kohei Miyazono. Roles of TGF-β family signaling in stem cell renewal and differentiation. *Cell Research* (2009) 19:103–115. doi: 10.1038/cr.2008.323.

[137] Yu, Young-Mi, Kurt M. Gibbs, Jonathan Davila,Neil Campbell, Simon, Sung,Tihomira I. Todorova, Seiji Otsuka, Hatem E. Sabaawy, Ronald P. Hart, Melitta Schachner. MicroRNA miR-133b is essential for functional recovery after spinal cord injury in adult zebrafish. *European journal of neuroscience*. 2011.

[138] Sofroniew, Michael V., Charles L. Howe, William C. Mobley. Nerve Growth Factor Signaling, Neuroprotection, and Neural Repair. *Annual Review of Neuroscience*. 2001.

[139] Jin, Kunlin, Lin Xie, Jocelyn Childs, Yunjuan Sun, Xiao Ou Mao, Anna Logvinova, and David A. Greenberg. Cerebral Neurogenesis Is Induced by Intranasal Administration of Growth Factors. *Ann Neurol* 2003;53: 405–409.

[140] Liu, Xin-Feng, John R. Fawcett, Robert G. Thorne, Terese A. DeFor, William H. Frey II. Intranasal administration of insulin-like growth factor-I bypasses the blood–brain barrier and protects against focal cerebral ischemic damage. *Journal of the Neurological Sciences*. 187 2001 91–97.

[141] Capsoni S., Ugolini, G., Comparini, A., Ruberti, F., Berardi, N. and Cattaneo, A. (2000), Alzheimer-like neurodegeneration in aged anti nerve growth factor transgenic mice. *Proc Natl Acad Sci USA* 97:6826–6831.

[142] Hanson, Leah R., Annina Roeytenberg, Paula M. Martinez, Valerie G. Coppes, Donald C. Sweet, Reshma J. Rao, Dianne L. Marti, John D. Hoekman, Rachel B. Matthews, William H. Frey II, and S. Scott Panter. Intranasal Deferoxamine Provides Increased Brain Exposure and Significant Protection in Rat Ischemic Stroke. *The journal of pharmacology and experimental therapeutics.* doi:10.1124/ jpet.108.149807.

[143] Balin, B.J., Broadwell RD, Salcman, M., E1-Kalliny, M. (1986): Avenues for entry of peripherally administered protein to the central nervous system in mouse, rat, and squirrel monkey. *J Comp Neurol* 251:260-280.

[144] Pietrowsky, Reinhard, Christel Strtiben, Matthias M611e, Horst L. Fehm, and Jan Born. Brain Intravenous for a Direct in Humans Potential Changes after Intranasal vs. *Administration of Vasopressin*: Evidence Nose-Brain Pathway for Peptide Effects.

[145] De Rosa, Roberta, Addys Ancheta Garcia, Chiara Braschi, Simona Capsoni, Lamberto Maffei, Nicoletta Berardi, and Antonino Cattane. Intranasal administration of nerve growth factor (NGF) rescues recognition memory deficits in AD11 anti-NGF transgenic mice. *PNAS.* March 8, 2005.

In: Stem Cells Origins of Cytokine Interactions
Editor: Prasad S. Koka
ISBN: 978-1-53615-302-6

Chapter 15

EFFECTS OF INTEGRATED APPROACH OF YOGA THERAPY (IAYT) ON CARDIAC VARIABLES AND MENTAL HEALTH IN HYPERTENSIVE PATIENTS WITH ELEVATED ANXIETY AND DEPRESSION

Kashinath G. Metri[1,*], BAMS, MD, PhD, Balaram Pradhan[1], PhD and H. R. Nagendra[1,2], PhD

[1]Department of Yoga and Life Sciences,
SVYAS University, Bengaluru, India
[2] Swami Vivekananda Yoga Anusandhana Samsthana, Bengaluru, India

ABSTRACT

Background: Hypertension significantly affects the mental health of the patient. Elevated anxiety and depression are commonly observed in hypertensive subjects compared to normotensives. Previous studies have demonstrated the positive impact of IAYT on mental health in many chronic health problems.

Objectives: To study the effects of Integrated Approach of Yoga Therapy (IAYT) on anxiety and depression in hypertensive subjects with elevated anxiety and depression.

Methodology: We screened 131 hypertensive patients within age range 40-60 years for anxiety and depression (using Hospital Anxiety Depression Scale (HADS)), 92 subjects found to have either elevated anxiety or depression or both. These potential subjects underwent one week residential yoga program at Holistic Health Care Centre, S-VYASA University, Bangalore-India. Subjects who were on antipsychotic medication, had recent surgery, previous exposure to any form of yoga in past one year, secondary hypertension, resistant hypertension were excluded from the study. Heart rate, systolic &

* Corresponding Author Email: kgmhetre@gmail.com.

diastolic blood pressure, anxiety, and depression were assessed at baseline and after one week of IAYT intervention. Data was analyzed using SPSS version 10.

Result: At the baseline, of 92 potential subjects, 47 (51.08%) had clinical anxiety, 26 (28.26%) had clinical depression, and 19 had both (20.65%). After one week of IAYT intervention significant decrease in pulse rate ($p < 0.001$), systolic ($p < 0.001$), & diastolic blood pressure ($p < 0.001$), anxiety ($p < 0.001$), and depression ($p < 0.001$) was observed compared to baseline.

Conclusion: Present study indicates the potential role of IAYT in improving mental health & cardiac variables in hypertension subjects. However, further randomized control trial should be performed to confirm the findings of the present study, which will be implemented in future studies.

Keywords: hypertension, anxiety, depression, yoga, IAYT

Introduction

Hypertension (HTN) is a major public health concern and a leading cause of morbidity and mortality across the globe [1]. It affects 25% of word population and it is estimated that, by the year 2025, 60% of world population will develop HTN [2]. HTN contributes to approximately 13% of yearly deaths worldwide [3, 4]. HTN is directly responsible for, 25% of deaths due to myocardial infarction, and 53% of death due to stroke. In a long run hypertension leads to several complications such as ischemic heart disease, stroke, glaucoma etc. [5]. Evidences from mind-body interaction studies showed a strong influence of mind on body physiology and vice-versa.

Studies demonstrated that emotions like, anger, anxiety, prolonged sadness (depression) have negative impact on heart rate, blood pressure, heart rate variability, basal metabolic rate etc. [6, 7, 8, 9]. Long term exposure to such emotions contributes the cause of many chronic ailments like, coronary artery disease, accelerated aging process, diabetes, hypertension, migraine etc. [10, 11]. Among these psychological factors anxiety and depression are most predominantly found in several chronic health conditions including hypertension. Depression and anxiety are involved in development, maintenance and complication of the hypertension [12]. Hypertensive people known to have elevated anxiety [13] and depression level compared to age and gender matched healthy subjects. Elevated anxiety in hypertension increases the progression of atherosclerosis [14]. Increased anxiety level in patients with coronary intervention increase the risk of re-vascular intervention [15]. In an observational study pre-pregnancy existence of anxiety and depression is associated with pregnancy induced hypertension [16].

Depression is associated with increased risk of hypertension [17]. A survey report showed that 44% of individuals with depression at baseline were found to have developed hypertension [18]. In a cross-sectional study hypertensive subjects showed higher chances of developing depression compared to normotensives [19]. Depression leads to many behavioral changes such as smoking, alcohol, reduced physical activity etc. which are associated which further increased risk of hypertension and CVD [20, 21, 22, 23].

Some studies showed the evidence for brain stem cells dysfunction in the pathology of mental ailments including depression. A strong association of brain stems cell dysfunction in depression is prominently reported in many studies. Hippocampus is a major part of the brain believed to be involved in the pathogenesis of depression. Persons with depressive disorder are known to have reduced hippocampus thickness [24]. The cause of hippocampus atrophy in depression is believed to be reduced adult neurogenes is that decreases production of functional neuronal cells in the hippocampus area of the brain. Some post-mortem studies of depression patients showed an increased neuronal apoptosis suggestive of decreased neuronal stem cell functioning in depression. Structural and functional neuronal cells in hippo-campus area are produced by neuronal stems cells present in subventricular zone [25]. Hence it thought that reduced hippocampus thickness is could be due to hypo-functioning of neuronal stems cell [26]. The anti-depressant interventions such as exercise [27], electroconvulsive therapy and some anti-depressant medications found to increase adult neurogenes [28].

YOGA

The science of yoga was proposed, developed, and practiced by ancient sages of Asian continent. Practice of yoga promotes physical and mental health leads to spiritual growth of the individual. Currently yoga is used as complementary and integrative medicine in many chronic health conditions such as, diabetes type-2, hypertension, coronary artery disease, chronic low back pain, Parkinson's disease, depression, schizophrenia, etc. [29, 30, 31] Various meta-analyses have confirmed the beneficial role of yoga in mental health. Practice of yoga is known to reduce anxiety [32] and depression [33]. A study by Rao et al. 2016 showed a significant improvement in psychological health of female teacher following one month of relaxation technique [34].

Integrated Approach of Yoga Therapy (IAYT)

IAYT at SVYASA University campus is a residential program consists of yoga based lifestyle modifications. This program includes yoga sessions specially designed for hypertension twice daily, *pranayama* (yogic breathing practices), meditation devotional session and yogic cleansing techniques twice in a week. This program also includes diet modification based on the health problem. The whole program is conducted in residential set up [35]. Previously scientific investigations on IAYT have shown beneficial effects in several health conditions [36, 67]. A study by Rao J et al. 2017 showed a significant improvement in constipation related sysmptoms following one week of IAYT. Another study by Padmini T et al. 2012 showed a significant improvement in anxiety and quality of in subjects with chronic low back pain.

NEED FOR THE STUDY

Previously, several cross-sectional and longitudinal studies have demonstrated that the psychological domains including anxiety and depression being contributing in factors cause of hypertension and complication of hypertension. Conventional medicine has limited role prevention and alleviation of anxiety and depression, anti-hypertensive medication does not reduce the anxiety and depression in hypertensive participants. Hence there is need of alternative intervention to control. Yoga is a non-pharmacological and cost effective intervention without side effects. Several studies have confirmed the positive role of yoga in reducing anxiety and depression. IAYT is a residential yoga based lifestyle intervention, conducted in SVYASA University. It is a comprehensive approach to deal modern non-communicable diseases. It includes yoga based diet, physical activity, stress management through, asanas, pranayama, meditation and devotional sessions. Several scientific investigation have shown the positive role of yoga in many health related conditions including diabetes [38], constipation [39], anxiety [40], hypertension [41] etc.

Methodology

Subjects

Hypertensive subjects within age range 40-60 years who visited to Arogyadhama, S-VYASA University, Bangalore for one week IAYT program for hypertension between January 2016 to June 2016, were considered for the study.

Inclusion Criteria

We included the Hypertensive subjects within age range 40 to 60 years, both gender, willing to participate the study, having HADS-A Score>11 or HADS-D Score>11 or both on only one antihypertensive medicine.

Exclusion Criteria

Subjects were excluded if they had: diabetes mellitus; resistant hypertension; secondary hyper-tension; taking antipsychotic medication; recent hospitalization for major health problem; pregnancy; previous exposure to any form of yoga in last one year; recent surgery and pregnant women were excluded from the study.

Procedure

We screened 131 subjects for clinical anxiety and depression, using Hospital Anxiety Depression Scale (HADS) for clinical anxiety and depression (using Hospital Anxiety Depression Scale (HADS) HADS-A Score>11 or HADS-D Score>11). Of 131 screened 92 participants found to have clinical caseness for anxiety and depression. Among 92 (70%), 47 found to have anxiety, 26 had depression and 19 had both.

Ethical Consideration

All the subjects enrolled in the study we explained the study protocol in their respective language and a duly signed written informed consent was obtained from each subjects before commencement of the study.

Study was evaluated and approved by Institutional Ethical Committee (IEC) of SVYASA University.

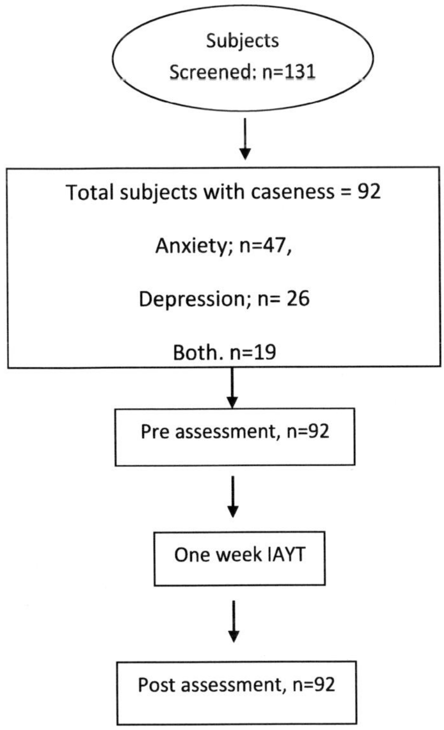

Assessments

Cardiac Variables

Heart rate, systolic and diastolic blood pressure were assessed using Omran BP monitoring device at baseline and after one week.

Psychological Variables

Anxiety and depression were assessed using Hospital anxiety and depression scale (HADS)

Hospital Anxiety and Depression Scale (HADS)

It considered as one of the reliable and valid tools to assess subjective depression and anxiety in the past week. This scale includes 14 items (7 items for anxiety and 7 items for depression). Score equal or more 11 indicates clinical case [42].

Intervention

Subjects in yoga group underwent one week IAYT intervention consisting of yogic three session, one each, specific yoga sessions for hypertension, three times a day consisting of physical practices, breathing practices, and relaxation techniques along with weekly twice different types of yogic cleansing techniques, daily ½ hr devotional session, counseling sessions and tuning to nature [Detailed in Tables 1 and 2].

Table 1. List of Yoga practices in yoga sessions

Type of Practices	
Loosening Practices (Sukshmavyayama)	Finger loosening, wrist loosening, elbow loosening, neck bending and rotation Twisting, hip rotation
Breathing Practices	Hands in & out breathing, hands stretch breathing, ankle stretch breathing, tiger breathing
Asanas (Physical Postures)	Tadasana, vrikshasana, ardhachakrasana, ardhakati chakrasana, trikonasana Bhunamanasana, ardhaustrasana, badhakonasana, setibandhasana, bhujanagasana, shalabhasana
Relaxation Techniques	Quick relaxation technique, deep relaxation technique, mind sound resonance technique, Cyclic meditation
Pranayama	Chandra anuloma vilom, nadishudhi pranayama, bhramari pranyama, sheetali pranayamas

Table 2. Schedule

Time	Activity	Time	Activity
05.00 AM	Ablution	03.00 PM	Cyclic meditation
05.30 AM	Prayer (Prathasmaran)	04.00 PM	Asana / Special yoga technique
06.00 AM	Asana / Special yoga technique	05.00 PM	Tuning to nature
07.15 AM	Friendship meet (Maitri milan)- Gita sloka chanting and discourse (Satsang)	06.00 PM	Devotional session (Bhajan)
08.00 AM	Breakfast	06.45 PM	MSRT (Mind sound resonance technique)
09.30 AM	Yogic Counselling	07.30 PM	Dinner
10.30 AM	Pranayama	08.30 PM	Happy assembly (yoga game session)/cultural program
11.30 AM	Milk or ayurvedic tea (Malt)	09.15 PM	Group discussion/self practice
12.05 PM	Special yoga techniques	10.00 PM	Lights off
01.00 PM	Lunch and rest		

Data Analysis

Descriptive data was showed in terms of mean and standard deviation. Data was found to be normally distributed by the Shapiro-Wilk test. Data prepost changes were assessed using paired sample t test. SPSS version 10.

RESULTS

Demographics

We screened 131 (68 males) of which 92 (70.21%) among 92 potential subjects, 47 (51.08%) had clinical anxiety, 26 (28.26%), had clinical depression, and 19 (20.65%) had anxiety and depression both (See Table 3). No significant change was found in mean age of both gender.

Correlation

We found a significant positive correlation between anxiety and depression ($p < 0.001$, Pearson's Corr. = 0.501).

Pre-Post Comparison

Subjects with Clinical Anxiety

We found a significant decrease in pulse rate ($p < 0.001$, -15.09%), systolic blood pressure ($p < 0.001$, -12.40%) and diastolic blood pressure ($p < 0.001$, -17.14%) along HADS A ($p < 0.001$, -54.98%) scores after one week IAYT intervention in subjects who had subjects clinical anxiety (Table 4).

Table 3. Demographics

Variables	Male (n = 68)	Female (n = 63)	Total (131)
Age (Mean ± SD)	53.45±11.54	55.61±10.91	54.99 ± 11.30
Clinical Anxiety	24	23	47 (51.08%)
Clinical Depression	9	17	26 (28.26%)
Both	12	7	19 (20.65%)

Table 1 shows the gender-wise age, number of subjects with clinical anxiety, depression and both. Average age of the subjects is 54.99. Among 131 subjects 92 subjects had found clinical caseness and Majority of the subjects had clinical anxiety (51) compared to depression.

Table 4. Pre-post changes in all the variables in patients with clinical anxiety

Variables	Pre [Mean ± SD]	Post [Mean ± SD]	% Change	df	P Value
Pulse (bpm)	82.40 ± 9.97	69.96 ± 7.36	-15.09	46	<0.001
Sys BP (mmHg)	144.68 ± 11.76	126.73 ± 7.22	-12.40	46	<0.001
Dia BP (mmHg)	93.48 ± 9.75	77.45 ± 5.19	-17.14	46	<0.001
Anxiety score	13.44 ± 21.66	6.05 ± 2.57	-54.98	46	<0.001

Table 2 shows the changes in blood pressure and anxiety scores following one week yoga intervention. Significant improvement is seen in all the variables. P value of 0.001 indicates the highly significant change Abbreviations: Sys BP-Systolic blood pressure, Dia BP-Diastolic blood pressure.

Table 5. Pre-post changes in all the variables in patients with clinical depression

Variables	Pre [Mean ± SD]	Post [Mean ± SD]	% Change	df	P Value
Pulse (bpm)	87.65 ± 7.33	68.30 ± 7.63	-22.08	25	<0.001
Sys BP (mmHg)	152.65 ±6.73	125.76 ± 4.73	-17.62	25	<0.001
Dia BP (mmHg)	94.87 ± 7.38	78.57 ± 4.81	-17.18	25	<0.001
Depression score	13.57 ± 1.66	6.48 ± 2.65	-52.21	25	<0.001

Table 2 shows the changes in blood pressure and depression scores following one week yoga intervention. Significant improvement is seen in all the variables. P value of 0.001 indicates the highly significant change. Abbreviations: Sys BP-Systolic blood pressure, Dia BP-Diastolic blood pressure.

Table 6. Pre-post changes in all the variables in patients with both clinical anxiety and depression

Variables	Pre	Post	% Change	df	P Value
Pulse (bpm)	85.63 ±7.76	69.72 ± 8.74	-18.58	18	<0.001
Sys BP (mmHg)	153.83 ± 6.61	123.66 ± 5.65	-19.61	18	<0.001
Dia BP (mmHg)	95.62 ± 6.79	78.62 ± 4.66	-17.78	18	<0.001
Anxiety score	13.82 ± 2.92	7.44 ± 31.82	-46.16	18	<0.001
Depression score	13.19 ± 1.32	6.57 ± 2.77	-50.19	18	<0.001

Table 6 shows the changes in blood pressure and anxiety & depression scores following one week yoga intervention. Significant improvement is seen in all the variables. P value of 0.001 indicates the highly significant change. Abbreviations: Sys BP-Systolic blood pressure, Dia BP-Diastolic blood pressure.

Subjects with Clinical Depression

There was significant decrease in pulse rate ($p < 0.001$, -22.08%), systolic blood pressure ($p < 0.001$, -17.62%) and diastolic blood pressure ($p < 0.001$, -17.18%) along with significant decrease in depression ($p < 0.001$, -52.21%) scores observed after one week IAYT intervention in subjects with clinical depression (Table 5).

Subjects with Both Clinical Anxiety and Depression

There was significant decrease in pulse rate ($p < 0.001$, -18.58%), systolic blood pressure ($p < 0.001$, -17.68%) and diastolic blood pressure ($p < 0.001$, -19.61%) along with significant decrease in anxiety ($p < 0.001$, -46.16%) and depression ($p < 0.001$, -50.19%) scores observed after one week IAYT intervention in subjects with clinical depression (Table 6).

DISCUSSION

This study intended to see the effects of one week of yoga based lifestyle intervention (IAYT) on cardiac and mental health parameters in hypertensive patients with elevated anxiety and depression. At the baseline among 92 potential subjects 47 (51.08%) participants had anxiety and 26 participants had clinical depression (28.26%) and 19 (20.65) had both anxiety and depression.

Though our study screened small sample of 131 subjects with hypertension, we noticed a high prevalence of anxiety and depression (70%), which is in consistent with previous studies [43].

After one week of IAYT intervention there was a significant decrease in anxiety, depression, systolic & diastolic blood pressure in subjects with either anxiety or depression or both.

We also observed a strong positive correlation between anxiety and depression. To the best of our knowledge this is the first to assess the impact of yoga intervention on mental health among hypertensive subjects.

IAYT is residential yoga based lifestyle intervention. Earlier studies of IAYT have confirmed its usefulness in several chronic diseases such as chronic low back pain [44], osteoarthritis, [45] menopausal syndrome etc.

Previously few studies have looked into effectiveness of IAYT on mental health in chronic health disorders. Among these a study by Padmini T found a significant decrease in anxiety associated with chronic low back pain after one week of IAYT. Another study by John E found a significant reduction in depression after one week of IAYT in participants with osteoarthritis.

Anxiety and depression are the common mental health problems associated with several chronic health conditions including hypertension. IAYT intervention shows a promising role in improving mental health by reducing anxiety and depression in hypertension.

The exact mechanism IAYT in current study is not known. Possible mechanisms can could be, practice of various *asanas, pranayama*, meditation and devotional sessions in IAYT lead to down regulation of hypothalamo-pituitary-adrenal axis, causing parasympathetic dominance which lead to decrease in pulse, blood pressure and anxiety [46, 47]. Change in the atmosphere, Interaction with therapist, group practice of yoga, and increased physical activity might have led to decrease in depression.

Further, neural stem cell proliferation is involved in various brain functions including, learning, stress induced neuropathology, & affective disorders [48]. There are evidences suggesting the strong involvement of neuronal stem cell proliferation in pathology of depression [49]. Studies reported decreased neural proliferation in hippocampus in depressed patients. And some post-mortem studies on depressed patients showed an evidence of increased apoptosis in hippocampus. These findings indicate the neural stem cell involvement in pathology of depression.

There are studies demonstrated learning, enriched environment and exercise are some of the factors promote neural stem cells proliferation in hippocampus [50]. Similarly in our study the residential atmosphere at SVYASA where the study has been conducted is enriched and the lot of yoga sessions was comprised of adequate physical activity. Probably this might have led to enhanced proliferation of neural stem cell in the hippocampus and improved mental health. However, future studies should attempt to explore the impact of yoga on neural stem cell proliferation in clinical population.

CONCLUSION

Present study indicates the potential role of IAYT in improving mental health & cardiac variables in hypertension subjects. However, further randomized control trial should be performed to confirm the findings of the present study, which will be implemented in future studies.

Strengths of this study are; 1) Good adherence rate 2) No adverse effects were observed during the intervention 3) It is first study to assess the effect of IAYT on psychological states in hypertensive participants 5) IAYT is a comprehensive treatment modality which includes diet, physical activity, meditation, pranyama and counseling

Limitations: This study has got several limitations; 1) Lack of control group 2) Small sample size, 4) Short term intervention 5) Lack of follow up.

FUTURE SUGGESTIONS

Future studies should include 1) Control group with Proper randomization 2) Control group should follow some international standard intervention 3) Objective variables like serum cortisol levels, heart rate variability etc. should also be assessed.

CONFLICT OF INTEREST

The authors declare no conflict of interest.

ACKNOWLEDGMENTS

We are thankful to all the therapist, doctors and staff of *Arogydhama* holistic health centre, SVYASA. We thank all the subjects who participated in this study.

REFERENCES

[1] Lim SS, Vos T, Flaxman AD, Danaei G, Shibuya K, Adair-Rohani H, AlMazroa MA, Amann M, Anderson HR, Andrews KG, Aryee M. A comparative risk assessment of burden of disease and injury attributable to 67 risk factors and risk factor clusters in 21 regions, 1990–2010: a systematic analysis for the Global Burden of Disease Study 2010. *The lancet*. 2013 Jan 4; 380(9859):2224-60.

[2] Adrogué HJ, Madias NE. Sodium and potassium in the pathogenesis of hypertension. *New England Journal of Medicine.* 2007 May 10;356(19):1966-78.

[3] World Health Organization. *The world health report 2002: reducing risks, promoting healthy life.* World Health Organization; 2002.

[4] World Health Organization, UNAIDS. *Prevention of cardiovascular disease*. World Health Organization; 2007.

[5] Chatterjee S, Chattopadhya S, Hope-Ross M, Lip PL. Hypertension and the eye: changing perspectives. *Journal of human hypertension.* 2002 Oct 1;16(10).

[6] Gentry DW, Chesney AP, Gary Jr HE, Hall RP, Harburg E. Habitual anger-coping styles: I. Effect on mean blood pressure and risk for essential hypertension. *Psychosomatic medicine*. 1982 May 1;44(2):195-202.

[7] Brosschot JF, Thayer JF. Heart rate response is longer after negative emotions than after positive emotions. *International Journal of Psychophysiology*. 2003 Nov 30;50(3):181-7.

[8] McCraty R, Atkinson M, Tiller WA, Rein G, Watkins AD. The effects of emotions on short-term power spectrum analysis of heart rate variability. *TheAmerican journal of cardiology*. 1995 Nov 15;76(14):1089-93.

[9] Thayer JF, Yamamoto SS, Brosschot JF. The relationship of autonomic imbalance, heart rate variability and cardiovascular disease risk factors. *International journal of cardiology*. 2010 May 28; 141(2):122-31.

[10] Kubzansky LD, Cole SR, Kawachi I, Vokonas P, Sparrow D. Shared and unique contributions of anger, anxiety, and depression to coronary heart disease: a prospective study in the normative aging study. *Annals of Behavioral Medicine*. 2006 Feb 1;31(1):21-9.

[11] Gross JJ, Carstensen LL, Pasupathi M, Tsai J, Götestam Skorpen C, Hsu AY. Emotion and aging: experience, expression, and control. *Psychology and aging*. 1997 Dec;12(4):590.

[12] Rutledge T, Hogan BE. A quantitative review of prospective evidence linking psychological factors with hypertension development. *Psychosomatic Medicine*. 2002 Sep 1;64(5):758-66.

[13] Wei TM, Wang L. Anxiety symptoms in patients with hypertension: a community-based study. *The International Journal of Psychiatry in Medicine*. 2006 Sep 1;36(3):315-22.

[14] Paterniti S, Zureik M, Ducimetière P, Touboul PJ, Fève JM, Alpérovitch A. Sustained anxiety and 4-year progression of carotid atherosclerosis. *Arteriosclerosis, Thrombosis, and Vascular Biology*. 2001 Jan 1;21(1): 136-41.

[15] Meyer T, Hussein S, Lange HW, Herrmann-Lingen C. Anxiety is associated with a reduction in both mortality and major adverse cardiovascular events five years after coronary stenting. *European journal of preventive cardiology*. 2015;22(1):75-82.

[16] Thombre MK, Talge NM, Holzman C. Association between pre-pregnancy depression/anxiety symptoms and hypertensive disorders of pregnancy. *Journal of Women's Health*. 2015 Mar 1;24(3):228-36.

[17] Meng L, Chen D, Yang Y, Zheng Y, Hui R. Depression increases the risk of hypertension incidence: a meta-analysis of prospective cohort studies. *Journal of Hypertension*. 2012 May 1;30(5):842-51.

[18] Gangwisch JE, Malaspina D, Posner K, Babiss LA, Heymsfield SB, Turner JB, Zammit GK, Pickering TG. Insomnia and sleep duration as mediators of the relationship between depression and hypertension incidence. *American journal of hypertension*. 2010 Jan 1;23(1):62-9.

[19] Bosworth HB, Bartash RM, Olsen MK, Steffens DC. The association of psychosocial factors and depression with hypertension among older adults. *International journal of geriatric psychiatry*. 2003 Dec 1;18(12): 1142-8.

[20] Allan JL, Johnston DW, Johnston M, Mant D. Depression and perceived behavioral control are independent predictors of future activity and fitness after coronary syndrome events. *Journal of Psychosomatic Research.* 2007 Nov 30;63(5):501-8.

[21] De Moor MH, Boomsma DI, Stubbe JH, Willemsen G, de Geus EJ. Testing causality in the association between regular exercise and symptoms of anxiety and depression. *Archives of General Psychiatry*. 2008 Aug 4;65(8):897-905.

[22] Appleton KM, Woodside JV, Yarnell JW, Arveiler D, Haas B, Amouyel P, Montaye M, Ferrières J, Ruidavets JB, Ducimetiere P, Bingham A. Depressed mood and dietary fish intake: direct relationship or indirect relationship as a result of diet and lifestyle?. *Journal of affective disorders.* 2007 Dec 31;104(1):217-23.

[23] Strine TW, Mokdad AH, Dube SR, Balluz LS, Gonzalez O, Berry JT, Manderscheid R, Kroenke K. The association of depression and anxiety with obesity and unhealthy behaviors among community-dwelling US adults. *General hospital psychiatry*. 2008 Apr 30;30(2):127-37.

[24] Geuze E, Vermetten E, Bremner JD. MR-based in vivo hippocampal volumetrics: 2. Findings in neuro-psychiatric disorders. *Mol Psychiatry* 2005; 10: 160–184.

[25] Song HJ, Stevens CF, Gage FH. Neural stem cells from adult hippocampus develop essential properties of functional CNS neurons. *Nat Neurosci* 2002; 5: 438–445.

[26] van Praag H, Schinder AF, Christie BR, Toni N, Palmer TD, Gage FH. Functional neurogenesis in the adult hippocampus. *Nature* 2002; 415: 1030–1034.

[27] van Praag H, Kempermann G, Gage FH. Running increases cell proliferation and neurogenesis in the adult mouse dentate gyrus. *Nat Neurosci* 1999; 2: 266–270.

[28] Madsen TM, Treschow A, Bengzon J, Bolwig TG, Lindvall O, Tingstrom A. Increased neurogenesis in a model of electroconvulsive therapy. *Biol Psychiatry* 2000; 47: 1043–1049.

[29] Sherman KJ, Cherkin DC, Erro J, Miglioretti DL, Deyo RA. Comparing yoga, exercise, and a self-care book for chronic low back pain: a randomized, controlled trial. *Annals of Internal Medicine.* 2005 Dec 20;143(12):849-56.

[30] Duraiswamy G, Thirthalli J, Nagendra HR, Gangadhar BN. Yoga therapy as an add-on treatment in the management of patients with schizophrenia–a randomized controlled trial. *Acta Psychiatrica Scandinavica.* 2007 Sep 1;116(3):226-32..

[31] Colgrove YS, Sharma N, Kluding P, Potter D, Imming K, VandeHoef J. Effect of yoga on motor function in people with Parkinson's disease: a randomized, controlled pilot study. *J Yoga Phys Ther*. 2012;2(2):112.

[32] Gupta N, Khera S, Vempati RP, Sharma R, Bijlani RL. Effect of yoga based lifestyle intervention on state and trait anxiety. *Indian journal of physiology and pharmacology*. 2006 Jan 21;50(1):41.

[33] Pilkington K, Kirkwood G, Rampes H, Richardson J. Yoga for depression: the research evidence. *Journal of affective disorders.* 2005 Dec 31;89(1):13-24.

[34] Manas Rao, Kashinath Metri, Nagaratna Raghuram, Nagendra R. Hongasandra. *Effects of Mind Sound Resonance Technique (Yogic Relaxation) on Psychological States, Sleep Quality, and Cognitive Functions in Female Teachers: A Randomized, Controlled Trial.*

[35] Nagendra HR. Integrated yoga therapy for mental illness. *Indian journal of psychiatry*. 2013 Jul;55(Suppl 3):S337.

[36] Tekur P, Singphow C, Nagendra HR, Raghuram N. Effect of short-term intensive yoga program on pain, functional disability and spinal flexibility in chronic low back pain: a randomized control study. *The journal of alternative and complementary medicine*. 2008 Jul 1;14(6):637-44.

[37] Ebnezar J, Nagarathna R, Bali Y, Nagendra HR. Effect of an integrated approach of yoga therapy on quality of life in osteoarthritis of the knee joint: A randomized control study. *International journal of yoga*. 2011 Jul 1;4(2):55.

[38] Mohanty S, Metri K, Nagaratna R, Nagendra HR. Immediate effect of mind sound resonance technique (msrt-a yogic relaxation technique) on cognitive functions in type 2 diabetes. *Voice of Research*. 2015 Jun;4(1):44-5.

[39] Rao J, Metri KG, Singh A, Nagaratna R. *Effect Of Integrated Approach Of Yoga Therapy On Chronic Constipation.* 2016 Jun.

[40] Dhansoia V, Bhargav H, Metri K. Immediate effect of mind sound resonance technique on state anxiety and cognitive functions in patients suffering from generalized anxiety disorder: A self-controlled pilot study. *International journal of yoga*. 2015 Jan;8(1):70.

[41] Metri KG, Pradhan B, Nagendra HR. Impact of Short-Term Residential Yoga-Based Lifestyle Intervention on Cognitive Function in Hypertensive Patients. *Journal of Stem Cells.* 2017;12(1):33.

[42] Mykletun A, Stordal E, Dahl AA. Hospital Anxiety and Depression (HAD) scale: factor structure, item analyses and internal consistency in a large population. *The British journal of psychiatry*. 2001 Dec 1;179(6):540-4.

[43] Ginty AT, Carroll D, Roseboom TJ, Phillips AC, De Rooij SR. Depression and anxiety are associated with a diagnosis of hypertension 5 years later in a cohort of late middle-aged men and women. *Journal of human hypertension.* 2013 Mar 1;27(3):187-90.

[44] Tekur P, Singphow C, Nagendra HR, Raghuram N. Effect of short-term intensive yoga program on pain, functional disability and spinal flexibility in chronic low back pain: a randomized control study. *The Journal of alternative and complementary medicine*. 2008 Jul 1;14(6):637-44.

[45] Ebnezar J, Nagarathna R, Bali Y, Nagendra HR. Effect of an integrated approach of yoga therapy on quality of life in osteoarthritis of the knee joint: A randomized control study. *International journal of yoga*. 2011 Jul 1;4(2):55.

[46] Ross A, Thomas S. The health benefits of yoga and exercise: a review of comparison studies. *The journal of alternative and complementary medicine.* 2010 Jan 1;16(1):3-12.

[47] Streeter CC, Gerbarg PL, Saper RB, Ciraulo DA, Brown RP. Effects of yoga on the autonomic nervous system, gamma-aminobutyric-acid, and allostasis in epilepsy, depression, and post-traumatic stress disorder. *Medical hypotheses.* 2012 May 31;78(5):571-9.

[48] Crews FT, Nixon K, Wilkie ME. Exercise reverses ethanol inhibition of neural stem cell proliferation. *Alcohol.* 2004 May 31;33(1):63-71.

[49] Carro, E., Nunez, A., Busiguina, S., & Torres-Aleman, I. (2000). Circulating insulin-like growth factor I mediates effects of exercise on the brain. *J Neurosci* 20, 2926–2933.

In: Stem Cells Origins of Cytokine Interactions
Editor: Prasad S. Koka
ISBN: 978-1-53615-302-6

Chapter 16

STEM CELL GENES SYNERGISM: ELUCIDATION OF THE MECHANISTIC BASIS OF RESPONSE INDUCED BY CYTOKINE COMBINATION (IL-3, FLT-3, AND SCF)

***Swati Sharma*[1], *Gurudutta U. Gangenahalli*[2,*] and *Upma Singh*[1,†]**

[1]Department of Applied Chemistry, School of Vocational Studies and Applied Sciences, Gautam Buddha University, Greater Noida, U. P., India
[2]Scientist 'G' FNABS, Fellow RS Medicine (UK), FRSB(UK), FRSC(UK), Head, Division of Stem Cell and Gene Therapy Research, Institute of Nuclear Medicine and Allied Sciences (INMAS), Defence Research and Development Organisation (DRDO), Delhi, India

ABSTRACT

Background: The signals received by Hematopoietic stem/ progenitor cells (HSCs) for proliferation and differentiation within the bone marrow environment are critical for understanding and characterization of their mechanisms, which likely to decipher coordinated expression of related functional genes & associated multiple pathways. This study was aimed to identify potential biomarkers of transient effect on Umbilical Cord Blood (UCB) derived CD34+ cells in response to the treatment of a particular combination of cytokines Interleukin-3 (IL-3), Fms-Like Tyrosine Kinase-3 Ligand (FLT-3), and Stem Cell Factor (SCF).

* Corresponding Author Email: gugdutta@rediffmail.com.
† Corresponding Author Email: drupmasingh@gmail.com.

Method: The microarray data with accession number GSE3003 was extracted from NCBI to identify global signatures and to construct & analyze Protein-Protein Interaction (PPI) networks eliciting different biochemical responses in molecular signaling.

Results: In this study, a total of 22,283 gene transcripts were identified. The most significant genes that are found to be responsible for HSC responses were 144 & 111-proliferation, 98 & 58-differentiation and 10 & 28 -self- renewal among the group of up-regulated and down-regulated genes respectively. Further, nine notable genes (RBM3, DHX32, PSMD9, SAR1A, SMYD3, ST6GAL1, CKAP4, CTR9) were identified having a crucial role in molecular signaling. Strikingly, among the gene-interacting partner associations, few combinations like RRAGC & PSMC2; CKAP4 & MANF on proliferation and CTR9 & CNTNAP2 on self-renewal properties have been found unexplored previously.

Conclusion: The novel combinations-RRAGC & PSMC2; CKAP4 & MANF and CTR9 & CNTNAP2 that have been found unexplored previously would be a stepping stone to unlock their long-touted therapeutic & synergistic gene therapy potential through siRNA-mediated knock down techniques.

Keywords: Hematopoietic Stem Cells (HSCs), gene therapy, proliferation, and self-renewal

1. INTRODUCTION

Hematopoietic stem cells (HSCs) are a rare population of well-characterized cells that are capable of replenishing all cellular blood components [1]. They play a crucial role during development, tissue regeneration and maintaining healthy homeostasis by switching between self-renewal, differentiation, quiescence, and dormancy. However, owing to complex yet dynamic crosstalk between HSCs and its microenvironment (niche) that directs the stem cell fate, it is difficult to maintain and expanded *in vitro* [2]. Spectrums of regulatory signals govern a complex molecular milieu, which regulates the molecular signaling and propagates the downstream events by stimulation of several intrinsic and extrinsic factors. Biochemical signaling involved in the cellular function plays a critical role in controlling cell growth and describes alteration resulting in the development of diverse pathologies. A great number of pathologies appeared as a consequence of the abnormal activity of a specific protein, and therefore their treatment would be based on suppressing this activity [1, 2].

HSCs transplantation has been increasingly adopted as a therapy for many malignant and non-malignant hematologic diseases, including leukemia, lymphoma, etc. Several modalities for this purpose have been employed, the most frequent being bone marrow transplantation and cytokine-mediated treatment. Recent studies in this field uncover the ways underlying expansion of Hematopoietic Stem Cells (HSCs) and progenitor cells by exploring the stem cell niche that presents a range of molecular cues and pathways critical to modulating stem cell decision. However, nowadays, a constructive approach is being employed with a combination of different cytokines for the treatment of malignancies and a specific combination of cytokines and hematopoietic growth factors; Interleukin-3 (IL-3), Fms-Like Tyrosine Kinase-3 Ligand (FLT-3), and Stem Cell Factor (SCF) have been found promising application potential.

The FLT-3 ligand has been identified as a regulatory factor that binds to the type-III tryptophan kinase receptors of hematopoietic stem and progenitor cells. It is shown that FL alone or in combination with other cytokines can effectively stimulate stem and progenitor cell differentiation. In addition to stimulating hematopoiesis, Flt3 also augments immune functions. It can prime $CD8^+$ T cells effectively in cancer patients and accelerate B- and T-cell reconstitution after myeloablation and bone marrow transplantation (BMT). While, IL-3, also known as multiple colony stimulating factor (multi-CSF), is secreted by activated T lymphocytes, enhances survival, stimulates proliferation of pluripotent stem cells, and promotes colony formation by multipotential progenitor cells. The experimental and clinical evidence indicates that IL-3 works on many of the cell types and shows its effect in the treatment of bone marrow failure, autologous bone marrow transplant, and radiation injury while causing fewer side effects. When combined with FL or stem cell factor (SCF), it amplifies the numbers of granulocytes, erythrocytes, and macrophages [3].

Further, various hematopoietic stem cell expansion experiment revealed the roles of different combination of cytokines through multifactorial analysis. For instance, it was demonstrated that high concentrations of FLT-3, SCF & IL-3 is necessary and sufficient to maximize the amplification of LTC-IC (>50-fold over input values within the first 10 days in culture) from starting populations of CD34'CD38- human bone marrow cells in first set of experiment [4]. However, second set of the same experiment investigated the influence of relative concentration of cytokines (FLT-3, SCF & IL 3) found to be effective in retaining (or losing) LTC-IC function [4]. The same combination of cytokines also showed favourable expansion results even for the expansion of SCID repopulating $CD34^-$ stem cells [5].

Moreover, the alterations in the transcriptional profile of certain cytokines reveal that the epigenetic modifications caused at protein level may acts as biomarkers when exposed to low-dose radiation and can stimulate multiple receptors and excite the repopulation of hematopoietic cells. Oswald et al. (2006) reported the significant qualitative changes in the gene-expression profile of HSCs expanded in a Collagen I Matrix, in the presence of IL-3, FLT-3, and SCF. However, there is a need to understand the response elicited by overlapping groups of genes and signaling pathways (regulating the stem cell fate in vitro along with the signaling pathways govern by IL3, FLT-3, and SCF). Together, these findings enable us to analyze the gene expression data (microarray data) with accession no. GSE3003 obtained from NCBI-GEO as reported [6] and to identify the novel proteomic biomarkers and their synergistic role in molecular signaling which would be of great therapeutic interest.

2. Methods

2.1. Microarray Data Source

The data used in the study for analysis was extracted from Gene Expression Omnibus (GEO, http://www.ncbi.gov/geo/) and accessible through GEO series accession number GSE3003. A background study reporting the significant qualitative changes in the gene-expression profile of cultivated HSCs conducted by Oswald et al. in 2006 has been referred in the current study [6]. The background study showed the Gene-expression profiling of $CD34^+$ hematopoietic cells expanded in a Collagen I Matrix, in the presence of IL-3, FLT-3, and SCF along with the submission of data GSE3003 in NCBI-GEO. The same data has been discussed in this study.

2.2. Protein-Protein Interaction (PPI) Data

A number of open access databases, such as The Biological General Repository for Interaction Datasets (BioGRID), Search Tool for the Retrieval of Interacting Genes (STRING), GeneMANIA, etc. which allowed us to retrieve and analyze the gene lists, gene function and prioritize genes based on functional assays [7]. These databases differ in the source of information used for computational prediction of protein interactions. We constructed protein-protein interaction network for some selected gene candidates responsible for HSCs cell proliferation and self-renewal.

2.3. Gene Expression Profile Analysis

The raw data extracted from GSE3003 series was normalized to eliminate the effects of non-biological variations using mean log-centering method assuming that mean $\log_2$ (expression ratio) should be equal to 0 for the gene-set. The normalization factor was calculated as:

$$N_{mlc} = \frac{\sum_{k=1}^{N_{gene-set}} \log_2\left(\frac{R_k}{G_k}\right)}{N_{gene-set}}$$

The intensities are then rescaled such that $G'_k = G_k \times (2^{N_{mlc}})$ and $R'_k = R_k$. The normalized expression ratio becomes:

$$T'_k = \frac{R'_k}{G'_k} = \frac{R_k}{G_k \times (2^{N_{mlc}})} = \frac{T_k}{2^{N_{mlc}}}$$

which is equivalent to:

$$\log_2(T'_k) = \log_2(T_k) - \log_2(2^{N_{mlc}}) = \log_2(T_k) - N_{mlc}$$

This adjusts the ratio such that the mean $\log_2$ (expression ratio) for the gene-set is equal to 0 [21].

After the background correction in microarray data, the normalized data was used for creating an expression matrix to enlist the Differentially Expressed Genes (DEGs) based on their fold change (FC) and p-values. We selected the threshold of 10 FC and $p \leq 0.05$ to identify the highly expressed genes. These genes were analyzed and clustered based on the biochemical function (proliferation and self-renewal), they elicit.

2.4. Analysis of Pathway Crosstalk

Further, the resulting up-regulated and down-regulated genes were mapped in a symmetric manner using computational tools such as DAVID, STRING etc. and detailed literature mining approaches. Each candidate gene of the up- and down-regulated gene list is literature searched combined with the computational analysis to make clusters based on its biological response, i.e., proliferation, differentiation, and self-renewal. Further, the detailed network-based and crosstalk analysis among the pathways have been investigated for proliferation and self-renewal.

3. RESULTS

3.1. Overview of Gene Expression Analysis

The microarray dataset with accession no. GSE3003 was selected for gene expression analysis [6]. The analysis was done after preprocessing of raw data primarily, to identify the Differentially Expressed Genes (DEGs) and the molecular signatures responsible for programming the stem cells function- proliferation, differentiation, and self-renewal. During data analysis, a total of 22,283 gene candidates were found expressed, which were normalized using a log-centering method based on their fold change (FC) and p-values. The initial screening and segregation of gene candidates resulted in 10,710 up-regulated and 11,573 down-regulated genes (*see supplementary file*).

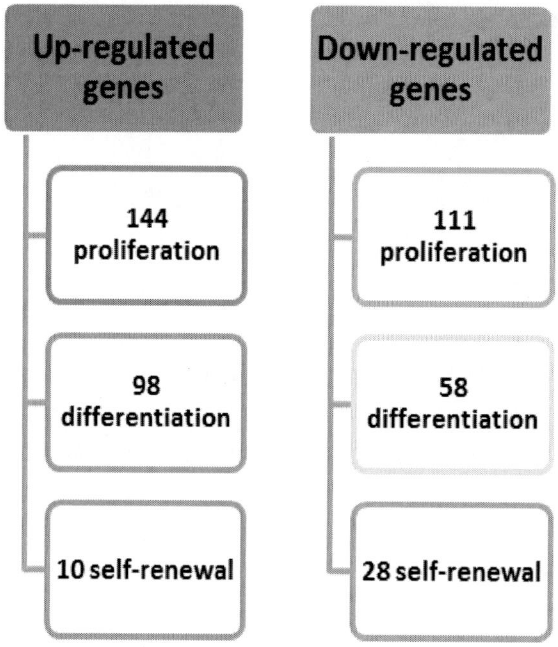

Figure 1. Significantly expressed genes (up- and down-regulated) responsible for proliferation, differentiation, and self-renewal.

A threshold of 10 fold change and $p \leq 0.05$ was selected to identify highly expressed genes (1802 up-regulated and 1618 down-regulated) that may play a crucial role in molecular signaling. For further analysis, both sets of selected genes were segregated into up- & down-regulated subsets and clustered based on their biochemical properties (proliferation, differentiation, and self-renewal) using computational tools and data mining approaches (as shown in Figure 1). However, the network-based analysis was employed, particularly for proliferation and self-renewal properties.

3.2. Identification of Molecular Signatures

The selected subset of genes that were differentially expressed in the subgroups of up- & down-regulated genes were hierarchically clustered based on the HSC functional properties. These clusters were then classified according to the function by using simplified gene ontologies (Table 1 & 2). The classification along with the clustering of DEGs and gene ontology analysis was done using computational tools like DAVID, GOrilla, etc. and databases like KEGG for pathway analysis combined with data mining through specific literature search for individual gene [7].

Table 1. Gene ontology of differentially expressed up-regulated genes

Go Term	Description	p-value	FDR q-Value	Enrichment (N, B, n, b)	Genes
GO:0006396	RNA processing	1.33E-7	3.98E-4	5.92 (142,16,15,10)	RRP1B, IMP3, HNRNPK, PABPC4, RBM3, WDR12, PRPF31, HNRNPC, HNRNPD, RPS19
GO:0044237	Cellular metabolic process	8.06E-7	1.2E-3	1.14 (142,99,118,94)	CDC20, BIRC5, CDK1, DDX3X, DHX32, STRA13, WDR12, TK1, RYBP, STK40, HMGB1, CDC25B, USP1, PDS5A, WWTR1, CTBP2, NAP1L1, MKI67, PRPF31, CNOT2, ILF2, EIF3I, ILF3, PHB, USP14, LPXN, SNRK, UBE2E3, KPNA2, CHD1L, ZMIZ2, GMPS, RRAGC, RAD23B, NELFCD, MCM2, PSMD9, EIF4G2, EIF5A, VCP, PAK2, TSC22D1, AURKA, CDC16, DDX39A, MCM7, DDX18, CENPF, RPS19, STAT6, EGR1, NFE2L2, HNRNPK, TOP2B, JAK1, TOP2A, DDX21, USP22, CEBPB, MAP2K1, MAP2K2, ZMIZ1, GATAD2A, THOC5, IMP3, FBXO7, ASF1B, UBR5, ADNP, CDK4, SMARCA4, SMARCD2, PABPC4, RBM3, EIF4E2, TBL1X, HNRNPC, EFTUD1, HNRNPD, PDS5B, SMARCA2, PHB2, CUL4B, RRP1B, WRAP53, DRG1, HSP90B1, PRKCD, ANXA1, NDUFS3, CNOT8, HSPA8
GO:0008152	Metabolic process	1.61E-6	1.6E-3	1.11 (142,103,126,101)	CDC20, BIRC5, DDX3X, CDK1, STRA13, STK40, CDC25B, WWTR1, PDS5A, DPP8, NAP1L1, MKI67, ILF2, ILF3, SNRK, CHD1L, GMPS, EIF4G2, PREPL, PSMD9, YBX3, EIF5A, PAK2, TSC22D1, CDC16, DDX39A, MCM7, DDX18, CENPF, EGR1, TOP2B, DDX21, USP22, TOP2A, SMARCE1, CEBPB, GATAD2A, THOC5, UBR5, ADNP, CDK4, SMARCA4, SMARCD2, TBL1X, EIF4E2, EFTUD1, PDS5B, SMARCA2, RP1B, WRAP53, HSP90B1, HSPA8, DHX32, WDR12, TK1, RYBP, HMGB1, USP1, DPAGT1, CTBP2, CNOT2, PRPF31, EIF3I, PHB, LPXN, USP14, SKP1, UBE2E3, KPNA2, ZMIZ2, RRAGC, FUT4, RAD23B, NELFCD, MCM2, DUSP3, VCP, AURKA, RPS19, STAT6, NFE2L2, HNRNPK, JAK1, MAP2K1, ZMIZ1, MAP2K2, IMP3, ASF1B, FBXO7, PABPC4, RBM3, HNRNPC, HNRNPD, PHB2, CUL4B, DRG1, ANXA2, PRKCD, ANXA1, CNOT8, NDUFS3
GO:0044260	Cellular macromolecule metabolic process	1.78E-6	1.33E-3	1.15 (142,95,117,90)	CDC20, BIRC5, CDK1, DDX3X, STRA13, DHX32, WDR12, TK1, STK40, RYBP, HMGB1, CDC25B, USP1, WWTR1, PDS5A, NAP1L1, CTBP2, MKI67, CNOT2, PRPF31, ILF2, ILF3, PHB, LPXN, USP14, SNRK, SKP1, UBE2E3, CHD1L, KPNA2, ZMIZ2, RRAGC, RAD23B, NELFCD, MCM2, PSMD9, EIF4G2, YBX3, EIF5A, PAK2, VCP, TSC22D1, AURKA, MCM7, DDX39A, DDX18, CENPF,

Table 1. (Continued)

Go Term	Description	p-value	FDR q-Value	Enrichment (N, B, n, b)	Genes
					RPS19, STAT6, EGR1, NFE2L2, HNRNPK, TOP2B, DDX21, USP22, TOP2A, JAK1, CEBPB, MAP2K1, GATAD2A, ZMIZ1, THOC5, MAP2K2, IMP3, ASF1B, UBR5, FBXO7, ADNP, CDK4, SMARCA4, SMARCD2, PABPC4, RBM3, TBL1X, HNRNPC, EFTUD1, HNRNPD, PDS5B, SMARCA2, PHB2, RRP1B, CUL4B, WRAP53, DRG1, HSP90B1, ANXA1, PRKCD, CNOT8, HSPA8
GO:0006364	rRNA processing	7.66E-6	2.54E-3	22.72 (142,5,5,4)	IMP3, RRP1B, WDR12, RPS19
GO:0016072	rRNA metabolic process	7.66E-6	2.28E-3	22.72 (142,5,5,4)	RRP1B, IMP3, WDR12, RPS19
GO:0043170	Macromolecule metabolic process	9.18E-6	2.49E-3	1.10 (142,99,126,97)	CDC20, BIRC5, CDK1, DDX3X, STRA13, STK40, CDC25B, PDS5A, WWTR1, DPP8, NAP1L1, MKI67, ILF2, ILF3, SNRK, CHD1L, PREPL, PSMD9, EIF4G2, YBX3, EIF5A, PAK2, TSC22D1, CDC16, DDX39A, MCM7, DDX18, CENPF, EGR1, TOP2B, TOP2A, DDX21, USP22, SMARCE1, CEBPB, THOC5, GATAD2A, UBR5, CDK4, ADNP, SMARCA4, SMARCD2, TBL1X, EFTUD1, PDS5B, SMARCA2, RRP1B, WRAP53, HSP90B1, HSPA8, DHX32, WDR12, TK1, RYBP, HMGB1, USP1, DPAGT1, CTBP2, PRPF31, CNOT2, PHB, LPXN, USP14, SKP1, UBE2E3, KPNA2, ZMIZ2, RRAGC, RAD23B, FUT4, NELFCD, MCM2, DUSP3, VCP, AURKA, RPS19, STAT6, NFE2L2, HNRNPK, JAK1, MAP2K1, ZMIZ1, MAP2K2, IMP3, ASF1B, FBXO7, PABPC4, RBM3, HNRNPC, HNRNPD, PHB2, CUL4B, DRG1, ANXA2, ANXA1, PRKCD, CNOT8
GO:0022402	Cell cycle process	2.14E-4	2.45E-2	1.65 (142,29,74,25)	CDC16, MCM7, CDC20, BIRC5, CDK1, STRA13, CENPF, XPO1, CDC25B, PDS5A, TOP2B, TOP2A, MAP2K1, CDK4, STMN1, SKP1, PDS5B, KIF20A, RRAGC, CKAP2, NUDC, MCM2, PSMD9, EIF4G2, AURKA
GO:0009893	Positive regulation of metabolic process	4.14E-4	4.12E-2	1.17 (142,62,117,60)	CACYBP, CDC20, DDX3X, CDK1, STRA13, NCL, HMGB1, CDC25B, WWTR1, CTBP2, ILF2, ILF3, PHB, MTDH, SKP1, ZMIZ2, CKAP2, PSMD9, YBX3, EIF5A, PAK2, VCP, RAB27A, AURKA, CDC16, RAP1A, RPS19, STAT6, EGR1, NFE2L2, HNRNPK, DDX21, TOP2A, USP22, RAP2B, CEBPB, MAP2K1, MAP2K2, THOC5, ZMIZ1, UBR5, FBXO7,

Go Term	Description	p-value	FDR q-Value	Enrichment (N, B, n, b)	Genes
					ADNP, CDK4, SMARCA4, RBM3, TBL1X, HNRNPC, HNRNPD, SMARCA2, PHB2, CUL4B, WRAP53, ANXA2, GNL3, PRKCD, ANXA1, CIRBP, CNOT8, HSPA8
GO:0010468	Regulation of gene expression	8.55E-4	5.8E-2	1.17 (142,67,114,63)	BIRC5, CDK1, DDX3X, NCL, STK40, RYBP, HMGB1, XPO1, WWTR1, CTBP2, CNOT2, YWHAQ, ILF2, ILF3, AFF1, PHB, MTDH, ZMIZ2, CKAP2, NELFCD, EIF4G2, PSMD9, YBX3, EIF5A, TSC22D1, RAB27A, DDX39A, CENPF, RPS19, STAT6, EGR1, NFE2L2, HNRNPK, USP22, DDX21, TOP2A, SMARCE1, CEBPB, MAP2K1, GATAD2A, MAP2K2, THOC5, ZMIZ1, ASF1B, UBR5, ADNP, CDK4, SMARCA4, SMARCD2, RBM3, EIF4E2, TBL1X, HNRNPC, HNRNPD, SMARCA2, PHB2, GNL3, PRKCD, CIRBP, CNOT8, HSPA8, TRIM44
GO:0006355	Regulation of transcription, DNA-templated (transcription factors)	3.9E-4	4.33E-2	1.20 (151,54,124,53)	BIRC5, DDX3X, NCL, RYBP, HMGB1, XPO1, WWTR1, ZFP36L2, CTBP2, KDM5B, CNOT2, ILF2, YWHAQ, ILF3, AFF1, PHB, LPXN, MTDH, ZMIZ2, CKAP2, NELFCD, PSMD9, YBX3, TSC22D1,H2AFZ, CENPF, STAT3, STAT6, EGR1,NFE2L2, HNRNPK, USP22, TOP2A, SMARCE1, CEBPB, MAP2K1, THOC5, ZMIZ1, MAP2K2, GATAD2A, ASF1B, ADNP, MARCA4, SMARCD2, TBL1X, DNMT1, HNRNPD, KAT6A, SMARCA2, PHB2, GNL3, CNOT8, HSPA8
GO:0016301	Kinase activity	4.51E-5	1.36E-3	2.68 (142,11,53,11)	CDK4, JAK1, CDK1, SNRK, PRKCD, TK1, STK40, PAK2, MAP2K1, MAP2K2, AURKA
GO:0016740	Transferase activity	3.76E-4	7.42E-3	2.01 (142,24,53,18)	USP14, UBR5, FBXO7, CDK4, CDK1, SKP1, SNRK, UBE2E3, TK1, STK40, CUL4B, USP22, JAK1, PRKCD, PAK2, MAP2K1, MAP2K2, AURKA
GO:0003824	Catalytic activity	7.97E-4	1.39E-2	1.52 (142,58,53,33)	MCM7, DDX39A, DDX18, CDK1, DDX3X, DHX32, TK1, STK40, USP1, TOP2B, JAK1, USP22, DDX21, TOP2A, MAP2K2, USP14, FBXO7, UBR5, CDK4, SNRK, SKP1, UBE2E3, CHD1L, KIF20A, CUL4B, GNL3, MCM2, PRKCD, PAK2, VCP, HSPA8, AURKA
GO:1990904	Ribonucleoprotein complex	5.96E-7	1.29E-4	4.26 (142,25,16,12)	RRP1B, IMP3, HNRNPK, PABPC4, RBM3, WDR12, NCL, CIRBP, HNRNPC, PRPF31, HNRNPD, RPS19

Table 1. (Continued)

Go Term	Description	p-value	FDR q-Value	Enrichment (N, B, n, b)	Genes
GO:0044428	Nuclear part (self-renewal)	6.25E-9	2.77E-6	1.15 (151,93,131,93)	CACYBP, CDC20, BIRC5, DDX3X, CDK1, DHX32, STRA13, WDR12, NCL, STK40, RYBP, HMGB1, XPO1, CDC25B, USP1, WWTR1, PDS5A, TEX10, KDM5B, MKI67, PRPF31, ILF2, ILF3, AFF1, PHB, LPXN, MTDH, SKP1, UBE2E3, CHD1L, KPNA2, ZMIZ2, RAD23B, NELFCD, MCM2, RAC2, PSMD9, DUSP3, EIF5A, VCP, AURKA, CDC16, DDX39A, MCM7, H2AFZ, DDX18, CUEDC2, CENPF, STAT3, RPS19, STAT6, EGR1, NFE2L2, HNRNPK, TOP2B, DDX21, TOP2A, USP22, SMARCE1, WDR6, CEBPB, GATAD2A, THOC5, ZMIZ1, IMP3, ASF1B, UBR5, CDK4, CSE1L, SMARCA4, SMARCD2, RBM3, TBL1X, HNRNPC, DNMT1, CLINT1, PDS5B, HNRNPD, KAT6A, SMARCA2, KIF20A, PHB2, RRP1B, CUL4B, WRAP53, DRG1, GNL3, NUDC, PRKCD, ANXA1, CIRBP, NDUFS3, HSPA8

The analysis of dynamic transcriptional profiles revealed *up-regulated gene signatures* (shown in Table 1) that were found falling into various categories of gene ontology including RNA processing (RRP1B, IMP3, HNRNPK, PABPC4, RBM3 RPS19 etc.), metabolic process (CDC20, BIRC5, CDK1, EIF4G2, ILF3, HSPA8, CTBP2, CNOT2, RRAGC, FBXO7, ANXA2, NDUFS3 etc.), cell cycle process (CDC16, MCM7, CDC20, BRIC5, CDK1, STRA13, CDK4, STMN1, RRAGC, CKAP2, ARKA etc.), regulation of gene expression (BIRC5, CDK1, DDX3X, NCL, STK40, XPO1, WWTR1, ILF2, CNOT2, AFF1, PHB, EGR1, HNRNPK, TOP2A, HNRNPC, SMARCA2, PHB2 etc.), kinase activity (CDK4, JAK1, CDK1, PRKCD, TK1, PAK2, MAP2K2, AURKA etc.), catalytic activity (MCM7, DDX39A, DDX18, CDK1, DHX32, TK1, USP1, TOP2B, JAK1, TOP2A, MAP2K2, CDK4, MCM2, PAK2, VCP etc.), regulation of transcription/ transcription factors (BIRC5, DDX3X, NCL, RYBP, HMGB1, XPO1, WWTR1, ZFP36L2, CTBP2, KDM5B, CNOT2, ILF2, YWHAQ, ILF3, MAP2K2, GATAD2A, DNMT1, HNRNPD etc.), cell proliferation (PAK2, CTBP2, RRAGC, FBXO7, RBM3, TOP2B, JAK1, CEBPB, SMARCA4, WDR12, PHB2, CNOT8, THOC5, SNRK etc.) & nuclear factor satisfying self-renewal properties (H2AFZ, ZFP36L2, STAT3, KDM5B, DNMT1, CD47, PTP4A1, PTP4A2, MYST3, TEX10 etc.).

Table 2. Gene ontology of differentially expressed down-regulated genes

Go Term	Description	p-value	FDR q-Value	Enrichment (N, B, n, b)	Genes
GO:0006396	RNA processing	2.43E-10	6.82E-7	5.52 (130,17,18,13)	RPL24, DDX5, RPL22, C1QBP, RBM5, RPS19, SKIV2L2, HNRNPK, RPS9, CELF1, CELF2, SF3B1, HSPA8
GO:0043603	Cellular amide metabolic process	1.23E-5	1.65E-3	17.33 (130,6,5,4)	RPL24, RPS9, RPL22, RPS19
GO:0016071	mRNA metabolic process	2.37E-9	3.34E-6	4.60 (130,22,18,14)	HSPA1A, RPL24, EIF4B, DDX5, RPL22, C1QBP, RBM5, RPS19, SKIV2L2, HNRNPK, RPS9, CELF1, SF3B1, HSPA8
GO:0044260	Cellular macromolecule metabolic process	8.24E-4	3.22E-2	1.17 (130,89,86,69)	ZNF24, CELF2, HOXA10, METAP2, USP3, TRIM33, C1QBP, RBM4, CDKN3, RUNX1, CSNK1A1, PIM1, PRKCB, MBD2, CNOT8, THRA, RUVBL2, HSPA8
GO:0006364	rRNA processing	1.23E-5	1.51E-3	17.33 (130,6,5,4)	RPL24, RPS9, RPL22, RPS19
GO:0016072	rRNA metabolic process	1.23E-5	1.45E-3	17.33 (130,6,5,4)	RPL24, RPS9, RPL22, RPS19
GO:0051252	Regulation of RNA metabolic process	5.51E-4	2.38E-2	1.32 (130,61,76,47)	HSPA1A, DDX5, RPS27A, EDF1, HMGN1, ELF4, TCF12, NUP107, RAF1, C14orf166, HSBP1, CTR9, NCOR1, WWTR1, CNOT7, HNRNPK, PLAGL1, ATF4, UXT, MEF2C, TOP2A, CTBP1, UBN1, KDM5B, NFE2, CELF1, ZNF24, HOXA10, ATF1, ILF3, MTDH, SMYD3, USP3, C1QBP, TRIM33, RBM4, RBM5, DYRK1A, RUNX1, CYTL1, PIM1, CBX4, PRKCB, MBD2, THRA, HSPA8, RUVBL2
GO:0010468	Regulation of gene expression	7.12E-4	2.94E-2	1.25 (130,75,76,55)	HSPA1A, RPL24, DDX5, RPL22, HMGN1, ELF4, TCF12, NUP107, HSBP1, EIF3C, WWTR1, NCOR1, ATF4, CTBP1, KDM5B, ILF3, ATF1, MTDH, EIF4B, SMYD3, RBM5, DYRK1A, CYTL1, CBX4, SF3B1, RPS27A, EDF1, C14orf166, RAF1, METAP1, CTR9, RPS19, HNRNPK, CNOT7, PLAGL1, RPS9, TOP2A, MEF2C, UXT, UBN1, NFE2, CELF1, ZNF24, HOXA10, USP3, TRIM33, C1QBP, RBM4, RUNX1, PIM1, PRKCB, MBD2, THRA, HSPA8, RUVBL2
GO:0000184	Nuclear-transcribed mRNA catabolic process, nonsense-mediated decay	3.69E-6	6.92E-4	20.80 (130,5,5,4)	RPL24, RPS9, RPL22, RPS19
GO:0003676	Nucleic acid binding (transcription factors)	2.49E-4	8.59E-3	1.79 (130,66,22,20)	HSPA1A, RPL24, EIF4B, DDX5, RPL22, G3BP1, C1QBP, RBM5, RAD51C, HMGN1, RPS19, SKIV2L2, EIF3C, HNRNPK, RPS9, CELF1, CELF2, SF3B1, HSPA8, RUVBL2
GO:0016887	ATPase activity	5.95E-5	3.42E-3	2.83 (130,10,46,10)	HSPA1A, VPS4B, LONP1, DDX5, TOP2A, G3BP1, RAD51C, HSPA8, SKIV2L2, RUVBL2
GO:1990904	Ribonucleoprotein complex	9.03E-8	3.36E-5	3.54 (130,29,19,15)	HSPA1A, RPL24, EIF4B, DDX5, RPL22, G3BP1, RBM5, RPS19, SKIV2L2, HNRNPK, CSNK1A1, RPS9, CELF1, SF3B1, HSPA8

Table 2. (Continued)

Go Term	Description	p-value	FDR q-Value	Enrichment (N, B, n, b)	Genes
GO:0044428	Nuclear part (self-renewal)	8.4E-5	4.46E-3	1.26 (130,75,81,59)	HSPA1A, ZC3H15, DDX5, HMGN1, ELF4, TCF12, TIMM17A, NUP107, HSBP1, WWTR1, NCOR1, ATF4, CTBP1, KDM5B, PCNP, ATF1, ILF3, PPP2R5C, MTDH, SMYD3, RBM5, DYRK1A, OTUB1, CBX4, COPS3, SF3B1, LONP1, MOB1A, RPS27A, EDF1, RAD51C, ANP32B, RAF1, C14orf166, CTR9, RPS19, SKIV2L2, CNOT7, HNRNPK, PLAGL1, RPS9, MEF2C, TOP2A, UBN1, NFE2, CELF1, USP3, TRIM33, C1QBP, RBM4, PCM1, RUNX1, CSNK1A1, PIM1, PRKCB, MBD2, THRA, HSPA8, RUVBL2

Similarly, gene lists specifically for a down-regulated group was also subjected to gene ontology analysis to determine their molecular functions (shown in Table 2). The analysis yielded genes belonging to different categories of functional groups which includes, RNA processing (DDX5, RPL22, C1QBP, RBM5, RPS19, SKIV2L2, HNRNPK, RPS9 etc.), cellular macromolecule metabolic process (ZNF24, CELF2, HOXA10, METAP2, USP3, TRIM33, C1QBP, RBM4, CDKN3, RUNX1, CSNK1A1, PIM1, PRKCB, etc.), mRNA metabolic process (RBM5, RPS19, SKIV2L2, HNRNPK, RPS9, CELF1, SF3B1, HSPA8 etc.), regulation of gene expression (HSPA1A, RPL24, DDX5, RPL22, HMGN1, ELF4, TCF12, NUP107, HSBP1, EIF3C, WWTR1, NCOR1, ATF4, USP3, TRIM33, C1QBP, RBM4, RUNX1, PIM1, PRKCB, MBD2, THRA, HSPA8, RUVBL2 etc.), nucleic acid binding/ transcription factors (HSPA1A, RPL24, EIF4B, DDX5, RPL22, G3BP1, C1QBP, RBM5, RAD51C, HMGN1, RPS19, SKIV2L2, EIF3C, HNRNPK, RPS etc.), cell proliferation (RPL22, RPS9, RPS19, RBM5, SF3B1, DDX5, RAD51C, RUVBL2, METAP1, RAF1, PRKCB, NEK9, PCM1, TSPO, COX5B, NFE2, TRIM33, ZNF24, CYTL1, MTDH, PTP4A2, VAMP8, PLP2, CKAP4, MS4A3, etc.), ATPase activity (HSPA1A, VPS4B, LONP1, DDX5, TOP2A, G3BP1, RAD51C, HSPA8, etc.), nuclear part satisfying self-renewal properties (RPL24, THOC2, BANF1, GPI, MAP3K4, TBPL1, PGK1, MKRN1, ATF4, YY1, CTBP2, CTR9, USP9X, ZMPSTE24 etc.).

3.3. Signatures of Proliferation and Self-Renewal

The genes found responsible for HSC proliferation and self-renewal responses in both up- & down-regulated sub-groups were further analyzed and comprehensively studied to explore their functional relevance.

In case of an *up-regulated genes* sub group, the genes exhibiting the fundamental property of HSCs- capacity and ability to proliferate are PAK2, CTBP2, RRAGC, FBXO7, RBM3, TOP2B, JAK1, CEBPB, SMARCA4, WDR12, PHB2, CNOT8, THOC5, SNRK, etc. The extracted information about the genes contributing to proliferation activities was validated with several biochemical studies existing in the literature. For instance, p21-activated kinases (Paks) has been reported to be a serine/threonine kinases that guide the different cellular activities such as cyto-skeletal remodeling, cell motility, cell proliferation, apoptosis, and mitosis. In a recent study, Pak2 disruption showed reduced survival and proliferation of hematopoietic progenitor cells (HPCs) based on a novel lentiviral vectors and a conditional Pak2-KO murine model in *in-vitro* and *in-vivo* conditions. This resulted into peripheral blood leukopenia and skewed granulocyte / monocyte along with differentiation / maturation defects in T and B cell [8].

RBM3, a protein from the heterogeneous nuclear ribonucleoprotein family, and NACA are associated with a GM-CSF–dependent proliferative response of the TF-1 cells. Both of these genes are reported to participate in hematopoiesis by showing a cytokine-specific activity and maintain the population of GM-CSF–dependent myeloid precursor cells during the stage of proliferation by preventing their differentiation [9].

It has also been established that Jak kinases (JAK1) are essential for the activation of STATs downstream of cytokine receptors. Also, the downstream to these signaling utilizes Jak kinases for activating other signaling molecules. The activation of STATs downstream of Jak kinases was also shown in a v-Abl–dependent manner such that v-Abl involves the activation of signaling pathways, including Ras, PI 3-K, and Akt. Interestingly, it was also shown that Jak activity is required by certain cytokines for activating Ras or PI 3-K. Hence, the expression of Jak kinases (Jak1) is highly significant for the activation of these signaling molecules upon v-Abl, which critically leads to cytokine-independent proliferation and transformation [10].

The HSC self-renewal is equally important for the process of HSC development and maintaining its life-long integrity. The genes whose expression is associated with self-renewal and multipotentiality are H2AFZ, ZFP36L2, STAT3, KDM5B, DNMT1, CD47, PTP4A1, PTP4A2, MYST3, TEX10, etc., which were also literature searched in support of the findings of present study. A molecular expression study confirmed H2A.Z, a variant of the nucleosomal core histone H2A, to be a prominent pattern-specific biomarker for identifying distributed stem cells (DSCs) in tissues involved in interlinked processes, i.e., deterministic asymmetric self-renewal and non-random segregation [11].

It is noteworthy to mention the presence of activated STAT3 in HSCs influences the self-renewal of transplanted bone marrow in-vivo. It was demonstrated that forced expression of a dnSTAT3, down-regulated the STAT3 activity in HSCs through scaling down of the competitive reconstituting ability of transduced HSCs upon transplantation of unmanipulated cells into lethally irradiated recipients. Conversely, up-regulation of

STAT3, which is achieved by transducing HSCs with a constitutively activated form STAT3 (STAT3-C), enhances the regenerative activity of small transplants by expanding transduced HSCs during the initial phase of hematologic recovery [12].

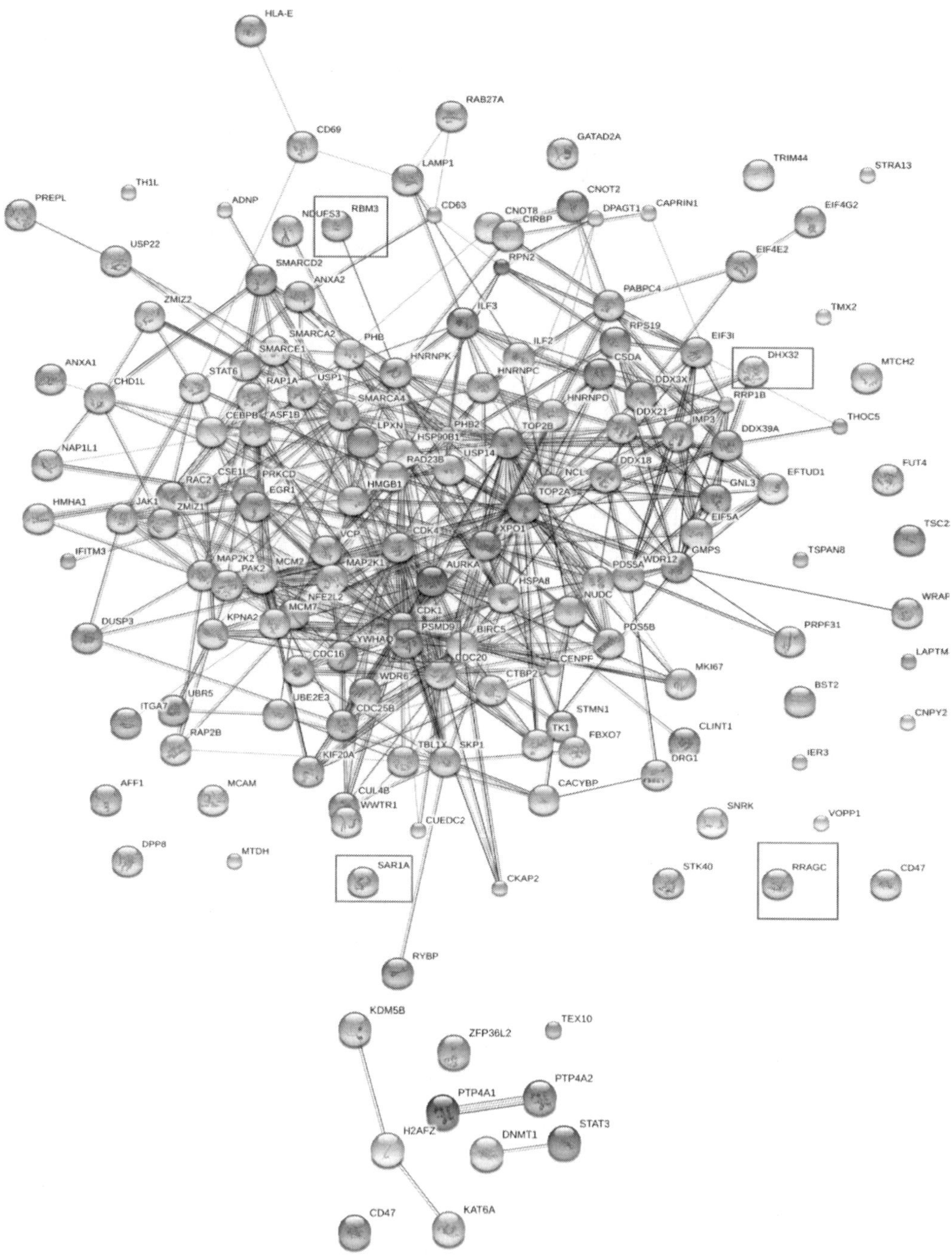

Figure 2. (a). PPI network Up-regulated genes eliciting proliferation; (b). PPI network Up-regulated genes eliciting self-renewal. Protein-protein interaction network of up-regulated genes responsible for (a) proliferation (b) self-renewal.

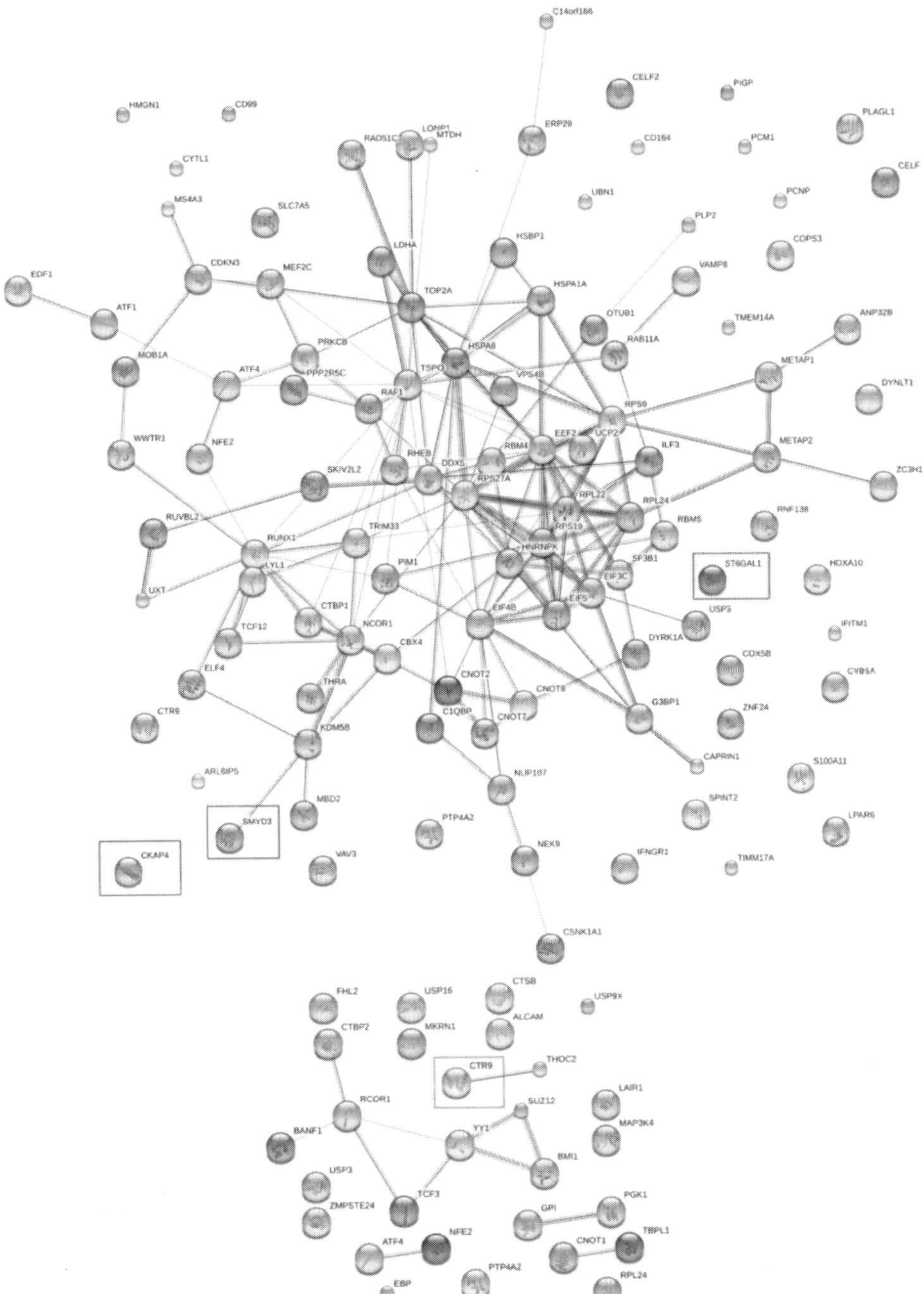

Figure 3. (a). PPI network down-regulated genes eliciting proliferation; (b). PPI network down-regulated genes eliciting self-renewal. Protein-protein interaction network of down-regulated genes eliciting (a) proliferation (b) self-renewal.

The expressed *down-regulated genes* associated with the proliferation of HSCs are RUVBL2, RPL22, RPS9, RPS19, RBM5, SF3B1, DDX5, RAD51C, METAP1, RAF1, PRKCB, NEK9, PCM1, TSPO, COX5B, NFE2, TRIM33, ZNF24, CYTL1, MTDH, PTP4A2, VAMP8, PLP2, CKAP4, MS4A3, etc. Largely, these findings were confirmed by the available research studies. For example, it has been illustrated that binding of GATA-3 to RuvB-like protein 2 (RUVBL2), represses the expression of a CDK inhibitor- cyclin-dependent kinase inhibitor 2c (*Cdkn2c*) which further facilitates the proliferation of Th2 cells. GATA-3 is known to directly bind to *Cdkn2c* locus in a Ruvbl2-dependent manner indicating that a critical role of Cdkn2c molecule in the GATA-3-mediated induction of Th2 cell proliferation [13].

A recent study using cell culture models reported that depletion of small 40S protein RPS9 negatively affects cell proliferation by activating p53-mediated cell cycle arrest [14]. Another study showed that Rpl22, a ubiquitously expressed RNA binding protein and a component of the 60S ribosomal subunit, is selectively required during early B cell development. It was also illustrated that elimination of Rpl22 resulted into decrease in the pro-B, pre-B and immature B cell subsets within the bone marrow and transitional B cell subsets in the spleen. In response to IL-7, which is essentially required for the pro-B cell stage, Rpl22-deficient pro-B cells do not show any survival or proliferation activity as its deficiency fails to alter IL-7 signaling and its downstream targets. Instead, the arrest of B-cell development is due to increased levels of p53 protein and mRNA of the PUMA and p21 which are found to be target genes of p53. The results also indicated an important role of Rpl22 as a mediator of B-cell development to regulate p53 at various developmental checkpoints [15].

With the similar approach, genes exhibiting self-renewal properties such as RPL24, THOC2, BANF1, GPI, MAP3K4, TBPL1, PGK1, MKRN1, ATF4, YY1, CTBP2, CTR9, USP9X, ZMPSTE24, etc. were also validated with recent experimental evidence. For example, it has been confirmed that deficiency or mutation of RNA binding protein-RPL24 impairs both HSC self-renewal and proliferation potential in HSCs in a cell-autonomous manner. It also indicated that the rate of protein synthesis is highly critical for regulating HSCs self-renewal [16].

Another study demonstrated that protein kinase MAP3K4, which has been found lacking in trophoblast stem (TS) cells, is essential for maintenance of both stemness and epithelial-mesenchymal transition (EMT). MAP3K4 was also reported to control the activity of the histone acetyltransferase CBP, which is required to maintain the epithelial phenotype by acetylation of histones H2A and H2B. Hence, the combined loss of MAP3K4/CBP activity suppresses the expression of epithelial genes and also allows TS cells to undergo EMT while maintaining their self-renewal and multipotency properties [17].

A well-established global regulator of oxidative stress response, NRF2 (or Nfe2l2) plays a significant role in HSC homeostasis. Lack of NRF2 at the expense of HSC quiescence and self-renewal resulted in an expansion of hematopoietic stem and progenitor cells (HSPCs) leading to cell-intrinsic hyperproliferation. NRF2 was further shown to mediate migration and retention of HSCs in their niche, through direct regulation of CXCR4 expression [18].

3.4. Protein-Protein Interaction Networks for Predicting Functional Domains

Furthermore, to delineate the role played by interaction between the extracted genes in all four subgroups: Up-regulated (proliferation & self-renewal) and down-regulated (proliferation and self-renewal) and their involvement in various other pathways and processes, indicated a need to comprehend the association among these genes. It guided us to construct gene interaction networks (Protein-Protein Interaction/PPI network) for above-mentioned all four subgroup of genes (Figure 2 & 3).

In this regard, four such PPI networks were retrieved using STRING database with evidence view display option, where each interaction edge is coloured differently and represented the different sources of the interaction data. Findings from these, lead us to the further selection of *9 notable genes/proteomic biomarkers* (highlighted in boxes in Figure 2 & 3) depicting 178 interaction links [19]. The interaction network is visualized as a graph with the protein molecule/gene corresponds to node and edge represents the biological relationships between the nodes. Most of the proteins are concentrated at the center of the network while few are arranged loosely at the periphery. The coloured lines of interactions with which the proteins are connected to onc another represents that these interactions were derived from more than one source of information. The input (gene names) provided to retrieve all of these PPI networks along with their interactions were comprehensively studied individually and literature searched. Our study emphasized on the 9 selected genes- RBM3, DHX32, PSMD9, RRAGC, SAR1A, SMYD3, ST6GAL1, CKAP4 and CTR9 (highlighted ones) found to be critical in regulation of hematopoiesis. However, there was lack of sufficient evidences showing their functional mechanism in the molecular signaling in the current background of study. Hence, these wcrc further chosen for analysis and validation for revealing their functional importance.

4. Discussion

The overall goal of this work was to identify the molecular signatures eliciting different biochemical responses in molecular signaling of $CD34^+$ HSCs in response to a cocktail of IL-3, FLT-3 and SCF using a network-based approach with the data extracted from GSE3003 [6]. With this microarray data, gene expression analysis was conducted to identify molecular signatures eliciting different biological responses. The data analysis and clustering of genes revealed the significant 1802 up-regulated and 1618 down-regulated gene candidates those were found responsible for proliferation, differentiation and self-renewal. Further, PPI network for up-regulated and down-regulated gene responsible for proliferation and self-renewal indicated the interactions among the genes and enabled us to select nine potential biomarkers for further analysis.

From the Figure 2 & 3, a comprehensively analysis of the PPI network (as shown in the boxed items)- about the nine genes/ proteins which may serve as a biomarker were retrieved, through several sources such as databases of physical interactions and databases of curated biological pathway knowledge. The targeted sets of genes/ proteins are up-regulated proliferation (RBM3, DHX32, PSMD9, RRAGC, SAR1A) and down-regulated proliferation (SMYD3, ST6GAL1, CKAP4) and self-renewal (CTR9).

Subsequently, each of these nine proteins was treated as an individual biomarker, and their potential interacting partners also found out using the GENEMANIA software tool [20] to analyze their functional organization (as shown in Table 3).

The analysis revealed that the numbers of interactions represented in each case of gene or protein in particular module are associated through text-mining, experimental and database evidence. Strikingly, few of them are found to be unexplored or if reported but with an unelucidated mechanism with the current background of the study. While gaining insights about each of the gene-interacting partner associations, the proteins which were found i) co-expressed and predicted (ii) physical interaction and predicted (shown in Table 4) would be considered for confirming the existing and establishing the novel potential interactions through biochemical experimentation (under process). Thus, these combinatorial approaches would reveal new findings of these interactions and their functional role in molecular signaling.

Table 3. Functional biomarkers of targeted genes and their interacting partners

Gene Symbol	Interacting Partners
Up-Regulated Proliferation	
RBM3	HNRNPA2B1, HNRNPK, ABI1, HNRNPA1, PSMA7, CHAF1A, SLC25A5, TOMM20, UXT, ABCB7, SLMO2, NACA, CDK2, PLP2, MSN, EEF1B2, NONO, RPS20, NAP1L1
DHX32	DHX33, TMEM231, DHX35, DHX34, NSUN7, DHRS7B, DHX9, DHX40, SAV1, DHX30, DHX29, EIF2B4, C10orf118, DHX38
PSMD9	PSMC3, PDX1, PSMC6, ACO2, ANAPC5, NSF, TCF3, PSMB1, PSMB6, TMOD3, PSMD10, PSMC5, RHOA, SLC25A3, UBE2I, PSMB5, ERP29, PSMD11
RRAGC	RRAGA, ATP2B2, ATP2B4, NOL8, COQ4, SUPT16H, RUVBL2, RUVBL1, PSMC2, WDR41, SGPP1, RRAGD, HPS1, PHLDA3, SNX6
SAR1A	TPD52, ATF6, TGFB2, SEC23A, TGFB3, SMAD7, TGFB1, DCTN1, TGFBR2, ARF4, ANXA7, PRDX4, DERL1, PSMA5, GDE1, MAGED1
Down-Regulated Proliferation	
SMYD3	HELZ, TGDS, FAR1, MEST, PBX2, TRIT1, MAGED1, EGLN1, HDAC1, CCNB1IP1, MAGED2, VKORC1, INSL6, EED
ST6GAL1	CD22, SEL1L3, ST3GAL6, CLINT1, BTG2, CXCR5, MS4A1, GCDH, ST6GALNAC5, ST3GAL4, ST3GAL2, DCAF11
CKAP4	ADRA1D, HDLBP, SEC61A1, KDELR2, RPN2, P4HB, MANF, HAMP, PSME4, TMED2, HSP90B1, FAM98A, GOLGA3, YIPF2, L1TD1
Down-Regulated Self-Renewal	
CTR9	CDC73, WDR61, PAF1, RTF1, CXCR5, CNTNAP2, BLK, FGG, SSRP1, MYH14, RNGTT, SUPT16H, CSNK2B, TMEM41B, PPM1B

Table 4. The gene/proteins which were found to be i) co-expressed and predicted (ii) physical interaction and predicted

S. No.	Predicted & Co-Expressed	
1	RBM3	PSMA7
2	RRAGC	NOL8
3	RRAGC	PSMC2
4	CKAP4	MANF
	Predicted and Physical Interaction	
5	SMYD3	MEST
6	CTR9	CNTNAP2
7	CTR9	BLK

Legends:

S. No. 1 and 5 depict the pair which might have relative significance in HSC functions.

S. No. 3, 4, and 6 depict the pair is unexplored for their relative significance in HSC functions.

S. No. 2 and 7 depict that the pair has evidently relative significance in HSC functions.

CONCLUSION

The clinical applications of HSCs have fueled enough excitement about prospects for various cell-based therapies. However, the focus has remained on exploring different ways of regulating the stem cell fate decisions such as survival, self-renewal, and differentiation on a molecular level. Recent studies have reportedly explored the different strategies with an aim to expand HSCs in number *ex-vivo* and also improving transplant efficiency. But, the limitation in purification and inferring the ex-vivo has renewed the interest of research exploring the key role of *in-vivo* microenvironment (stem cell niche) in molecular functions using different combinations of cytokines. Consistently, a combination of cytokine (IL-3, FLT-3 & SCF) has been used to understand their impact on HSC fate decisions- proliferation, self-renewal and differentiation in the present study. Additionally, the gene expression analysis enabled us to identify nine potential biomarkers (RBM3, DHX32, PSMD9, SAR1A, SMYD3, ST6GAL1, CKAP4, CTR9) of transient effect which would unlock long-touted therapeutic potential. Further, the combinations of protein with and their potential interacting partners investigated, which were (i) co-expressed and predicted (ii) physical interaction and predicted). The interaction among the combinations- RRAGC & PSMC2; CKAP4 & MANF and CTR9 & CNTNAP2 in the current background of the study was found to be unexplored previously. However, siRNA-mediated knock down of these proteins in combination, which is underway, would elucidate their synergistic role in hematopoiesis.

ACKNOWLEDGMENTS

The help of Dr. Yogesh Kumar Verma for providing technical comments and editing of the manuscript is acknowledged.

CONFLICT OF INTEREST

The author declares no conflict of interest exists.

SUPPLEMENTARY MATERIAL

Supplementary material includes the Microarray data (Accession No. GSE3003), normalized data, complete list of DEGs- up regulated and down-regulated and separate lists of up-regulated & down-regulated genes contributing in proliferation, differentiation and self-renewal.

- S-1: Raw Micro Array data (Accession No.) GSE3003
- S-2: Normalized data
- S-3: Up-regulated gene list
- S-4: Down-regulated gene list
- S-5: Selected Up-regulated genes (threshold value: Fc>10 and $p < 0.05$)
- S-6: Selected Down-regulated genes (threshold value: Fc>10 and $p < 0.05$)
- S-7: Up-regulated selected genes involved in proliferation
- S-8: Up-regulated selected genes involved in differentiation
- S-9: Up-regulated selected genes involved in self-renewal
- S-10: Down-regulated selected genes involved in proliferation
- S-11: Down-regulated selected genes involved in differentiation
- S-12: Down-regulated selected genes involved in self-renewal.

REFERENCES

[1] Han T, Yang CS et al. Identification of novel genes and networks governing hematopoietic stem cell development. *EMBO reports* 2016; e201642395. doi: 10.15252/embr.201642395.

[2] Celso CL & Scadden DT. Journal of Cell Science 2011; 124, 3529-3535. Published by The Company of *Biologists* Ltd. doi: 10.1242/jcs.074112.

[3] Seita J, Weissman IL. Hematopoietic Stem Cell: Self-renewal versus Differentiation. Wiley interdisciplinary reviews *Systems biology and medicine*. 2010;2(6):640-653. doi: 10.1002/wsbm.86.

[4] Eaves C, Zandstra P, Conneally E et al. Changes in the Cytokine Regulation of Stem Cell Self-Renewal during Ontogeny. *Stem Cells*. 1998;16 Suppl 1:177-84.

[5] Bhatia M, Bonnet D, Murdoch B, Gan OI et al. A newly discovered class of human hematopoietic cells with SCID-repopulating activity. *Nat Med.* 1998;4(9):1038-45.

[6] Oswald J, Steudel S, Salchert et al. Gene-Expression Profiling of CD34+ Hematopoietic Cells Expanded in a Collagen I Matrix. *Stem Cells* 2006; 24: 494–500. doi: 10.1634/stemcells.2005-0276.

[7] Taye B, Vaz C, Tanavde V et al. Benchmarking selected computational gene network growing tools in context of virus-host interactions, *Scientific Reports* 2017; 7(5805). doi:10.1038/s41598-017-06020-6.

[8] Zeng Y, Broxmeyer HE, Staser K, et al. Pak2 regulates hematopoietic progenitor cell proliferation, survival and differentiation. *Stem cells* (Dayton, Ohio) 2015; 33(5): 1630-41. doi:10.1002/stem.1951.

[9] Baghdoyan S, Dubreuil P, Eberle F, Gomez S. Capture of cytokine-responsive genes (NACA and RBM3) using a gene trap approach. *Blood* 2000; 95(12): 3750-7.

[10] Oki S, Limnander A, Danial NN and Rothman PB. Functional involvement of Akt signaling downstream of Jak1 in v-Abl–induced activation of hematopoietic cells. *Blood* 2002; 100: 966-73. doi: https://doi.org/10.1182/blood.V100.3.966.

[11] Huh YH, Noh M, Burden FR, Chen JC, Winkler DA, Sherley JL. Sparse Feature Selection Identifies H2A.Z as a novel, pattern-specific biomarker for asymmetrically self-renewing distributed stem cells. *Stem cell research* 2015; 14(2): 144-54. doi:10.1016/ j.scr.2014.12.007.

[12] Chung YJ, Park BB, Kang YJ, Kim T, Eaves CJ and Oh IH. Unique effects of Stat3 on the early phase of hematopoietic stem cell regeneration. *Blood* 2006; 108: 1208-15. doi: https://doi.org/10.1182/blood-2006-01-01 0199.

[13] Hosokawa H, Tanaka T, Kato M, et al. Gata3/Ruvbl2 complex regulates T helper 2 cell proliferation via repression of Cdkn2c expression. *Proceedings of the National Academy of Sciences of the United States of America 2013*; 110(46): 18626-31. doi: 10.1073/pnas.1311100110.

[14] Deisenroth C, Zhang Y. The Ribosomal Protein-Mdm2-p53 Pathway and Energy Metabolism: Bridging the Gap between Feast and Famine. Levine AJ, ed. *Genes & Cancer* 2011; 2(4): 392-403. doi: 10.1177/19476019114 09737.

[15] Fahl SP, Harris B, Coffey F, Wiest DL. Rpl22 loss impairs the development of B lymphocytes by activating a p53-dependent checkpoint. *Journal of immunology* (Baltimore, Md : 1950) 2015; 194(1): 200-9. doi:10.4049/jimmunol.1402242.

[16] Gao R and Liu Y. Protein Synthesis: More Than a House-Keeping Function in Hematopoietic Stem Cells. *J Stem Cell Res Transplant* 2014; 1(1): 1003.

[17] Abell AN, Jordan NV, Huang W, et al. MAP3K4/CBP Regulated H2B Acetylation Controls Epithelial-Mesenchymal Transition in Trophoblast Stem Cells. *Cell stem cell* 2011; 8(5): 525-37. doi: 10.1016/j.stem. 2011.03.008.

[18] Tsai JJ, Dudakov JA, Takahashi K, et al. Nrf2 regulates haematopoietic stem cell function. *Nature cell biology* 2013; 15(3): 309-16. doi: 10.1038/ncb2699.

[19] Szklarczyk D, Franceschini A, Wyder S, et al. STRING v10: protein–protein interaction networks, integrated over the tree of life. *Nucleic Acids Research* 2015; 43(Database issue): D447-52. doi:10.1093/nar/gku1003.

[20] Warde FD, Donaldson SL, Comes O, et al. The GeneMANIA prediction server: Biological network integration for gene prioritization and predicting gene function. *Nucleic Acids Research* 2010; 38(Web Server issue): W214-W220. doi:10.1093/nar/gkq537.

[21] Babu MM. *An introduction to Microarray data analysis. Computational Genomics* (Ed: R. Grant), Horizon Press, UK 2004; 225-49.

In: Stem Cells Origins of Cytokine Interactions
Editor: Prasad S. Koka
ISBN: 978-1-53615-302-6

Chapter 17

EFFECT OF CULTURE CONDITIONS ON COLONY-FORMING ABILITY OF STEM CELLS FROM HUMAN EXFOLIATED DECIDUOUS TEETH

***Chareerut Phruksaniyom*[1], *Thanasup Gonmanee*[2], *Kutkao Vongsavan*[3], *Taweepong Arayapisit*[4], *Petcharat Kraivaphan*[1] and *Hathaitip Sritanaudomchai*[5,*]**

[1]Department of Pharmacology, Faculty of Dentistry,
[2]Department of Anatomy, Faculty of Science,
[3]Department of Pediatric Dentistry, Faculty of Dentistry,
[4]Department of Anatomy, Faculty of Dentistry,
[5]Department of Oral Biology, Faculty of Dentistry,
Mahidol University, Bangkok, Thailand

ABSTRACT

Stem cells from human exfoliated deciduous teeth (SHED) are originally referred to as fibroblastoid-colony-forming-cells because one of their characteristics is related to proliferation and differentiation potential. In this study, *we developed an efficient protocol for colony-forming-unit (CFU) assay for SHED.* The culture conditions including culturing time, seeding cell density and modifications of media supplements were optimized. The CFUs were determined on the proportion of colony types, sizes and the numbers of colony. Isolated SHED expressed CD44, CD73, CD105 and showed multipotent ability. Higher yield was obtained when SHED was cultured at the density of 500 cells per well of a 6-well plate for 14 days in Dulbecco's modified Eagle medium (DMEM) with 10% fetal bovine serum (FBS). The number and size of CFU was

[*] Corresponding Author Email: hathaitip.sri@mahidol.ac.th.

increased with the increase of FBS concentration in which the transit-amplifying cell (TAC) colonies were the most abundant, followed by the stem cell (SC) and differentiated cell (DC) colonies, respectively. In DMEM with 20% FBS and 100 μM L-ascorbic acid-2-phosphate, the CFU rate was highest with increasing the proportion of SC colonies. The improvement of the culture conditions described herein could be adopted to easily acquire more functional SHED for applying in cell therapy.

INTRODUCTION

A dental pulp tissue appears to be an excellent source of adult mesenchymal stem cells (MSCs). Clinically, interesting populations of pulp cells have also been isolated from deciduous teeth [1]. The stem cells from human exfoliated deciduous teeth (SHED) with self-renewal and multilineage differentiation capacities have been found to grow more rapidly than those from other sources. It is believed that this is because they may be less mature than other stem cells found in the body. Moreover, the stem cells from exfoliating deciduous teeth are readily available providing that they are stored properly until they are needed later in life and are used with a minimally invasive process that results in minimal trauma [2].

One of the most prominent properties of MSCs is their ability to generate fibroblastoid colony-forming cells, termed colony forming unit-fibroblasts (CFU-Fs) after they are plated at low density [3, 4]. The CFU-Fs are representative of the long term self-renewable and rapid proliferation properties in MSCs [5]. The efficiency with which they form colonies still remains an important assay to determine the quality of cell preparations. From several studies, there are generally three types of colonies, called holoclones, meroclones from transit-amplifying cells and paraclones [6-8]. The holoclones are large, contain 20,000–50,000 cells and consist mainly of small cells. The meroclones also have considerable proliferative capacity but are not as large as holoclones and have a greater range of cell size. When meroclones are subcloned, more than 5% but less than 100% of the colonies produced are terminal. The paraclones are small, highly irregular colonies with a short lifespan (less than 15 cell generations), consist mainly of large cells and do not form colonies on subcloning [6-8]. The holoclones are likely to be derived from stem cells (SC), meroclones from transit-amplifying cells (TAC) and paraclones from late stage precursors which quickly differentiate to become terminal or terminally differentiated cell colonies (DC) [6-8]. On the basis of isolation and expansion protocols, it is expected that the CFU-Fs originated from MSCs, as a result of clonal evolution, would rapidly become fast growing with a high clonogenic capacity. However, they actually have different proliferative and differentiation potentials which may be either subtle or significant [9, 10]. There is large variation in CFU-Fs among the cell lines [11]. Nevertheless, the CFU-Fs and the growth rate of most individual cell lines seem to be stable over many generations. Isolation facilities and protocols, including medium, plastic, seeding density, growth factors, and chemicals, influence the expansion, differentiation, and immunogenic properties of MSCs [12]. The optimal conditions that influence the CFU-Fs in SHED and the effects on the proliferative capacity of individual colonies have not been clearly studied. In order to increase the efficiency of colony-forming unit fibroblast formation in SHED culture, we examine the effects of culturing time, seeding cell density and modification of media supplements, serum and ascorbic substrates, on colony morphology and production colonies in an attempt to better optimize the isolation and generation of CFU-Fs from SHED.

MATERIALS AND METHODS

Tooth Sample Collection

Human exfoliated deciduous teeth were acquired as discarded biological samples from children (5-9 years of age) at Faculty of Dentistry, Mahidol University following the approval of the ethical committee of the Faculty of Dentistry, Mahidol University (IRB 2014/044.2710).

Isolation and Culturing of SHED

After disinfection of the tooth surface, the teeth were mechanically fractured and the dental pulps were gently isolated with forceps. The pulp tissue was rinsed in a growth medium [Dulbecco's modified Eagle's medium (DMEM, Hyclone, Logan, UT, USA) supplemented with 10% fetal bovine serum (FBS) from Biochrom GmbH, Berlin, Germany or from Sigma St. Louis, MO, USA and 1% penicillin/streptomycin (Gibco, Rockville, MD, USA)], after which it was cut into small pieces of 1 – 2 mm^3. The small tissue pieces were cultured in 25 mm^2 culture flasks in growth medium at 37°C in a humidified atmosphere containing 5% CO_2. The medium was changed every 2 days and the cell cultures were monitored regularly with an inverted phase-contrast microscope (Nikon, Eclipse TS100, Tokyo, Japan). When reaching 70% – 80% confluence, cells were harvested by using 0.05% Trypsin/EDTA (Sigma) and sub-cultured for further experiments. The cells between passage 3 and 5 were used. For each assay, the stem cells of at least three different donors were used.

Immunophenotypic Characterization of SHED

For the cell surface antigen analysis, SHED were harvested, washed in cold PBS supplemented with 0.5% Bovine Serum Albumin (BSA, Sigma), and aliquots of 5 x 10^5 cells were labeled (for 30 min in the dark at 4°C) with monoclonal antibodies specific for human markers associated with mesenchymal and hematopoietic lineages. Namely, mouse anti-human antibodies against the following antigens were used: CD34 (FITC conjugated), CD44 (FITC conjugated), CD73 (APC conjugated), and CD105 (PE conjugated) (all purchased from BioLegend, San Diego, CA, USA). To determine the level of nonspecific binding, fluorochrome conjugated isotype control antibodies were used. Flow cytometry was performed using a FACSCanto flow cytometer (BD Biosciences, San Jose, CA, USA).

Differentiation

After 28 days in complete commercial medium (HyClone™ AdvanceSTEM™ Osteogenic Differenti-ation Kit, Hyclone) for osteogenesis, calcium depositions were demonstrated in the extracellular matrix. The cells were fixed in 10% paraformalde-hyde for 15 min at room temperature (RT). Cultures were then stained with 40 mM Alizarin Red S solution (Sigma), pH 4.1, for 20 min at RT, followed by rinsing three times with deionized water.

After 28 days in complete medium (HyClone™ AdvanceSTEM™ Adipogenic Differentiation Kit, Hyclone) for adipogenesis, cells were fixed in 4% paraformaldehyde for 15 min at RT and stained with a fresh 0.35% Oil Red O solution for 10 min, followed by washing twice with deionized water. The differentiated cells were observed and imaged using an inverted microscope.

After 21 days in D-MEM, 10% FBS, 10 ng/ml TGF-β1 (sigma), 100 nmol/L dexamethasone (sigma), 6.25 μg/ml insulin (sigma), 110 mg/L sodium pyruvate (sigma) and 1% penicillin/streptomycin for chondrocyte differentiation, proteoglycans of differentiated cells were examined by staining with Alcian blue (Sigma), pH 2.5, for 20 min at RT, followed by three washes in tap water to remove unspecific binding of the dye. The differentiated cells were observed using an inverted microscope and images were taken.

After 14 days in DMEM/F12 (Gibco), 20 ng/ml neurotrophins brain-derived neurotrophic factor (BDNF), 20 ng/mL neurotrophin-3 (NT-3), 20 ng/mL glial cell-derived neurotrophic factor (GDNF) (all from Sigma), supplemented with 2% N-2 (Gibco), 2% B-27 supplement, 100 U/mL Penicillin, and 100 μm/mL Streptomycin for neuron differenti-ation, the differentiated cells were stained with β-tubulin III (Tuj1; BioLegend, San Diego, CA, USA) and Tropomyosin receptor kinase B (TrkB; Santa

Cruz Biotechnology, Inc, Dallas, TX, USA). Nuclei were counterstained with 20 µg/mL 4',6-diamidino-2-phenylindole (DAPI; Sigma).

Colony-Forming Unit-Fibroblast (CFU-F) Assay

Culturing Time

After trypsinization with 0.05% Trypsin/EDTA, cells were resuspended in growth medium to disaggregate the clumps and produce a single cell suspension. Viability, cell number, and proportion of single cells were determined by Trypan Blue (Sigma). The cell suspension was seeded into 6-well culture plates at 250 cells per well. Cells were incubated for 12-14 days at 37°C in 5% CO_2 with 2 ml growth medium per well which was changed every 2 days.

Seeding Density

The SHED expanded in culture to 80 - 90% confluency were harvested with Trypsin/EDTA treatment and counted using a haemocytometer. A single cell suspension was diluted in growth medium, and they were plated into 6-well culture plates at various densities. After incubation at 37°C in 5% CO_2, the cells were washed with PBS and assayed for CFU-F.

FBS Supplement

Cells were seeded in triplicate into each well of a 6-well plate containing growth medium supplemented with 1% penicillin/streptomycin and 10% or 20% FBS Biochrom GmbH (Catalog no. S 0613/S 0615: FBS(Bio)) or Sigma St (Catalog no. F7524: FBS(Sig)). Cells were incubated at 37°C in 5% CO_2 with 2 ml growth medium per well which was changed every 2 days.

Ascorbic Acid and L-Ascorbic acid 2-Phosphate Supplement

A single cell suspension was seeded into a 6-well plate containing growth medium supplemented with 1% penicillin/streptomycin, the best concentration of FBS from the previous experiment and 0, 50, 100, 200 µM ascorbic acid (AA; Sigma St) or L-ascorbic acid 2-phosphate (A2P; Sigma St). Both AA and A2P were dissolved in SHED growth media. Cells were incubated at 37°C in 5% CO_2 with 2 ml growth medium per well which was changed every 2 days.

Analysis

After 12-14 days, culture medium was removed and the cells were fixed for 20 min with 1% paraformaldehyde (Sigma St) in PBS buffer and stained with 0.1% toluidine blue O (Certistain Merck, Frankfurter, Germany) dissolved in 1% paraformaldehyde. The cells were washed with distilled water and colony morphology was inspected under a light microscope. The colonies larger than 2 mm in diameter were enumerated using ImageProPlus 7.0 program (Media Cybernetics, Rockville, MD, USA).

Statistical Analysis

All experiments in this study were performed 3 times. The data shown are representative results. Values were calculated as the means and standard deviations. The differences in colony numbers and sizes were analyzed by one-way ANOVA of variance using SPSS software (Version 10.0; SPSS, Chicago, IL, USA). The values at $p < 0.05$ were deemed significant.

Results

Isolation and Characterization of SHED

To isolate and identify MSCs in dental pulp tissue, the cells were extracted from the remnant pulp of exfoliated deciduous teeth after outgrowth method. The fibroblast-like cells began to migrate from pulp tissue into the flask by almost 7 days after seeding (Figure 1a) and reached 70-80% confluence by 20 days in growth medium with 10% FBS (Figure 1b). At passages 3 - 4, the quality of cells acquired was assessed using flow cytometry analysis for the expression of MSC surface markers. All cells were highly positive for the surface antigens CD44, CD73, and CD105 (Figure 1c). Moreover, a CD34 was not expressed on the cell population examined (Figure 1c). The differentiation potential of isolated cells is important considering their potential to regenerate specified tissues (Figure 1d). After 28 days of cultivation in osteogenic medium, the presence of calcium mineral was observed and confirmed by positive Alizarin red staining (Figure 1d). Moreover, the adipogenic differentiation capacity of cells was confirmed by the formation of lipid-filled vacuoles stained with Oil red O after 28 days of incubation in an adipogenic differentiation medium. To demonstrate chondrogenesis, the SHED were placed into chondrogenic medium for 21 days. SHED lines had the capacity to differentiate into chondrocytes (Figure 1d). The differentiated cells showed cartilage-specific proteoglycan and were positive for Alcian blue staining. Finally, fourteen days after transfer into neurons induction medium, the morphology of SHED changed from elongated to extension of the cell body into neuron-like processes and a spherical soma. The SHED expressed mature neuronal marker Tuj1 and all cell nuclei were positive for DAPI (Figure 1d).

Optimal Culture Time for Colony Forming Unit-Fibroblast (CFU-F) Assay

To determine the optimal time that is best suited for the formation of stem cell colonies, three SHED lines (SHED #1, #2, and #3) from exponential growth were plated at a density of 250 cells/well of 6-well plate and cultured for 12-14 days in DMEM supplemented with 10% FBS (Table 1). All counting colonies were larger than 2 mm in diameter. The average number of SHED colonies (from 3 wells of 6-well plate) on day 12 was 3 ± 0.33. The colony numbers were higher on day 13 (5 ± 0.77) and day 14 (17 ± 0.69) than on day 12 (Table 1). These data suggested that the optimal time point for SHED colony formation assay was 14 days.

Effect of Cell Seeding Density on the Formation of SHED Colonies

To determine the cell seeding density for optimal CFU-F assay, three lines of SHED were plated at densities of 100, 250, 500 and 1,000 cells/well of 6-well plates and cultured for 14 days in medium supplemented with 10% FBS (Table 2). All SHED lines attached to the plastic plates and had the capability to form fibroblasts-like colonies (Figure 2a). The easily distinguishable colonies greater than 2 mm in diameter were measured the colony size (Figure 2b) and the number of colonies (Figure 2c) using ImageProPlus 7.0 program. In normal growth medium, SHED cells formed three types of colonies. The holoclones or SC colonies were large consisting of many small, tightly packed cells that stained darkly (Figure 2d). The meroclones or TAC colonies were a mixture of cell types, small, tightly packed cells at the center and larger cells at the edges (Figure 2e). The paraclones or DC colonies were the smallest and consisted mainly of large, flattened and scatter cells (Figure 2f). In all densities, the TAC colonies were most abundant, followed by SC and DC colonies, respectively (Figure 2c). Increasing SHED seeding number (100, 250, and 500 cells/well) in 6-well plates resulted in a strong increase statistically significant in colony numbers (4 ± 0.19, 18 ± 0.33 and 26 ± 1.45 colonies, respectively) (Table 2). On day 14, the SHED seeding of 1,000 cells/well grew to a very high density and the colony number could not be counted (Table 2). These data suggested that the best cell seeding density for SHED colony assay was 500 cells/well of 6-well plate within 14 days.

Table 1. Number of SHED colonies in day 12-14

Day	The mean numbers of colonies (per 3 wells)			
	SHED #1	**SHED #2**	**SHED #3**	**Mean ± SD**
12	2.67	3.33	3.00	3.00 ± 0.33
13	4.33	5.67	5.67	5.22 ± 0.77
14	16.00	17.33	16.33	16.56 ± 0.69

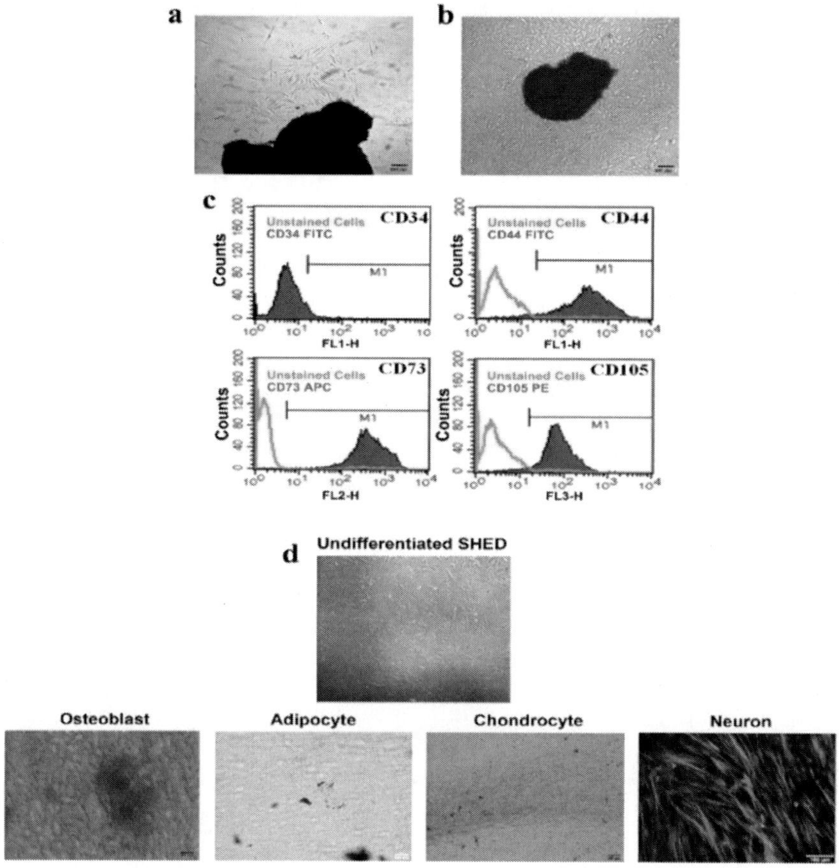

Figure 1. Isolation, morphological observation and characterization of stem cells from pulp tissue of human deciduous teeth. (a) Outgrowth of stem cells from dental pulp tissue of human exfoliated deciduous teeth. (b) Phase-contrast images showing the fibroblast-like morphology of the *in vitro* expanded SHEDs. (c) Flow cytometric analysis of the expression of mesenchymal stem cell markers CD44, CD73, CD105 and the hematopoietic marker CD34 on SHEDs. Gate M1 are deemed to be as positive. (d) Multi-differentiation of undifferentiated SHED, cells staining with Alizarin Red S for osteogenic differentiation, Oil Red O for adipogenic differentiation, Alcian blue for chondrogenic differentiation, and the mature neuronal marker Tuj1 (green) for neuronal differentiation. Cell nuclei were counterstained with DAPI (blue). Scale bars: 100 μm in a, b and d (chondrogenic and neuronal differentiation), 5 μm in d (undifferentiated SHED), 50 μm in d (Osteogenic differentiation), and 200 μm in d (adipogenic differentiation). SHED, stem cell from human exfoliated deciduous teeth, Tuj1, β-tubulin III, and DAPI, 4',6-diamidino-2-phenylindole.

Table 2. Seeding density and colony number of SHED

Seeding densities (Cells)	**The mean numbers of colonies (per 3 wells)**			
	SHED #1	**SHED #2**	**SHED #3**	**Mean ± SD**
100	4.33	4.00	4.33	4.22 ± 0.19
250	18.33	18.00	18.67	18.33 ± 0.33
500	28.00	25.67	25.33	26.33 ± 1.45
1000	ND	ND	ND	ND

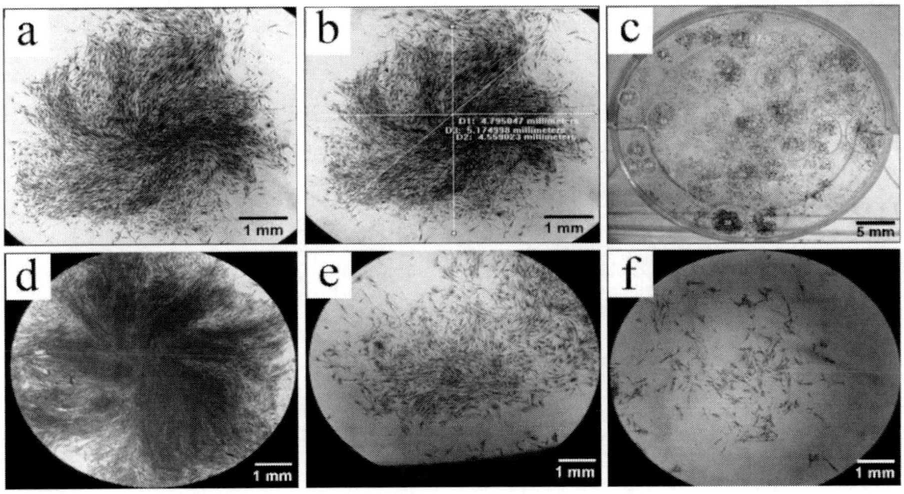

Figure 2. Morphology of SHED colonies stained by toluidine blue. (a) Phase-contrast images showing the morphology of SHED colony. (b) Assessing colony size using ImageProPlus 7.0 program. (c) Measurement of colony number by ImageProPlus 7.0 program which showing in green circle. SHED form colonies of three distinct morphologies; (d) stem cell, (e) transit-amplifying and (f) differentiated colonies. Scale bars: 1 mm in a, b, d, e, and f, and 5 mm in c.

Selection of Optimal Supplemented Substrates for CFU-F Assay

To define the optimal supplemented substrate for the colony-forming unit assay, the SHED seeding density at 250 and 500 cells/well of 6-well plate were further chosen since these densities provided the higher numbers of colonies. The SHED were cultured for 14 days in the presence of different medium supplements as described in Materials and Methods.

Fetal Bovine Serum

In this study, the FBS from 2 companies, Biochrom GmbH and Sigma St called FBS(Bio) and FBS(Sig), were selected for the evaluation of CFUs. Since all cell lines exhibited similar results, SHED #1 was used for further experiments. The experiments were repeated 6 times. The FBS concentration was first experimented for the effect on CFU-F (Figure 3). The TAC colonies were observed at the highest proportion of both 10% & 20% serum concentrations of Biochrom GmbH company (Figure 3a). In contrast, the colonies formed only DC colony type, which were found in the presence of 20% FBS(Sig) from Sigma St company (Figure 3a). The results were similar when seeding cells at a density of 250 & 500 cells/well in the presence of both FBS(Bio) and FBS(Sig) (Figure 3a).

When seeding cells at a density of 250 cells/well in the presence of 10% FBS(Bio), the number and size of CFU-F were lower compared to 20% FBS(Bio) with cell seeding at 500 cells/well (Figure 3b-c). The number of CFU-F-derived SHED were dependent on the concentration of FBS(Bio). For FBS from Biochrom GmbH, CFU-F in medium containing 20% serum had significantly higher number than 10% serum in both cell seeding densities. However, very few colonies (1 - 3 colonies) were observed at both the start of seeding SHED of 250 and 500 cells/well in the presence of FBS(Sig) (Figure 3b).

For colony diameter (Figure 3c), colonies in 20% FBS(Bio) were significantly larger than at 10% FBS(Bio) concentration when starting the cells density at 500 cells/well. While seeding density of 250 cells/well in the presence of FBS(Sig), the SHED colonies with average 2.60 mm in diameter was rarely found. The range of colonies size was increased with a diameter of 3.1 - 3.3 nm with cell seeding at 500 cells/well (Figure 3c). The largest number and size of CFU-F were obtained in 20% FBS(Bio) after cell seeding at 500 cells/well (Figure 3b-c). Therefore, we used only 20% FBS(Bio) from Biochrom GmbH company in subsequent experiments.

Ascorbic Acid and L-Ascorbic Acid 2-Phosphate

To observe the effect of ascorbic acid and it derivative on the forming of colonies, the SHED #1 was cultured in DMEM supplemented with 20% FBS(Bio) and 0, 50, 100, or 200 μM ascorbic acid or 0, 50, 100, or 200 μM L-ascorbic acid 2-phosphate for 14 days (Figure 4). The concentrations of ascorbic acid (vitamin C; AA) had large impact on the CFU-F of SHED (Figure 4). From the 250 cells/well seeding in culture medium supplemented with AA at 50 μM, there was a marked increase in the proportion of SC colonies (Figure 4a; upper row). However, the SHED seeding at 500 cells/well became over confluent and did not result in well-separated colonies (data not shown). All tested concentration of L-ascorbic acid 2-phosphate (A2P), a long-acting vitamin C derivative, also resulted in a reduction in the proportion of TAC colonies and an increase in the proportion of SC colonies. There was no DC colonies in medium supplemented with A2P (Figure 4a; lower row). Similar to the observation in ascorbic acid, there were no isolated colonies in SHED seeding densities of 500 cells/well (data not shown).

SHED in culture medium supplemented with AA at 50 μM, the number and size of colonies were increased compared to the medium with no AA (Figure 4b-c). The ascorbic acid at 100 and 200 μM had significant toxic effects toward SHED and very few colonies formed in both cell seeding densities (Figure 4a-b). On the other hand, the A2P had no toxic effect on SHED proliferation (Figure 4a, 4c). The A2P also did influence the colony size (Figure 4c) but small changes in colony number when using SHED seeding at 250 cells/well (Figure 4b). The sizes of colonies were gradually increased with increasing concentrations of A2P, with the biggest colonies observed at 100 μM (4.5-5.5 mm in diameter). The decline in a colony size was found in culture medium with 200 μM A2P (Figure 4c). The A2P was superior to AA as a supplement for SHED colony formation *in vitro* cultures (Figure 4). The data suggested that the highest CFU-F efficiency was obtained when SHED was cultured in growth media supplemented with 100 μM L-ascorbic acid-2-phosphate.

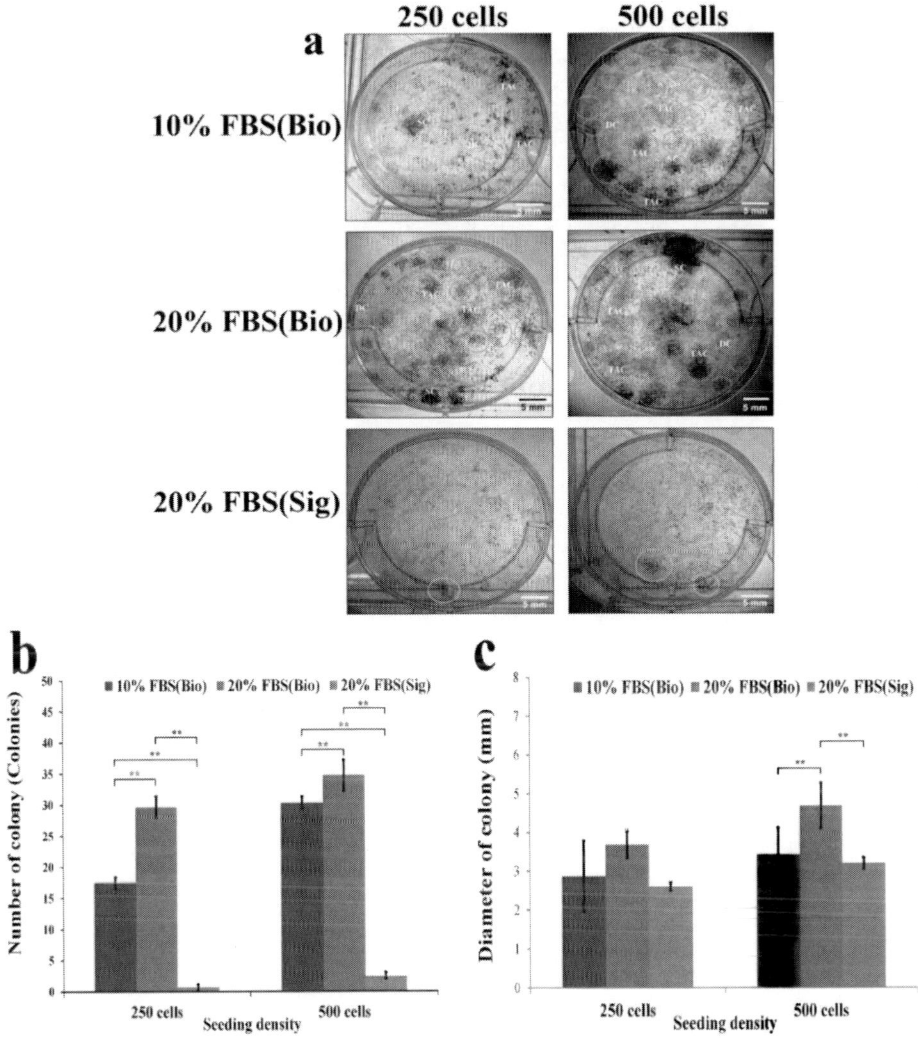

Figure 3. Influence of FBS on morphology, number, and size of SHED colonies. SHED at 250 cells and 500 cells per well of 6-well plated were cultured with 10% and 20% FBS from two companies (Biochrom GmbH; FBS(Bio) and Sigma St; FBS(Sig)). (a) Representative images of the colonies stained by toluidine blue after 14 days of cultured in 10% and 20% FBS from Biochrom GmbH company and 20% FBS from Sigma St company. Each colony in green circle. Scale bars: 5 mm. (b) The total number of colonies counted within each group. (c) The mean diameters per colony of SHED. The data are expressed as the mean ± SD of 6 replicate cultures (**$p < 0.05$). SC, stem cell; TAC, transit-amplifying; and DC, differentiated colonies; SHED, stem cell from human exfoliated deciduous teeth.

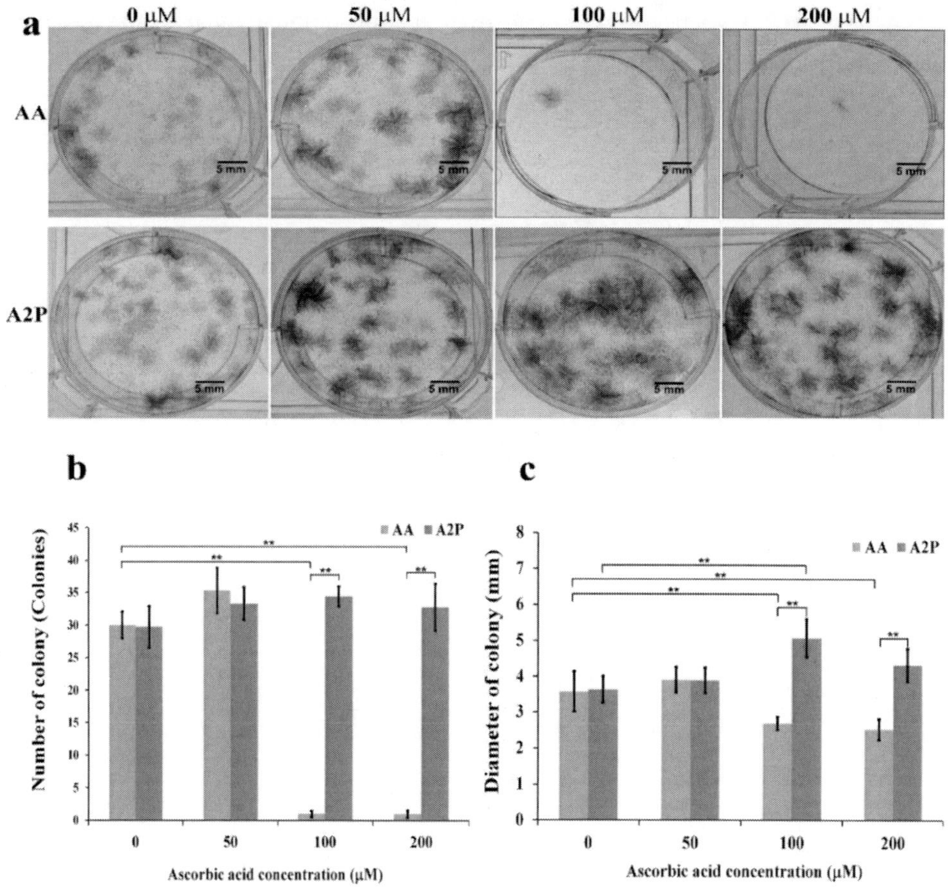

Figure 4. Influence of ascorbic acid and ascorbic acid 2-phosphate on morphology, number, and size of SHED colonies. SHED at 250 cells per well of 6-well plated were cultured in a range of concentrations between 0 and 200 μM. (a) Representative images of the colonies stained by toluidine blue after 14 days of cultured in 0-200 μM ascorbic acid (upper row) and 0-200 μM ascorbic acid 2-phosphate (lower row). Scale bars: 5 mm. (b) The total number of colonies counted within each group. (c) The mean diameters per colony of SHED. The data are expressed as the mean ± SD of 6 replicate cultures (**$p < 0.05$). AA, ascorbic acid; A2P, ascorbic acid 2-phosphate; SHED, stem cell from human exfoliated deciduous teeth.

Discussion

The SHED holds promise as unique resource for dental therapies including autologous stem-cell transplantation and tissue engineering and does not carry the ethical burden of embryonic stem cells. The CFU-F with the crucial character of stem cells enables it a long term self-renewal and rapid proliferation properties [3, 13] and an ability to differentiate [14]. The low initial number of CFU-F is a disadvantage. On the other hand, the accurate culture conditions of CFU-F are able to support proliferation and differentiation of stem cells for their future therapeutic uses. However, several factors were reported to have impact on the CFU-F assay including the time of culturing, cell-seeding density, and medium composition [15-17]. In this study, the highest CFU-F formation was obtained when SHED was cultured at the density of 500 cells per well of a 6-well plate for 14 days in Dulbecco's modified Eagle medium (DMEM) with 20% FBS and 100 μM L-ascorbic acid-2-phosphate.

The influence of culturing time was first evaluated for the formation of CFU-F in SHED. Culturing time for CFU-F formation in stem cells from exfoliated deciduous teeth is still a controversial issue. It occurs between day 10 and day 14 [18-20], our work found that 14 days were adequate for the colony formation of SHED.

The CFU-F colonies exhibit heterogeneous proliferation and differentiation potentials, which are correlated with their heterogeneous morphologies [21-24]. Under normal growth culture condition, the SHED formed colonies of three morphological types: holoclones or stem cells (SC), meroclones or transit-amplifying colonies (TAC) and paraclones or differentiated colonies (DC) as previously described [6, 15]. The high number of cells in the SC colonies corresponds to the high proliferative potential and possess differentiation potential, whereas the DC colonies have a limited remaining proliferative capacity and are fully differentiated [6]. The TAC are in between these two colony types in terms of size and number of cells within the colony and have a mixture of characteristics. Under standard culture conditions, the colony morphology formed varies according to the cells line [15]. Here, the colony formed by SHED was shown to be TAC colonies with relatively high prevalence, followed by SC and DC colonies, respectively.

Previous reports about the critical parameters for human bone marrow MSC expansion have reported that the plating MSC density affects the proliferation and stemness preservation [5, 25, 26]. In addition, the starting population is crucial for the final numbers of cells to be obtained albeit the proliferative capacity of MSCs is very high [12]. Our results from the testing of four different plating densities, initial plating density of 250 cells and 500 cells/well of 6-well plate, gave much higher colony numbers of the starting SHED adherent population. Better yet, the SHED colony morphology was not critically altered when different cell densities are seeded as the highest colony's proportion was TAC.

The FBS plays a pivotal role in cell culture [16]. The concentration of serum affects the MSC propagation. Without the serum, the mouse bone marrow stem cells show senescence signs and died in a few days [27]. In 10% FBS media, mouse MSC proliferation is visibly increased and 20% FBS media has stronger effects on cell growth than 10% FBS media [16, 28]. This study showed that the number and size of SHED colonies in 20% FBS were significantly higher than those in 10% FBS. However, both FBS concentrations did not affect the morphology of colonies and the TAC colonies were still the most prevalence. The content of protein components such as bovine serum albumin and growth factors can vary widely between lots of the sera and possibly reduce or increase the CFU-F formation and cellular growth [29]. In our experiments, the FBS from two different companies (Biochrom GmbH and Sigma St) were tested and observed the colony formation after 14-day. FBS from Sigma St company significantly reduced the ability to promote the formation of SHED colonies and the morphology appeared to be more like DC colonies. Thus the commercial FBSs with variation in the concentrations and composition of components such as albumin, globulins, hemoglobin and total proteins can affect the proliferation and morphology of SHED colonies. Inconsistent cell growth can be due to the different lots of FBSs used which has varying concentrations of various proteins particularly the growth stimulatory factors, growth inhibitory factors and other growth related proteins [30].

In addition, we chose to study the ascorbic acid (AA) and ascorbic acid 2-phosphate (A2P) which are essential supplements in a variety of cultures [31-34]. One of the essential functions of AA is its role as a cofactor for the hydroxylation of proline and lysine residues in collagen, which is the most abundant protein in the body [35]. The *in vitro* proliferation and differentiation of MSCs into skeletal tissues requires ascorbic acid as an essential medium component [34]. In some cell culture conditions, the ascorbic acid concentrations of 0.01 – 0.5 mM inhibit or stimulate the proliferation of various kinds of cells including HEP2, KB, ascites, bone marrow and melanoma cells [34, 36-38]. The AA can induce morphological changes and apoptosis in a cell culture of chondrocytes after 18 h of cultivation [39]. Moreover, the numbers of leukemic cell colonies grown in culture are reduced by the addition of L-ascorbic acid (0.3 mM) to the culture medium [40]. When 50 μM AA was added in the cultivation of SHED, it stimulated the SHED colony formation and growth and increased in the proportion of SC colonies. On the contrary, the AA at 100 and 200 μM repressed nearly 100% of SHED colony formation. This inhibitory effect may be due to several possibility. There may be the accumulation of H_2O_2 and the formation of lipid peroxide in mitochondria that lead to cellular damages [5, 40-43]. The AA may induce an increasing level of cyclic adenosine 3':5'-monophosphate that inhibits cell growth [44]. Moreover, the AA is unstable in solution, especially under the normal culture conditions of neutral pH and 37°C. Instead of AA, an ascorbic acid 2-phosphate (A2P), a phosphate derivative of AA, has been used as a supplement in a variety of cultures [31, 45, 46]. In human fibroblasts culture, the A2P has a cofactor activity for collagen biosynthesis [47]. Moreover, it is very stable under the culture conditions [47]. Similar to osteoblast cells [33], the A2P up to 200 μM has no inhibitory effect on SHED growth. The size of SHED colonies and the proportion of SC colonies were increased by the A2P. The A2P is also resistant to oxidation by oxygen. It releases free AA by the action of alkaline phosphatase in the cell membranes of most animal cells. Our data suggested that the A2P-supplement was an effective way to help increase the formation of undifferentiated SHED colonies.

Collectively, this study provides the general guidelines for the establishment of a standardized protocol aiming at the preparation of high-quality adherent dental stem cells from human exfoliated deciduous teeth clusters (or colony forming units-fibroblastic) for use as source for preclinical and clinical studies.

ACKNOWLEDGMENTS

The authors would like to thank L. Amornwichitwech, S. Jaiton, S. Srisawan and P. Wiphakphongpakorn from Faculty of Dentistry mahidol University for their helpful contribution and technical assistance. We are grateful to Professor Vichien Rimpanitchayakit for helpful discussions and critical reading of the manuscript.

REFERENCES

[1] d'Aquino, R; De Rosa, A; Laino, G; Caruso, F; Guida, L; Rullo, R; Checchi, V; Laino, L; Tirino, V; Papaccio, G. Human dental pulp stem cells: from biology to clinical applications. *Journal of experimental zoology Part B, Molecular and developmental evolution*, 2009, 312B(5), 408-415.

[2] Miura, M; Gronthos, S; Zhao, M; Lu, B; Fisher, LW; Robey, PG; Shi, S. SHED: stem cells from human exfoliated deciduous teeth. *Proceedings of the National Academy of Sciences of the United States of America*, 2003, 100(10), 5807-5812.

[3] Friedenstein, AJ; Chailakhjan, RK; Lalykina, KS. The development of fibroblast colonies in monolayer cultures of guinea-pig bone marrow and spleen cells. *Cell and tissue kinetics*, 1970, 3(4), 393-403.

[4] Prockop, DJ. Marrow stromal cells as stem cells for nonhematopoietic tissues. *Science*, 1997, 276(5309), 71-74.

[5] Colter, DC; Sekiya, I; Prockop, DJ. Identification of a subpopulation of rapidly self-renewing and multipotential adult stem cells in colonies of human marrow stromal cells. *Proceedings of the National Academy of Sciences of the United States of America*, 2001, 98(14), 7841-7845.

[6] Barrandon, Y; Green, H. Three clonal types of keratinocyte with different capacities for multiplication. *Proceedings of the National Academy of Sciences of the United States of America*, 1987, 84(8), 2302-2306.

[7] Rochat, A; Kobayashi, K; Barrandon, Y. Location of stem cells of human hair follicles by clonal analysis. *Cell*, 1994, 76(6), 1063-1073.

[8] Pellegrini, G; Golisano, O; Paterna, P; Lambiase, A; Bonini, S; Rama, P; De Luca, M. Location and clonal analysis of stem cells and their differentiated progeny in the human ocular surface. *The Journal of cell biology*, 1999, 145(4), 769-782.

[9] Kogler, G; Sensken, S; Airey, JA; Trapp, T; Muschen, M; Feldhahn, N; Liedtke, S; Sorg, RV; Fischer, J; Rosenbaum, C; et al. A new human somatic stem cell from placental cord blood with intrinsic pluripotent differentiation potential. *The Journal of experimental medicine*, 2004, 200(2), 123-135.

[10] Jiang, Y; Jahagirdar, BN; Reinhardt, RL; Schwartz, RE; Keene, CD; Ortiz-Gonzalez, XR; Reyes, M; Lenvik, T; Lund, T; Blackstad, M; et al. Pluripotency of mesen-chymal stem cells derived from adult marrow. *Nature*, 2002, 418(6893), 41-49.

[11] Masters, JR; Thomson, JA; Daly-Burns, B; Reid, YA; Dirks, WG; Packer, P; Toji, LH; Ohno, T; Tanabe, H; Arlett, CF; et al. Short tandem repeat profiling provides an international reference standard for human cell lines. *Proceedings of the National Academy of Sciences of the United States of America*, 2001, 98(14), 8012-8017.

[12] Sotiropoulou, PA; Perez, SA; Salagianni, M; Baxevanis, CN; Papamichail, M. Characterization of the optimal culture conditions for clinical scale production of human mesenchymal stem cells. *Stem cells*, 2006, 24(2), 462-471.

[13] Bianco, P; Riminucci, M; Gronthos, S; Robey, PG. Bone marrow stromal stem cells: nature, biology, and potential applications. *Stem Cells*, 2001, 19(3), 180-192.

[14] Nombela-Arrieta, C; Ritz, J; Silberstein, LE. The elusive nature and function of mesenchymal stem cells. *Nature reviews Molecular cell biology*, 2011, 12(2), 126-131.

[15] Beaver, CM. The effect of culture conditions on colony morphology and proliferative capacity in human prostate cancer cell lines. *Cell biology and toxicology*, 2012, 28(5), 291-301.

[16] Quan, H; Kim, SK; Heo, SJ; Koak, JY; Lee, JH. Optimization of growth inducing factors for colony forming and attachment of bone marrow-derived mesenchymal stem cells regarding bioengineering application. *The journal of advanced prosthodontics*, 2014, 6(5), 379-386.

[17] Kuznetsov, SA; Mankani, MH; Bianco, P; Robey, PG. Enumeration of the colony-forming units-fibroblast from mouse and human bone marrow in normal and pathological conditions. *Stem cell research*, 2009, 2(1), 83-94.

[18] Wang, X; Sha, XJ; Li, GH; Yang, FS; Ji, K; Wen, LY; Liu, SY; Chen, L; Ding, Y; Xuan, K. Comparative characterization of stem cells from human exfoliated deciduous teeth and dental pulp stem cells. *Archives of oral biology*, 2012, 57(9), 1231-1240.

[19] Nakamura, S; Yamada, Y; Katagiri, W; Sugito, T; Ito, K; Ueda, M. Stem cell proliferation pathways comparison between human exfoliated deciduous teeth and dental pulp stem cells by gene expression profile from promising dental pulp. *Journal of endodontics*, 2009, 35(11), 1536-1542.

[20] Govindasamy, V; Abdullah, AN; Ronald, VS; Musa, S; Ab Aziz, ZA; Zain, RB; Totey, S; Bhonde, RR; Abu Kasim, NH. Inherent differential propensity of dental pulp stem cells derived from human deciduous and permanent teeth. *Journal of endodontics*, 2010, 36(9), 1504-1515.

[21] Rhodes, NP; Srivastava, JK; Smith, RF; Longinotti, C. Heterogeneity in proliferative potential of ovine mesenchymal stem cell colonies. *Journal of materials science Materials in medicine*, 2004, 15(4), 397-402.

[22] Russell, KC; Phinney, DG; Lacey, MR; Barrilleaux, BL; Meyertholen, KE; O'Connor, KC. *In vitro* high-capacity assay to quantify the clonal heterogeneity in trilineage potential of mesenchymal stem cells reveals a complex hierarchy of lineage commitment. *Stem cells*, 2010, 28(4), 788-798.

[23] Sengers, BG; Dawson, JI; Oreffo, RO. Characterisation of human bone marrow stromal cell heterogeneity for skeletal regeneration strategies using a two-stage colony assay and computational modelling. *Bone*, 2010, 46(2), 496-503.

[24] Gothard, D; Dawson, JI; Oreffo, RO. Assessing the potential of colony morphology for dissecting the CFU-F population from human bone marrow stromal cells. *Cell and tissue research*, 2013, 352(2), 237-247.

[25] Colter, DC; Class, R; DiGirolamo, CM; Prockop, DJ. Rapid expansion of recycling stem cells in cultures of plastic-adherent cells from human bone marrow. *Proceedings of the National Academy of Sciences of the United States of America*, 2000, 97(7), 3213-3218.

[26] Sekiya, I; Larson, BL; Smith, JR; Pochampally, R; Cui, JG; Prockop, DJ. Expansion of human adult stem cells from bone marrow stroma: conditions that maximize the yields of early progenitors and evaluate their quality. *Stem cells*, 2002, 20(6), 530-541.

[27] Tropel, P; Noel, D; Platet, N; Legrand, P; Benabid, AL; Berger, F. Isolation and characterisation of mesen-chymal stem cells from adult mouse bone marrow. *Experimental cell research*, 2004, 295(2), 395-406.

[28] Li, X; Zhang, Y; Qi, G. Evaluation of isolation methods and culture conditions for rat bone marrow mesen-chymal stem cells. *Cytotechnology*, 2013, 65(3), 323-334.

[29] Talbot, NC; Powell, AM; Caperna, TJ. Comparison of colony-formation efficiency of bovine fetal fibroblast cell lines cultured with low oxygen, hydrocortisone, L-carnosine, bFGF, or different levels of FBS. *Cloning and stem cells*, 2004, 6(1), 37-47.

[30] Zheng, X; Baker, H; Hancock, WS; Fawaz, F; McCaman, M; Pungor, E. Jr. Proteomic analysis for the assessment of different lots of fetal bovine serum as a raw material for cell culture. Part IV. Application of proteomics to the manufacture of biological drugs. *Biotechnol Prog*, 2006, 22(5), 1294-1300.

[31] Fermor, B; Gundle, R; Evans, M; Emerton, M; Pocock, A; Murray, D. Primary human osteoblast proliferation and prostaglandin E2 release in response to mechanical strain *in vitro*. *Bone*, 1998, 22(6), 637-643.

[32] Gallagher, JA; Gundle, R; Beresford, JN. Isolation and culture of bone-forming cells (osteoblasts) from human bone. *Methods in molecular medicine*, 1996, 2, 233-262.

[33] Takamizawa, S; Maehata, Y; Imai, K; Senoo, H; Sato, S; Hata, R. Effects of ascorbic acid and ascorbic acid 2-phosphate, a long-acting vitamin C derivative, on the proliferation and differentiation of human osteoblast-like cells. *Cell biology international*, 2004, 28(4), 255-265.

[34] Choi, KM; Seo, YK; Yoon, HH; Song, KY; Kwon, SY; Lee, HS; Park, JK. Effect of ascorbic acid on bone marrow-derived mesenchymal stem cell proliferation and differentiation. *J Biosci Bioeng*, 2008, 105(6), 586-594.

[35] Prockop, DJ; Kivirikko, KI. Collagens: molecular biology, diseases, and potentials for therapy. *Annual review of biochemistry*, 1995, 64, 403-434.

[36] Leung, PY; Miyashita, K; Young, M; Tsao, CS. Cytotoxic effect of ascorbate and its derivatives on cultured malignant and nonmalignant cell lines. *Anticancer research*, 1993, 13(2), 475-480.

[37] Gardiner, NS; Duncan, JR. Enhanced prostaglandin synthesis as a mechanism for inhibition of melanoma cell growth by ascorbic acid. *Prostaglandins, leukotrienes, and essential fatty acids*, 1988, 34(2), 119-126.

[38] Brigelius-Flohe, R; Flohe, L. Ascorbic acid, cell proliferation, and cell differentiation in culture. *Sub-cellular biochemistry*, 1996, 25, 83-107.

[39] Malicev, E; Woyniak, G; Knezevic, M; Radosavljevic, D; Jeras, M. Vitamin C induced apoptosis in human articular chondrocytes. *Pflugers Archiv: European journal of physiology*, 2000, 440(5 Suppl), R46-48.

[40] Park, CH; Amare, M; Savin, MA; Hoogstraten, B. Growth suppression of human leukemic cells *in vivo* by L-ascorbic acid. *Cancer research*, 1980, 40(4), 1062-1065.

[41] Josephy, PD; Palcic, B; Skarsgard, LD. Ascorbate-enhanced cytotoxicity of misonidazole. *Nature*, 1978, 271(5643), 370-372.

[42] Koch, CJ; Biaglow, JE. Toxicity, radiation sensitivity modification, and metabolic effects of dehydroascorbate and ascorbate in mammalian cells. *Journal of cellular physiology*, 1978, 94(3), 299-306.

[43] Wills, ED; Wilkinson, AE. Release of enzymes from lysosomes by irradiation and the relation of lipid peroxide formation to enzyme release. *The Biochemical journal*, 1966, 99(3), 657-666.

[44] Gericke, D; Chandra, P. Inhibition of tumor growth by nucleoside cyclic 3'-5'-monophosphates. *Hoppe-Seyler's Zeitschrift fur physiologische Chemie*, 1969, 350(11), 1469-1471.

[45] Hitomi, K; Torii, Y; Tsukagoshi, N. Increase in the activity of alkaline phosphatase by L-ascorbic acid 2-phosphate in a human osteoblast cell line, HuO-3N1. *Journal of nutritional science and vitaminology*, 1992, 38(6), 535-544.

[46] Mizutani, A; Sugiyama, I; Kuno, E; Matsunaga, S; Tsukagoshi, N. Expression of matrix metalloproteinases during ascorbate-induced differentiation of osteoblastic MC3T3-E1 cells. *Journal of bone and mineral research: the official journal of the American Society for Bone and Mineral Research*, 2001, 16(11), 2043-2049.

[47] Hata, R; Senoo, H. L-ascorbic acid 2-phosphate stimulates collagen accumulation, cell proliferation, and formation of a three-dimensional tissuelike substance by skin fibroblasts. *Journal of cellular physiology*, 1989, 138(1), 8-16.

In: Stem Cells Origins of Cytokine Interactions
Editor: Prasad S. Koka

ISBN: 978-1-53615-302-6

Chapter 18

IN VITRO PROLIFERATION OF HUMAN HAEMATOPOIETIC STEM CELLS BY *ANDROGRAPHIS PANICULATA* (BURM.F.) NEES EXTRACTS

Thiruvengadasamy Kannan[1], Manne Mudhu Sunitha[2], Lokanathan Srikanth[2], Pasupuleti Santhosh Kumar[2], Chodimella Chandrasekhar[3] and Potukuchi Venkata Gurunadha Krishna Sarma[2,†]

[1]Department of Medical Oncology
[2]Department of Biotechnology
[3]Department of Haematology
Sri Venkateswara Institute of Medical Sciences,
Tirupati, Andhra Pradesh, India

ABSTRACT

Andrographis paniculata (Burm.f.) Nees (Nelavembu) a natural herb extract largely used to increase the platelet count in thrombocytopenia condition. Normally, Hematopoietic stem cells (HSCs) are key cell in the formation of platelets and depletion of platelets results in thrombocyte-penia. The present study is aimed to understand the effect of *Andrographis paniculata* (Burm.f.) Nees extract (APE) on proliferation of HSCs that eventually exhibit anti-thrombocytopenic effect. Andrograph-olide a prominent compound in APE is known to stimulate thrombopoietin from liver which showed greater affinity towards thrombo-poietin receptor compared to eltrombopag and pegacaristim (compounds used for generation of platelets). The exertion of APE on HSCs proliferation showed a significant effect on cell count with increasing concentrations from 1 μM to 2

[†] Corresponding Author Email: sarmasvims@gmail.com.

mM. The results were concurrent with the in vitro IL-3 and G-CSF treated HSCs. These findings explain that administration of APE results in the HSCs proliferation and eventual differentiation into platelets.

Keywords: Nelavembu, *Andrographis panicualta*, haematopoietic stem cells, platelets

1. Introduction

Thrombocytopenia (Low platelet count) a serious life threatening condition arises due to ineffective production of platelets from bone marrow or increased destruction of platelets or it may occur by platelet splenic sequestration [1]. The decrease in normal platelet count might be caused by genetic disorders or it may be viral infectious like dengue fever, chicken pox, parvo virus, rubella and some bacterial infections like tuberculosis [2]. Whereas increased destruction of platelets were observed in thrombotic microangiopathies, disseminated intra-vascular coagulation (DIC) and immune thrombocytopenia (ITP). While platelet sequestration was noticed in congestive splenomegaly occurs due to portal hypertension caused by cardiac failure, hepatic vein thrombosis or vena cava thrombosis, cirrhosis [3]. In addition, increased destruction and decreased production of platelets are also observed in drug-induced thrombocytopenia and hepatitis C virus (HCV) and human immunodeficiency virus (HIV)-induced thrombocytopenia [4].

The therapeutic approaches of thrombocytopenia include platelet and recombinant human TPO (rHUTPO) transfusion leads to infections by generating auto-antibodies to these molecules and also develop cross reactivity with endogenous TPO, resulting in severe thrombocytopenia [5, 6]. These limitations develop an alternative synthetic drug such as Romiplostim (a peptide) an intravenously administrative drug and an Eltrombopag, a small molecule non-peptide oral drug for the treatment of thrombocytopenia. The major disadvantage of Romi-plostim was the development of thromboembolism, myelofibrosis and reversible reticulin deposition in bone marrow [7]. In addition, Eltrombopag activates the TPO receptor in non-competitive manner and act as a stabilizer to endogenous TPO [8]. Julian et al. reported that oral administration of eltrombopag a chemical compound to thrombocytopenia patients showed increased platelet count in a dose dependent manner [9]. However, Eltrombopag also develops adverse effects like nausea, pain, insomnia etc. As synthetic drugs posse's major drawbacks which direct the development of novel compounds which has capacity to cure thrombocytopenia without any adverse effects. In this regard, researchers are concentrated on the development of plant based medicines which are safer and healthier to the human than synthetic drugs.

Nowadays, herbal medicines also play a major role in treatment of many diseases in humans and animals with unique approach to safe, effective and inexpensive new remedies for variety of diseases as they do not have any adverse effects [10]. Therefore, studies on herbal medicine constituents for the treatment of thrombocytopenia are increasing significantly. Several studies have reported that extracts from different parts of the plant exhibited anti thrombocytopenic activity and also involved in mega-karyocyte differentiation as compared to synthetic analogues [11-13].

In this view the present study is concentrated on the use of *Andrographis paniculata* extract (APE) in the development of platelets. *Andrographis paniculata* belongs to the family acanthaceae is used conventionally for treatment of multiple diseases. It constitutes various chemical compounds like lactones, terpenes, diterpenoids, flavanoids etc and exhibiting anti-diabetic, anti-ulcer, anti-inflammatory, anti-viral properties [14]. Methanolic extract of *A. paniculata* reveal anti-dengue property particularly towards DENV-1 serotype [15]. This plant exhibited platelet augmentation activity through andrographolide in male Sprague-Dawley rats by inducing thrombocyte-penia with cyclophosphamide [16]. The *Andrographis paniculata* extract showed increased amount of andrographolide known to induce the production of thrombopoietin from liver thus plays vital role in the development of platelets. In addition to this the APE promotes a proliferation of human epidermal stem cells (EpSCs), through the up-regulation of the integrin β1 and VEGF expression [17].

Platelets are produced from megakaryocytic process by the action of thrombopoietin hormone [18]. Thrombopoietin is a glycoprotein hormone that regulates the production of platelets by binding to its receptor CD110 present on the surface of bone marrow haematopoietic stem cells (HSCs). These HSCs exist in the specialized bone marrow micro-environment (niches) under hypoxic conditions where the stem cell exhibits its quiescence nature [19, 20]. Human haematopoietic stem cells are characterized by the presence of its markers $CD34^{+}$. HSCs are multipotent cells that can either self-renew or differentiate into various blood lineages including the common lymphoid and myeloid progenitor. These myeloid progenitors further differentiate into the granulocyte/ macrophage progenitor and the mega-karyocyte-erythrocyte progenitors (MEP), which are responsible for the production of Megakaryocytes (MKs) further these cells upon maturation differentiate into platelets and thus formed platelets do not proliferate to form more platelets as they lack nucleus and their life span is about 10 days [21]. The present study was undertaken to evaluate the effect of *Andrographis paniculata* extract (APE) on the proliferation of HSCs and binding affinity of andrographolide a major constituent of APE toward thrombopoietin receptor.

2. Materials and Methods

2.1. Preparation of Medium

1g of powdered Dulbecco's Modified Eagle's Medium (DMEM) (Himedia), 370 mg of sodium bicarbonate was weighed and dissolved with 100mL of milliQwater and the pH of the medium was adjusted to 7.0. To this medium 100 μg/mL ampicillin, 25 μg/mL Streptomycin, 100 μg/mL Gentamycin, 25 μg/mL Tetracyclin and 50 μg/mL Amphotericin-B was added. Then the medium was filtered using 0.22 μm membrane filter under vacuum condition [22]. This study was carried out in the Department of Biotechnology, SVIMS University after obtaining approval from institutional ethics committee (IEC No: 31/01/2006 and 419/27-01-2015).

2.2. Thawing of Cryopreserved Cells

The cryopreserved HSCs were rapidly thawed in a water bath at 37°C. Cells were transferred carefully to 25 cm^2 tissue culture flasks (Corning) containing 4mL of DMEM with 10% of Fetal Bovine Serum (FBS) (Himedia). Cells were then incubated at 37°C with 5% CO_2 for 2 days to form a monolayer culture [23].

The Peripheral blood mononuclear cells (MNCs) which served as control were isolated from the peripheral blood following the method Riedhammer et al. [24].

2.3. Morphological Identification of Cultured Cells through Leishman's Staining

A thin smear of cultured PBSCs was prepared on a sterile glass slide and the smear was stained with 0.15% Leishman's stain for 2 min then observed under light microscope (Olympus, India). Viability of the cultured $CD34^+$ cells was identified by using dye exclusion method. In this procedure, equal volume of $CD34^+$ cell suspension and 0.4% (w/v) Trypan blue dye was mixed and counted in hemocytometer and calculated as [22].

Total number of viable cells = Total number of cells – Total number of dead cells

$$\% \text{ of Viability} = \frac{\text{Total number of viable cells} \times 100}{\text{Total number of cells}}$$

2.4. Identification of CD34 Surface Marker on HSCs

The presence of cell specific markers for grown HSCs ($CD34^+$) was assessed by ICC using specific antibodies. In this procedure, a smear of cultured $CD34^+$ cells were seeded onto a sterile slide. The cells were then incubated at 37°C for 1h in 4% paraformaldehyde containing PBS. The internal peroxidase activity was blocked by using 3% H_2O_2 dissolved in PBS and non-specific sites blocked with 2% gelatin. Then the cells were treated with cell specific primary antibodies for 1hr at 37°C. Bounded primary antibodies were detected by using HRP-conjugates as secondary antibodies. The HRP activity was revealed by adding 0.3% H_2O_2 and 1.5 mM diaminobenzidine tetra hydrochloride dissolved in 10 mM Tris–HCl (pH 8) for 30 min at room temperature. After washing with distilled water, the slides were observed under phase contrast microscope at 200× magnification (Olympus, India) [23].

2.5. Flowcytometry Analysis of HSCs

The assessment of human $CD34^+$ cells were carried out by using FACS (Beckman coulter Cytomics FC500 Flow Cytometry analyser, USA) by the stain-wash-no lyse method. 50 μL of generated $CD34^+$ culture and CD34-PE/CD45-FITC was added to single use tubes and vortexed thoroughly for 30 sec, incubated for 30 min at room temperature (the tubes are protected from direct light). Following incubation, the sample was loaded onto Flow cytometry for analysis. Absolute viable $CD34^+$ was determined using anti-human CD34 monoclonal antibodies conjugated with phycoerythrin and mouse Anti-human CD45-FITC antibody. The dead $CD34^+$ cells were excluded using 7-Amino-actinomycin D. The results were acquired for up to 75,000 events and maintained a co-efficient of variation at <2% and the absolute number of viable $CD34^+$ cells/μL was calculated as follow [25].

$$\frac{\#\ CD34^+\ \text{cells} \times \text{bead concentration} \times \text{dilution factor}}{\#\ \text{Single beads}}$$

2.6. Scanning Electron Microscopy

The morphology of the cultured $CD34^+$ stem cells was confirmed through scanning electron microscopic (SEM) analysis. Cells were fixed on a glass slide with 3% (v/v) glutaraldehyde in 0.1 M phosphate buffer (pH 7.4) at 37°C for 1h. The slides were rinsed with PBS (pH 7.2) for three times and then treated with 1% osmium tetraoxide for 1h. The sides were washed with PBS for five times and were gradually dehydrated using ethanol series (30–100%) for 10 min each, till critical point of drying was achieved. The samples were mounted on aluminum stubs with two-sided adhesive tape, sputter-coated with 2 nm gold/palladium (Au/Pd) and were examined with a field emission SEM (EVO MA15, Carl Zeiss, Germany) then the images were captured with Zeiss EVO MA 15 microscope at a magnification of 500X [23].

2.7. Preparation of Nelavembu Extracts

Stock of 2% solution was prepared by dissolving 20 mg of powdered dried Nelavembu extract in 1 mL of PBS. The stock solution was filtered using (0.22 μm pore membrane filters) and further diluted with culture medium to the required concentration for the assays.

2.8. *In Vitro* Proliferation Assay of HSCs

The *in vitro* proliferation activity was assessed by seeding 1×10^3 cells/well into each well of 6-well plate containing DMEM and the plate were incubated at 37°C with 5% CO_2 for 24 hours. After 24 hours, the medium was removed and the cells were treated with diluted stock of APE at a concentration ranging from 0.05 to 2.5 mg/mL and the cells were incubated at 37°C, 5% CO_2 for 2 days. In addition, 1×10^3 HSCs were also treated with the DMEM containing 5 μg of G-CSF, 10 ng of IL-3and were incubated at 37°C, 5% CO_2 for 2 days along with the cell control (cells + DMEM) and was observed under phase contrast microscope at 200X magnifications (Olympus, India). The same was carried on MNCs.

2.9. Preparation of TPO Receptor Structure

The protein sequence of TPO receptor obtained from NCBI was submitted to BLASTP for similarity search to find out its homolog's. By using ClustalX (1.83) Chain A of Erythropoietin receptor (protein data bank ID: 1CN4) showed maximum identity with TPO receptor therefore, 1CN4_A used as a template to build the structure of TPO receptor by using modeler9v9 software where it generates 20 possible models. Amongst best model was selected based on the least DOPE score (Table 1) and thus obtained model was used in Autodock.

Table 1. Represents 20 model of receptor structures for Thrombopoietin build with modeler9v9

S No	Summary of successfully produced models			
	Filename	molpdf	DOPE score	GA341 score
1	Target.B99990001.pdb	36122.10156	-23123.54297	0.18173
2	Target.B99990002.pdb	43250.31250	-23607.25195	0.10925
3	Target.B99990003.pdb	35018.17188	-23334.89453	0.08168
4	Target.B99990004.pdb	43327.78516	-22451.62109	0.07566
5	Target.B99990005.pdb	34661.49219	-23641.45898	0.16017
6	Target.B99990006.pdb	35053.33984	-24044.11523	0.22461
7	Target.B99990007.pdb	31707.76172	-21840.97656	0.04039
8	Target.B99990008.pdb	35066.90625	-21859.67773	0.06654
9	Target.B99990009.pdb	39716.65234	-24983.07617	0.20408
10	Target.B99990010.pdb	50880.82422	-22788.07813	0.18661
11	Target.B99990011.pdb	32933.69141	-21742.04688	0.05985
12	Target.B99990012.pdb	38588.20703	-22424.37305	0.07673
13	Target.B99990013.pdb	36635.85938	-22142.49414	0.05025
14	Target.B99990014.pdb	32859.32422	-22609.37500	0.03998
15	Target.B99990015.pdb	37335.66797	-23527.53516	0.08301
16	Target.B99990016.pdb	32916.05469	-23851.20703	0.06325
17	Target.B99990017.pdb	41510.19141	-21084.48828	0.07144
18	Target.B99990018.pdb	32631.05273	-22525.30664	0.15549
19	Target.B99990019.pdb	34822.76953	-21742.40820	0.05503
20	Target.B99990020.pdb	36376.56250	-23401.49023	0.08092

2.10. Preparation of Ligand

The 3D structures of andrographolide (And), eltrombopag (Elt) and pegacaristim (Peg) was retrieved from the pubchem as SD files and converted in to PDB files.

2.11. Docking

By using Autodock4.2 free software Molecular docking was carried out between the best receptor model9 and with ligands andrographolide (And), eltrombopag (Elt) and pegacaristim (Peg) to know their binding affinity by keeping constant grid position for all the three ligands by placing center grid box with dimensions of X center: -25.662, Y center: 37.587, Z center: 103.034. Auto grid records the interaction energy of a ligand atom of a specific element at each point in a 3-D grid around the rigid receptor in the corresponding gridmap file. During the AutoDock calculation the energetics of a particular ligand configuration is evaluated using the values from the gridmaps. Thus, obtained ten possible conformations were ranked based on their binding energy to determine the best ligand-receptor complex. The best complex was explored in Accelrys Discovery Studio 2.5 software for visualizing ligand binding interactions.

3. RESULTS

3.1. Characterization of $CD34^+$ Cells

The HSCs cells grown in the tissue culture flasks with DMEM supplemented with 10% FBS showed the monolayer formation of cells (Figure 1a). The morphology of the cells was clearly visualized by Leishman's staining, demonstrates round large nucleated cells with less cytoplasmic to nucleus ratio (Figure 1b). The viability of the cells through trypan blue dye exclusion method showed 92% cells are viable appeared as colorless (Figure 1c). SEM analysis further confirmed the morphology of $CD34^+$ stem cells showing round cells (Figure 1d). The isolated stem cells were purified using the limiting dilution method and the presence of CD34 cell surface glycoprotein was validated by anti-human CD34 monoclonal antibodies in ICC method demonstrated that the cells were positive for CD34, a HSC marker (Figure 1e). The flowcytometry analysis enforces the preceding results where CD34 stem cells were growing in the flasks showed 99.5% viability with 7AAD (Figure 1f-g).

3.2. *In Vitro* Assay

In order to screen the proliferation potential of HSCs by APE *in vitro* proliferation assay was performed. Increased HSCs proliferation was observed upon APE treatment from 1 μM to 2 mM (cell count = $2 \times 10^3 - 23 \times 10^6$ cells/ml). Further increase in the concentration of the drug there was no appreciable increase in cell count (Figure 2).

These results were concurrent with the results of IL-3, G-CSF treated HSCs which demonstrate an increased proliferation of HSCs were observed (5×10^6 cells) when compared to untreated control cells (1×10^3 cells) (Figure 3). The proliferation of MNCs in presence of APE and compared with MNCs in presence of GCSF. The MNCs in the presence of APE did not show any appreciable growth compared with HSCs while both MNCs and HSCs proliferation increased appreciably in the presence of GCSF.

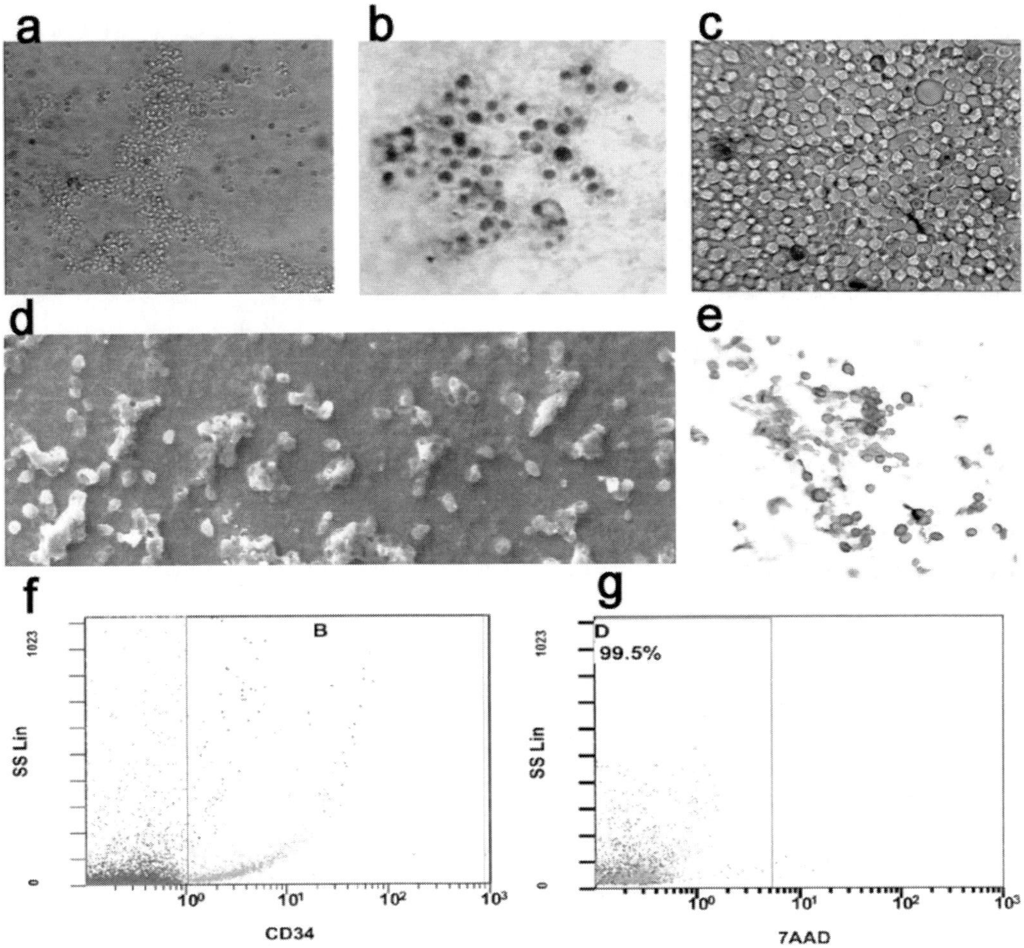

Figure 1. Characterization of $CD34^+$ stem cells a. Monolayer culture of human HSCs in DMEM b. Leishman's staining of cultured $CD34^+$ cells showing prominent large nucleus c. Trypan blue staining of cells showing blue coloured cells are dead cells and viable cells are colourless. d. SEM analysis demonstrating the morphology of individual $CD34^+$ cells appearing round cells e. ICC staining of cultured cells confirming the positivity to $CD34^+$ surface marker f. Flowcytometry analysis showing positive for CD34 marker in the gated region of 'B' g. Percent viability of $CD34^+$ cells with 7-AAD represents at gated region 'D' showing 99.5% cells are viable.

3.3. *In Silico* Analysis

Among 20 models of receptor build through modeller9v9, model 9 has least DOPE score of -24983.07617 and it was chosen as a best model for docking (Table 1). Docking results of mode l9 with andrographolide, eltrombopag and Pegacaristim showed -7.23 kcal/mol, 7.00 kcal/mol and -1.63 kcal/mol indicates that andrographolide and eltrombopag have close homology between the binding affinities (Table 2). The best docked complexes of three different ligands with its receptor were visualized by using Discovery studio software and inspect edits ligand binding amino acid sites in the receptor (Figure 4, Table 3).

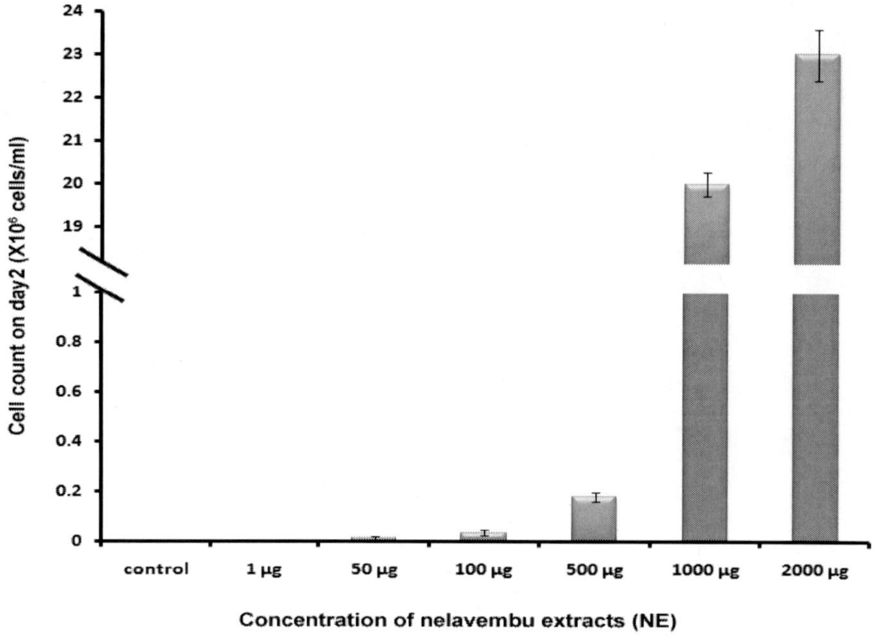

Figure 2. Effect of nelavembu on HSCs. Histogram represent that there is an increase in cell count of HSCs upon addition of the nelavembu extract till 2 mg/ml concentration. Further upon increasing the concentration of the drug revealed there is no significant effect on HSCs count.

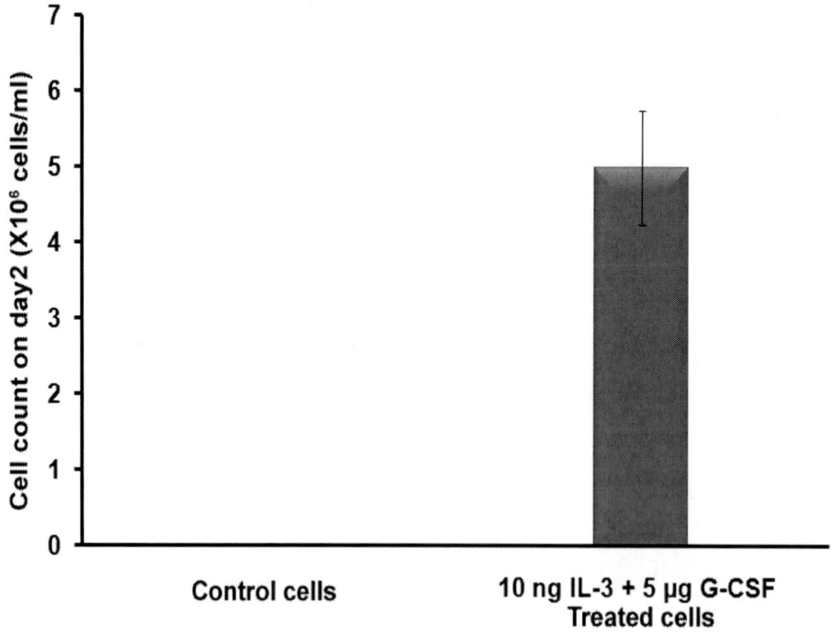

Figure 3. Effect of IL-3, G-CSF on HSCs. Histogram represents that there is an increase in cell count of HSCs were observed upon addition of IL-3 and G-CSF compared to untreated control HSCs.

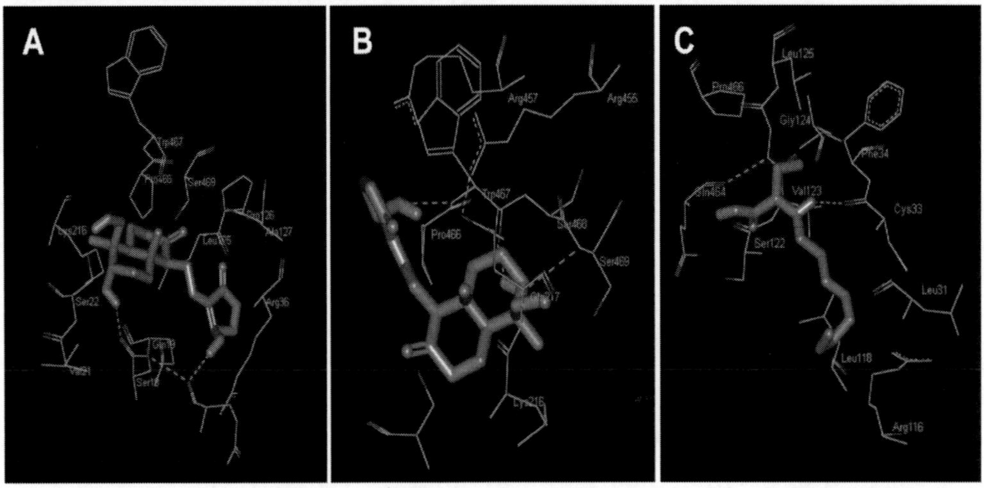

Figure 4. Visualising of Best docked complexes at ligand binding interactions sites of receptor amino acids A Andrographolide B Eltrombopag C Pegacaristim.

Table 2. Autodock results of receptor model 9 with andrographolide, eltrombopag and Pegacaristim

A. Binding energies of Andrographolide ranked based on their energy values

S No	Rank	Sub Rank	Run	Binding Energy	Cluster RMSD	Reference RMSD
1	1	1	1	-7.23	0.00	116.72
2	2	1	10	-6.90	0.00	115.46
3	2	2	9	-6.42	1.64	115.96
4	3	1	4	-6.44	0.00	116.04
5	4	1	3	-6.41	0.00	117.71
6	5	1	6	-6.28	0.00	106.65
7	6	1	7	-6.02	0.00	104.50
8	7	1	2	-5.96	0.00	106.09
9	7	2	5	-5.85	1.18	106.25
10	8	1	8	-5.41	0.00	121.84

B. Binding energies of Eltrombopag ranked based on their energy values

S No	Rank	Sub Rank	Run	Binding Energy	Cluster RMSD	Reference RMSD
1	1	1	7	-7.00	0.00	116.13
2	2	1	10	-6.82	0.00	118.00
3	3	1	5	-6.46	0.00	115.88
4	4	1	3	-6.38	0.00	116.10
5	5	1	2	-6.23	0.00	114.14
6	6	1	9	-6.16	0.00	118.00
7	7	1	6	-6.06	0.00	105.64
8	8	1	1	-6.03	0.00	115.15
9	9	1	8	-5.99	0.00	116.88
10	10	1	4	-5.77	0.00	115.41

C. Binding energies of Pegacaristim ranked based on their energy values

S No	Rank	Sub Rank	Run	Binding Energy	Cluster RMSD	Reference RMSD
1	1	1	1	-1.63	0.00	105.71
2	2	1	8	-1.48	0.00	118.54
3	3	1	6	-1.45	0.00	118.35
4	4	1	4	-1.44	0.00	116.90
5	5	1	2	-1.36	0.00	110.29
6	6	1	10	-1.23	0.00	105.47
7	7	1	5	-1.13	0.00	120.15
8	8	1	9	-0.78	0.00	116.21
9	9	1	3	-0.57	0.00	118.29
10	10	1	7	-0.53	0.00	110.90

Table 3. Molecular docking of Model 9 with its ligands

A. Andrographolide

Receptor amino acids and its positions interacting with ligand	No of hydrogen bonds
Typ467	-
Pro466	-
Ser469	-
Pro126	-
Ala127	-
Leu125	-
Arg36	-
Gln19	2
Ser18	-
Val21	-
Ser22	1
Lys216	-

B. Eltrombopag

Receptor amino acids and its positions interacting with ligand	No of hydrogen bonds
Arg457	-
Arg455	-
Trp467	2
Ser468	-
Ser469	-
Pro466	-
Gln217	2
Lys216	-
Val21	-
Leu125	-
Pro466	-
Gly124	-

C. Pegacaristim

Receptor amino acids and its positions interacting with ligand	No of hydrogen bonds
Leu125	-
Pro466	-
Gly124	-
Phe34	-
Val123	1
Gln464	1
Ser122	-
Cys33	-
Leu31	-
Leu118	-
Arg116	-

4. DISCUSSION

Andrographis paniculata contains various chemical constituents like lactones, diterpenoids, diterpene glycosides and flavanoids. The active diterpenoids in *Andrographis paniculata* analysed through various techniques showed the presence of highest concentration of andrographolide (AP1) in leaves and stem of *Andrographis paniculata* than other diterpenoids like 14-deoxy- 11,12-didehydroandrographolide (AP3) and neoandrograph-olide (AP4) [26]. It has multiple pharmacological actions like anti-inflammatory, anti-diabetic, anti-ulcer, anti-cancer, anti-angiogenic, anti-microbial and anti-oxidant activitiesdue to the presence of diterpens [27]. AP1 exhibits higher anti-inflammatory and hepatoprotective activities against galactosamine and paracetamol intoxication, whereas, AP3 produces a more potent hypotensive effect in anaesthetised rats and isolated right atria, while AP4 shows greater activity against malaria [26].

Under normal healthy conditions platelets are produced from megakaryocytes within 4 to 6 days and these platelets are stored in spleen. The stored platelets are released into the circulation through smooth muscle contraction of the spleen [28]. The presence of andrographolide constituent in the *A. Paniculata* extracts exhibit megakarypoietic stimulatory activity with the release of thrombo-poietin from liver playing a role in thrombopoiesis. The presence of andrographolide in APE not only exhibited haepatoprotective activity but also triggers the synthesis of thrombopoietin from liver [29]. Thus, released thrombopoietin binds to CD110 (TPO receptor) present on the surface of HSCs in the bone marrow and triggers the JAK/STAT signalling pathway in HSCs to initiate megakaryocytic process to produce platelets (Figure. 5).

In addition, the constituents of the APE also stimulate the proliferation of HSCs and there are no significant morphological changes of HSCs were observed upon APE treatment since premature megakaryocytes resembles alike HSCs (Figure 1). Therefore, upon administration of APE in thrombo-cytepenia patient enhances the release of thrombo-poietin from liver that induces the differentiation of bone marrow HSCs to megakarypoietic process which eventually forms platelets.

The APE possesses an andrographolide, a diterpe-noid which is an active compound of *Andrographis paniculata* that has the ability to initiate the proliferation of HSCs similar to G-CSF and IL-3 cytokines that supports the expansion of $CD34^{+}$ cells (HSCs) in the bone marrow by inducing HSC cycling followed by quiescence, and decreases HSC long-term repopulating activity through the activation of Toll like receptor (TLR) signaling in HSCs [30, 31]. In this view, the addition of G-CSF and IL-3 in cultured HSCs showed increased proliferation of HSCs like APE by stimulating intrinsic TLR signaling pathway (Figure 3) unlike GCSF, APE doesn't showed any proliferative activity on MNCs [32].

The *In silico* binding affinities of andrographo-lide, eltrombopag and pegacaristim to the structure build receptor was studied in order to understand the binding affinity native *in vivo* compound pegacaristim and medicinal drug eltrombopag. The binding affinity will be greater with least binding energy where andrographolide (-7.23 kcal/mol) has very close similarity withthe binding energy of eltrombopag (-7.00 kcal/mol) than pegacaristim (-1.23 kcal/mol) (Table 2). These data explain that natural drug andrographolide exhibits least binding energy with more affinity towards its receptor like medicinal drug eltrombopag for the activation of HSCs towards platelets generation.

These results explain that APE has anti-thrombo-cytopenic effect by the activation of HSCs and the increased population of HSCs are pushed more towards the formation of platelets as andrographolide induces release of thrombopoietin from liver. This APE is commonly used as drug in many countries for the treatment of thrombocytopenia.

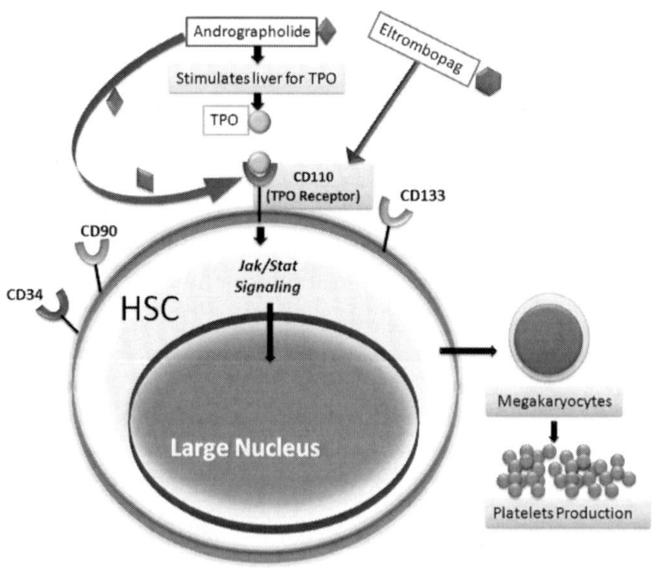

Figure 5. Mechanism of Andrographolide and Eltrombopag for the generation of platelets by binding to CD110 or thrombopoietin receptor.

CONCLUSION

A. paniculata extract (APE) showed HSCs proliferative property which is comparable with IL-3 and G-CSF stimulation. The andrographolide present in APE exhibited highest binding affinities compared to eltrombopag and Pegacaristim thrombopoietin receptor agonists. Also, the andrographolide present in APE stimulates the release of thrombopoietin from liver and thus differentiates HSCs in to mega-karyocytes further to platelet formation. This is the most probable molecular mechanism of APE used as drug to treat thrombocytopenia condition in many countries especially southern part of India.

ACKNOWLEDGMENTS

We sincerely acknowledge Sri Venkateswara Institute of Medical Sciences, Tirupati, Andhra Pradesh, India for providing facilities to carry out the work.

REFERENCES

[1] Visentin GP, Liu CY (2007) Drug induced thrombo-cytopenia. *Hematol Oncol Cin North AM* 21(4): 685-696.

[2] Greenberg EM, kaled ES (2013) *Thrombocytopenia*. Critical care nursing clinics of north America 25(4): 427-434.

[3] Aster RH (1966) Pooling of platelets in the spleen: role in the pathogenesis of "hypersplenic" thrombocyte-penia. *J Clin Invest* 45:645-57.

[4] Provan D, Stasi R, Newland AC, Blanchette VS, Bolton-Maggs P, Bussel JB, Chong BH, Cines DB, Gernsheimer TB, Godeau B, Grainger J, Greer I, Hunt BJ, Imbach PA, Lyons G, McMillan R, Rodeghiero F, Sanz MA, Tarantino M, Watson S, Young J, Kuter DJ (2010) International consensus report on the investigation and management of primary immune thrombocytopenia. *Blood* 115:168-86.

[5] Vadhan-Raj S, Verschraegen CF, Bueso-Ramos C, Broxmeyer HE, Kudelkà AP, Freedman RS et al. (2000) Recombinant human thrombopoietin attenuates carbo-platin-induced severe thrombocytopenia and the need for platelet transfusions in patients with gynecologic cancer. *Ann Intern Med* 132(5):364-8.

[6] Peeters K, Stassen JM, Collen D, Geet CV, Freeson K (2008) Emerging treatments for thrombocytopenia: Increasing platelet production. *Drug Discovery Today* 13:798-806.

[7] Stasi R (2012) How to approach thrombocytopenia," *Hematology Am Soc Hematol Educ Program*. 191-7.

[8] Erickson-Miller CL, Delorme E, Tian SS, Hopson CB, Landis AJ, Valoret EI, et al. (2009) Preclinical activity of eltrombopag (SB-497115), an oral, nonpeptide thrombo-poietin receptor agonist. *Stem Cells*. 27(2): 424-30.

[9] Julian MJ, Daphne W, Yanli D, Joanne U, Valerie K, David C, Connie L (2007) Erickson M. Phase 1 clinical study of eltrombopag, an oral, nonpeptide thrombo-poietin receptor agonist. *The American Society of Hematology* 109(11):4739-4741.

[10] Rates SMK (2001) Plants as a source of drugs. *Toxicon* 39:603-613.

[11] Ahmad N, Fazal H, Ayaz M, Abbasi BH, Mohammad I, Fazal L (2011) Dengue fever treatment with Carica papaya leaves extracts. *Asian Pac J Trop Biomed* 1(4):330-3.

[12] Beneytout JL, Nappez C, Leboutet MJ, Malinvaud G (1995) A plant steroid, diosgenin, a new megakaryo-cytic differentiation inducer of HEL cells. *Biochem Biophys Res Commun*. 207(1):398-404.

[13] Gao X, Wu j, Zou w, Dai Y (2014) Two ellagic acids isolated from roots of Sanguisorba officinalis L. promote hematopoietic progenitor cellproliferation and megakaryocyte differentiation. *Molecules* 19(4):5448-58.

[14] Akbar S (2011) Andrographis paniculata: a review of pharmacological activities and clinical effects. *Alternative medicine Review* 16(1):66-77.

[15] Leon IC, Anna PK, Rhun Y et al. (2012) Screening of anti-dengue activity in methanolic extracts of medicinal plants, *BMC complementary and Alternative medicine* 12(3):1-10

[16] Venkataraman N, Pamukuntala S, Banoth J, Boini P, Sampathi S et al. (2015) Platelet agumentation activity of andrographis paniculata extracts andrographolide against cyclophosphamide induced thrombocytopenia in rats. *Pharm Pharmacol Int J* 2(4): 00029.

[17] Jiyoung Y, Kyung-Baeg R, Zidan L, Guangrong L, Jian T, Seoungwoo S, Deokhoon P, Eunsun J (2015) The Antiaging Properties of Andrographis paniculata by Activation Epidermal Cell Stemness. *Molecules* 20, 17557-17569.

[18] Eissa LA, Gad LS, Rabie AM, El-Gayar AM (2008) Thrombopoietin level in patient with chronic liver diseases. *Ann Hepatol* 7(3):235-244.

[19] Srikanth L, Sunitha MM, Venkatesh K, Santhosh P, Chandrasekhar C, Vengamma B, Sarma PVGK (2015) Anaerobic glycolysis and HIF1a expression in haemato-poietic stem cells explains its quiescence nature. *J Stem Cells* 10: 97–106.

[20] Sunitha MM, Srikanth L, Santhosh Kumar P, Chandrasekhar C, Sarma PV (2016) Role of STPK in Maintaining the Quiescent Nature of Human CD34+ Stem Cells. *J Stem Cells* 11: 125–133.

[21] Guo T, Wang X, Qu Y, Yin Y, Jing T, Zhang Q (2015) Megakaryopoiesis and platelet production: insight into hematopoietic stem cell proliferation and differentiation. *Stem Cell Investig* 2:3.

[22] Sarma PV, Subramanyam G (2008) *In vitro* cardio-genesis can be initiated in human $CD34^+$ cells. *Indian Heart J* 60: 95–100.

[23] Sunitha MM, Srikanth L, Santhosh Kumar P, Chandrasekhar C, Sarma PV (2008) *In vitro* differentiation potential of human haematopoietic CD34+ cells towards pancreatic β-cells. *Cell Biol Int* 40: 1084–1093.

[24] Riedhammer C, Halbritter D, Weissert R (2016) Peripheral Blood Mononuclear Cells: Isolation, Freezing, Thawing, and Culture. *Methods Mol Biol* 1304:53-61.

[25] Luider J, Cyfra M, Johnson P, Auer I (2004) Impact of the new Beckman Coulter Cytomics FC 500 5-color flow cytometer on a regional flow cytometry clinical laboratory service. *Lab Hematol* 10(2):102-8.

[26] Nanthanit P, Nuchanart R, Sanit T, Somsak R, Mathuros R, Jutamaad S (2004) Determination and Variation of Three Active Diterpenoids in Andrographis paniculata (Burm.f.) Nees. *Phytochem. Anal* 15, 365–371.

[27] Agbonlahor O, Joyce EF, Osayemwenre E, Vincent I, Abiodun F, Peter Langer (2014) Harnessing the medicinal properties of Andrographis paniculata for diseases and beyond: a review of its phytochemistry and pharmacology. *Asian Pac J Trop Dis* 4(3): 213-222.

[28] Choi ES, Nichol JL, Hokom MM, Hornkohl AC, Hunt P (1995) Platelets generated *in vitro* from pro-platelet displaying human megakaryocytes are functional. *Blood* 85(2): 402-413.

[29] Qadrie ZL, Rajkapoor, Kavimani S (2015) Hepato-protective medicinal herbs and animal models for their screening - a review. *IJPS* 6(12): 5006-5028.

[30] Laura G. Schuettpelz1, Joshua N. Borgerding2, Matthew J Christopher2, Priya K. Gopalan3, Molly P. Romine1, Angela C. Herman1, Jill R. Woloszynek2, Adam M. Greenbaum2, and Daniel C. Link (2014) G-CSF regulates hematopoietic stem cell activity, in part, through activation of toll-like receptor signaling. *Leukemia* 28(9): 1851–1860.

[31] Nitsche A, Junghahn I, Thulke S, Aumann J, Radonić A, Fichtner I, Siegert W (2003) Interleukin-3 Promotes Proliferation and Differentiation of Human Hematopoietic Stem Cells but Reduces Their Repopulation Potential in NOD/SCID Mice. *Stem Cells* 21:236-244.

[32] Michelsen KS, Wong MH, Ko B, Thomas LS, Dhall D, Targan SR (2013) HMPL-004 (Andrographis paniculata extract) prevents development of murine colitis by inhibiting T-cell proliferation and TH1/TH17 responses. *Inflamm Bowel Dis* 19(1):151-64.

In: Stem Cells Origins of Cytokine Interactions
Editor: Prasad S. Koka
ISBN: 978-1-53615-302-6

Chapter 19

INTRINSIC VS. EXTRINSIC REGULATORY MECHANISMS OF LUNG STEM CELL BIOLOGY AND BEHAVIOR

Ahmed H. K. El-Hashash*, PhD
The University of Edinburgh-Zhejiang International Campus (UoE-ZJU),
Centre of Stem Cell and Regenerative Medicine
and Molecular Medicine Research Centre, Schools of Medicine
and of Basic Medicine, Zhejiang University, and Edinburgh Medical School,
University of Edinburgh, UK

The balance of stem cell self-renewal and differentiation maintains tissue homeostasis in different organs. Excess of stem cell self-renewal may lead to tissue hyperplasia and/ or tumorigenesis; while excess of differentiation may lead to tissue degeneration and/ or tissue aging. Asymmetric stem cell division, producing one stem cell and one differentiating cell, is a simple way to maintain the balance between stem cell and differentiated cell populations.

Several recent studies suggest that the biology and behavior of lung stem and progenitor cells is controlled by both extrinsic signals and intrinsic factors. The intrinsic aspects of stem cell control provide the context in which extrinsic signals are interpreted [1, 2, 3, 4].

During embryogenesis, stem and progenitor cells undergo a mixture of symmetric and asymmetric divisions [5, 6, 7]. It is currently difficult to distinguish between these two types of divisions at the cellular level. One possibility to do this is by looking at

* Corresponding Author Email: hashash05@yahoo.co.uk.

differences in the spindle orientation or differential inheritance of cytoplasmic or membrane-bound proteins such as the cell fate determinant Numb and atypical PKC zeta [7, 8, 9, 10, 11]. Cells divide asymmetrically in response to either extrinsic or intrinsic fate determinants. In the case of extrinsic fate determinants (i.e., the microenvironment), the daughter cells are placed in different micro-environments, thus the two daughters take on different fates. In the case of intrinsic fate determinants, cytoplasmic cell fate determinants (e.g., Numb) are asymmetrically localized within a cell and subsequently segregate differentially into the two daughter cells, thus the two daughters take on different fates [12, 13]. Yet, some molecules and signals have been suggested to function indirectly in the promotion of different aspects of stem/progenitor cell behavior such as the self-renewal or differentiation based on comparing changes in the number of these stem/progenitor cells between mutant and control lungs [14].

The proper balance between self-renewal and differentiation of lung-specific stem and progenitor cells at the distal epithelial tips is absolutely required for normal lung morphogenesis [3, 4, 7, 10, 11]. Both cell polarity and the orientation of mitotic spindles are crucial for proper self-renewal and differentiation of tissue-specific epithelial cells. In addition, proper cell polarity and mitotic spindle orientation can clearly influence various types of normal physiological processes such as epithelial branching morphogenesis, and both epithelial cell proliferation and differentiation. Therefore, understanding the behavior of distal epithelial stem and progenitor cells of the embryonic lung could identify innovative solutions to restoring normal lung morphogenesis. Little is currently known about the establishment of both proper cell polarity and mitotic spindle orientation, as well as the segregation and inheritance of cell fate determinant factors in the distal epithelial stem and progenitor cells of the embryonic lung. Findings from our laboratory showed that embryonic lung distal epithelial stem and progenitor cells are polarized and highly mitotic with characteristic perpendicular cell divisions [2, 3, 10, 11]. In addition, we have reported that the cell fate determinant Numb has a characteristic asymmetric distribution at the apical side of distal lung epithelial stem and progenitor cells. Numb was found to segregate to one daughter cell in most mitotic distal lung epithelial stem and progenitor cells. This suggests that these cells have an asymmetric mode of cell division [1, 3, 5, 6, 10, 11, 13].

Several signaling molecules are known to control lung growth and, therefore, probably affect stem/ progenitor cell proliferation. For instance, the expression of the transcription factor, thyroid transcription factor 1 (Ttf-1/Nk x 2.1) marks the lung lineage commitment in the early embryo. Ttf-1 is critical for the development of distal lung progenitor cells [15]. In addition, Sox9, which is HMG box transcription factor, is abundant in the distal lung epithelial stem and progenitor cells from E11.5 to E16.5 in the mouse embryo [16]. However, lung-specific conditional deletion of Sox9 has no effect on epithelial stem/ progenitor cell behavior [17]. Moreover, N-myc transcription factor functions to maintain the distal lung epithelial stem/progenitor cell population undifferentiated, and probably promotes the self-renewal of these stem/progenitor cells [18].

The forkhead/winged helix (fox) transcription factors have mutant lung phenotypes, and may maintain and promote lung epithelial progenitor cell self-renewing divisions. A conditional deletion of both *foxa1* and *foxa2* genes in the lung results in small lungs with decreased rates of cell division [19]. A similar phenotype was reported after conditional deletion of both *foxp1* and *foxp2* genes, which are enriched in the distal stem/epithelial progenitor cells. Thus, in *foxp22/2*; *foxp11/2* double mutants, the lungs are hypoplastic, with an inhibited proliferation, but normal proximal–distal patterning [20].

A network of growth factors is essential for both development and function of the lung and other organs. There are five major signaling pathways that control many aspects of embryogenesis: fibroblast growth factor (FGF), Hedgehog, Notch, Wnt and transforming growth factor-beta (TGF-beta) family. These signaling pathways can also control the self-renewal/proliferation of distal lung epithelial stem and progenitor cells. These signaling pathways are needed, at some level and at some time, to drive differentiation in many different stem cell lines, but it is still unknown what exactly it is that drives specific differentiation into the multiple cell types that have distinct roles in the lung.

Wnt2/2b and beta-catenin signaling are necessary and sufficient to specify lung progenitor cells in the foregut [21]. Mutant mice for Wnt2/2b have a complete lung agenesis. The expression of the earliest marker of the lung endoderm; Nkx 2.1, cannot be detected in these mice. This phenotype is recapitulated by an endoderm-restricted deletion of beta-catenin, which suggests that Wnt2/2b signaling through the canonical Wnt pathway is required to specify lung endoderm progenitors within the foregut [21, 22]. Wnt5a, a prominent Wnt ligand, is also highly expressed in and around the distal lung epithelial tips. *Wnt5a*$^{-/-}$ mutant lungs have a general increase in the levels of cell proliferation, and an additional branch of the conducting airways [23].

FGF signaling is critical for the specification of the lung lineages distal to the trachea. FGF10 is expressed by the lung mesenchyme and functions as a chemotactic factor during branching morphogenesis. Overexpression of FGF10 maintains epithelial stem/progenitor cell proliferation, and leads to a metaplastic differentiation of goblet cells [24]. In addition, FGF10 coordinates both alveolar smooth muscle cell formation and vascular development [25]. Retinoic acid signaling targets the gene expression of *Fgf10* through TGF-beta pathway and, therefore, is critical for both lung stem/progenitor cell expansion and the primary lung bud formation [26].

SHH is expressed in the distal epithelium and may control lung progenitor cell proliferation and branching [27]. Ectopic overexpression of Gli2, which is a mediator of SHH activity, in the lung mesenchyme results in an increase in both SHH activity and cell proliferation through regulation of cyclin expression in the lung [28]. Furthermore, autocrine BMP signaling is likewise important for proliferation of the distal epithelial progenitor cell compartment.

HuR is RNA-binding protein (RBP) that regulates lung mesenchymal cell responses during epithelial branching morphogenesis. HuR regulates both *Fgf10* and *Tbx4* transcripts, and its deletion can lead to a morphogenesis arrest of distal bronchial branches during the pseudo-glandular stage of lung development [29]. In addition, mesenchymal nuclear factor I (NFI) B can regulate lung cell proliferation and differentiation during morphogenesis [30]. Furthermore, Notch signaling pathway controls lung cell fates [31].

A member of the micro-RNAs family, miR-17-92, is critical for a proper lung development. miR-17-92 may induce the self-renewal of lung epithelial progenitor cells at the expense of their differentiation fate [32]. Downregulation of miR-17 and its paralogs, miR-20a/miR-106b in lung epithelial explants grown in culture remarkably changes the effect of key regulators of lung epithelial cell development, including FGF10 and beta-catenin signaling [33]. miR-17-92 expression is negatively regulated by a histone deacetylase 3 (Hdac3)-mediated epigenetic pathway, which is important for the correct remodeling of the developing lung saccules to form primitive alveoli during lung development [34].

In conclusion, the proper balance between stem cell self-renewal and differentiation in the lung is absolutely required for normal lung formation, development and function. Accumulated recent studies have demonstrated that the processes of stem cell self-renewal, differentiation and apoptosis in the lung are tightly regulated by several signaling pathways, molecules and transcription factors that have been discussed in this short review. Further studies are still needed to determine how these signaling pathways or molecules interact to control lung stem cell self-renewal, differentiation and apoptosis as well as to regulate the delicate balance between these processes that may help to identify new targets for the therapies of different lung diseases.

REFERENCES

[1] El-Hashash A (2013). Lung Stem Cells: Mechanisms of Behavior, Development and Regeneration. *Journal of Anatomy and Physiology* 3:119.

[2] Ibrahim A and El-Hashash A. Lung stem cell behavior in development and regeneration. *Edorium Journal of Stem Cell Research and Therapy*. 2015; 1:1-13.

[3] Ku J, El-Hashash AH. Molecular Control of the Mode of Cell Division: A View from Mammalian Lung Epithelial Stem Cells. A. *Journal of Anatomy* 2016; 3:3-6.

[4] Berika M, Ku J Huang, H, El-Hashash, AH. "Gene and Signals Regulating Stem Cell Fate." In *Developmental and Stem Cell Biology in Health and Disease*. Bentham Science Publisher, USA 2016. Vol. 1, 36-48.

[5] Berika M, Elgayyar M and El-Hashash A (2014). Asymmetric cell divisions of stem cells in the lung and other systems. *Front. Cell Dev. Biol.* (Stem Cell Treatments) 2:33.

[6] Elshahawy S, Ibrahim A, Soliman, et al. "Behavior and Asymmetric Cell Divisions of Stem Cells." In *Developmental and Stem Cell Biology in Health and Disease.* Bentham Science Publisher, US 2016. Vol. 1, 81-104.

[7] Chen F, Fine A. Understanding Lung Development, Injury, and Repair. *The Am. J. Path.* 2016; 186 (10): 2518.

[8] Morrison SJ, Kimble J. Asymmetric and symmetric stem cell divisions in development and cancer. *Nature.* 2006; 441:1068–1074.

[9] Wang Y, Maciejewski BS, Soto-Reyes D, et al. Mechanical stretch promotes fetal type II epithelial cell differentiation via shedding of HB-EGF and TGF-alpha. *J Physiol.* 2009; 587(Pt 8):1739–53.

[10] El-Hashash A, Warburton D. Cell polarity and spindle orientation in the distal epithelium of embryonic lung. *Dev Dyn.* 2011; 240(2):441-5.

[11] El-Hashash A, Warburton D. Numb expression and asymmetric versus symmetric cell division in embryonic distal lung epithelium. *J. Histochem.* Cytchem. 2012; 60 (9): 675-682.

[12] Yamashita Y· Asymmetric stem cell division and pathology: insights from Drosophila stem cell systems. *J Pathol.* 2009; 217(2):181-5.

[13] El-Hashash A (2014). New insights into the regulation and functional significance of Numb in lung stem cells during organogenesis. *Austin Journal of Anatomy* 1(4):2.

[14] Rawlins EL. Lung epithelial progenitor cells: lessons from development. *Proc Am Thorac Soc* 2008; 5:675–681.

[15] Kimura S, Ward JD, Minoo P. Thyroid-specific enhancer-binding protein/ transcription factor 1 is not required for the initial specification of the thyroid and lung primordia. *Biochemie.* 1999; 81: 321-328.

[16] Liu Y, Hogan BL. Differential gene expression in the distal tip endoderm of the embryonic mouse lung. *Gene Expr. Patterns.* 2002; 2: 229-233.

[17] Perl A-K, Kist R, Shan Z, et al. Normal lung development and function after *Sox9* in activation in the respiratory epithelium. *Genesis.* 2005; 41(1):23–32.

[18] Okubo T, Knoepfler PS, Eisenman RN et al. Nmyc plays an essential role during lung development as a dosage-sensitive regulator of progenitor cell proliferation and differentiation. *Development* 2005; 132(6):1363-74.

[19] Wan H, Dingle S, Xu Y, et al. Compensatory roles of Foxa1 and Foxa2 during lung morphogenesis. *J. Biol. Chem.* 2005; 280:13809-13816.

[20] Shu W, Lu MM, Zhang Y, et al. Foxp2 and Foxp1 cooperatively regulate lung and esophagus development. *Development.* 2007; 134:1991–2000.

[21] Goss AM, Tian Y, Tsukiyama T, et al. Wnt2/2b and beta-catenin signaling are necessary and sufficient to specify lung progenitors in the foregut. *Dev. Cell.* 2009; 17: 290–298.

[22] Harris-Johnson KS, Domyan ET, Vezina CM, et al. beta-Catenin promotes respiratory progenitor identity in mouse foregut. *Proc. Natl. Acad. Sci. USA*. 2009; 106: 16287–16292.

[23] Li C, Xiao J, Hormi K, et al. Wnt5a participates in distal lung morphogenesis. *Dev Biol.* 2002; 248, 68-81.

[24] Nyeng P, Norgaard GA, Kobberup S. et al. FGF10 maintains distal lung bud epithelium and excessive signaling leads to progenitor state arrest, distalization, and goblet cell metaplasia. *BMC Dev. Biol.* 2008; 8:2.

[25] Ramasamy SK, Mailleux AA, Gupte VV, et al. Fgf10 dosage is critical for the amplification of epithelial cell progenitors and for the formation of multiple mesenchymal lineages during lung development. *Dev. Biol.* 2007; 307:237-247.

[26] Chen F, Desai TG, Qian J, et al. Inhibition of Tgf beta signaling by endogenous retinoic acid is essential for primary lung bud induction. *Development.* 2007; 134: 2969–2979.

[27] Pepicelli CV, Lewis PM, McMahon AP. Sonic hedgehog regulates branching morphogenesis in the mammalian lung. *Curr Biol.* 1998; 8:1083–1086.

[28] Rutter M, Wang J, Huang Z, et al. Gli2 influences proliferation in the developing lung through regulation of cyclin expression. *Am J Respir Cell Mol Biol.* 2010; 42(5):615-25.

[29] Sgantzis N, Yiakouvaki A, Remboutsika E, et al. HuR controls lung branching morphogenesis and mesen-chymal FGF networks. *Dev Biol.* 2011; 354(2): 267-79.

[30] Hsu YC, Osinski J, Campbell CE, et al. Mesenchymal nuclear factor I B regulates cell proliferation and epithelial differentiation during lung maturation. *Dev Biol.* 2011; 354(2):242-52.

[31] Morimoto M, Liu Z, Cheng HT, et al. Canonical Notch signaling in the developing lung is required for determination of arterial smooth muscle cells and selection of Clara versus ciliated cell fate. *J Cell Sci*. 2010; 123(Pt 2):213-24.

[32] Lu Y, Thomson JM, Wong HY, et al. Transgenic over-expression of the microRNA miR-17-92 cluster promotes proliferation and inhibits differentiation of lung epithelial progenitor cells. *Dev. Biol*. 2007; 310: 442–453.

[33] Carraro G, El-Hashash A, Guidolin D, et al. miR-17 family of microRNAs controls FGF10-mediated embryonic lung epithelial branching morphogenesis through MAPK14 and STAT3 regulation of E-Cadherin distribution. *Dev. Biol.* 2009; 333(2): 238-50.

[34] Wang Y, Frank D, Morley MP. et al. HDAC3-Dependent Epigenetic Pathway Controls Lung Alveolar Epithelial Cell Remodeling and Spreading via miR-17-92 and TGF-β Signaling Regulation. *Dev. Cell.* 2016; 36: 303–315.

In: Stem Cells Origins of Cytokine Interactions
Editor: Prasad S. Koka
ISBN: 978-1-53615-302-6

Chapter 20

QUALITY MANAGEMENT IN THE MANUFACTURE OF STEM-CELL-BASED-PRODUCTS

Chandra Viswanathan, **Sushilkumar Ramdasi**†,*
***Rohit Kulkarni**‡ and **Abhijit Bopardikar**§*
Research and Development Team,
ReeLabs Private Limited, Mumbai, India

ABSTRACT

Stem cells have been classified as drugs and so will necessitate compliance with regulations. Cell therapy initially started in the form of a bone marrow transplant for hematopoietic reconstitution to produce blood, a complex liquid tissue. Several years later, this experience has culminated in the administration of specialized cells with one or more intended therapeutic effects; this is a hallmark of cellular therapy.

We are currently witnessing the emergence of this 'new age medicine' which uses a variety of stem cells for several unmet medical needs. Stem cells function by one or more mechanisms to regenerate the damaged tissue / organs by stimulating tissue-resident stem cells and by secreting different cytokines and growth factors. They also possess the potential to recruit themselves in the damaged part of the body and play a pivotal role to reinstate degenerated area through neoangiogenesis and neovascularization.

While it is no secret that cells have been used as drugs for decades in one or the other forms, testing a specific therapeutic effect of cells objectively in the form of clinical trials is relatively recent. Lessons in the establishment of quality systems have been learnt as a part of establishment of blood banking and subsequently, cord blood banking, with emphasis on key components in the assurance of the safety and potency of cell therapy

* Email: chandra.viswanathan1@gmail.com.
† Corresponding Author Email: sushbiotech@gmail.com.
‡ Email: rrk@reelabs.com.
§ Email: asb@reelabs.com.

preparations. As more stakeholders in the academic medical centers and private companies are vying to be part of this new age medicine, there is a compelling need to get to the granular details of the development of the infrastructure, method developments, research requirements, analytical tools to assess and quantify efficacy, before it becomes part of main stream medicine. In this article, we discuss the various challenges specific to this industry, with reference to establishment of a quality management program.

Keywords: stem cells, quality assurance, quality management system, cell therapy, AABB

Background

Considerable challenges are to be overcome before you can think of a launch of stem cell products for clinical use. Development and implementation of quality management is not just limited to development of a quality vision, and a quality policy, but includes a host of other aspects like biological raw material sourcing, making of bio-repositories, specifics of product manufacturing, cryopreservation, shipping and logistics, validation plans, implementation and finally, standardization of each of these aspects [1]. This learning is more relevant today than ever before, because we observe sketchy regulatory framework in a few countries, while a few others have some stem cell guidelines in place. It is also worth noting that many countries including India are entering into stem cell based clinical trials that makes it imperative for us to review and update ourselves with best practices available elsewhere. Having a deeper understanding of the quality aspects will compel management and leadership to invest suitably so that it poses no surprise at a later stage. We strongly feel that a, comprehensive and robust quality strategy must be available right from the stage of conception to avoid disappointment, loss of resources, time and opportunity. A good strategy not only attracts a larger customer base but also safeguards the organization from medico-legal battles.

This could promote trade and exchange of cellular therapy products across borders if there is harmonization of understanding across teams and collaborators, especially now, as many researchers are moving their clinical trials to countries where the regulatory environment is more conducive and speedier. Through this paper, we would also insist on the establishment of universal biological quality control methods on the basis of high precision and accuracy with special emphasis on participant safety.

Basic Concepts of Quality Management System

Quality Management System (QMS) is defined as a comprehensive set of policies, processes and procedures with a well laid down plan for, execution and implementation of quality standards for the products, services and product development [2, 3]. QMS when aligned with organizational goals is expected to impart confidence with respect to policies, processes, procedures, customer focus, quality commitment, regulatory compliance and contribute to the safety of the product, environment, customers and workers. The Stem cell industry-specific QMS is expected to evolve along with research progress, very similar to what has been witnessed in other areas.

Quality, safety and efficacy are the pivotal traits of products and services on the basis of which any life science organization builds its reputation [4]. Out of these three traits, safety and efficacy are inherent to any drug or cell type. It is only the quality trait which is extrinsic and largely dependent on many variables such as machines, raw materials, skilled personnel, manufacturing conditions, infrastructure, storage conditions and testing specifications. There are several parameters that are specific to stem cells, and the recognition of this is increasing. The overall awareness is on the rise given the fact that there are about 24000 stem cell publications in the year 2016 as compared to about 15000 in the year of 2010 [5].

The QMS Components

QMS encompasses quality assurance, quality control, and quality improvement / process improvement of all activities that lead to a product or service. Quality assurance (QA) function involves all activities designed to ensure that the policies, processes and procedures are made effective in line with national and/or international regulatory and/or available quality guidelines to maintain the quality of products and services [1, 6]. QA also takes active part in designing new products and services to align them with the organizational goals but within the regulatory framework. It also serves to maintain the integrity of policies, processes and procedures. The International Standards Organization (ISO) developed standards for QMS which are based on seven basic 'Quality Management Principles' (QMPs) which can be used as a reference point [2]. However, Cellular therapies being relatively new, several 'not for profit' organizations like the AABB (American Association of Blood Banks) and the FACT (Foundation for the Accreditation of Cellular Therapy) took the lead in shaping the quality program [1]. In the absence of a firm regulatory pathway, this turned out to be a blessing to the stem cell community. It has allowed a lot of flexibility within the framework with ample opportunities for improvement and learning. A beginner in cellular therapy can get a fair idea of how various operational and policy aspects are to be understood and evaluated. Some of these are, but not limited to the following:

Vision and Mission of the Organization

The vision and the mission of an organization are critical aspects that define the motive, direction and ethics of an organization. The Vision statement can be revised constantly as and when the set goals are achieved, however, the mission statement remains the same. Hence due care should be taken while designing the vision and mission statement to align them with the organizational objectives [2]. The Regenerative Medicine space is very huge and varied; players can pick and choose what they want to excel in. Some organizations discuss their value systems in details, while others are clear on which domain they will work on, for e.g., their desire to focus on adult cells, or iPS (Induced Pluripotent Stem Cells) or embryonic, or mesenchymal stem cells, while few choose to focus in the clinical trials' segment or in reagent manufacture and so on. This is a very important step for your identity in the domain.

Quality Manual and Quality Policy of the Organization

This is a well narrated document, with a certain specific expiry, and a definite review frequency, that outlines all the policies that company's top management will apply, educate and enforce through the company's quality team. The Quality Manual (QM) & Quality Policy (QP) document outlines policies, processes and procedures on handling manpower, policies for discard of biowaste, raw material procurement (bone marrow, fetal cells, cord tissue, embryo etc.), reordering levels (based on shelf life and cost). Interim and final storage, (whether store at one location or more) documentation practices (electronic or otherwise), customer interface (marketing needs), software development and validation, management review meetings (frequency and format), format for SOPs, collaboration, partnerships and so on. This list is very exhaustive. All these form a part of the quality document [1].

Customer Focus

Regenerative Medicine and cell based therapies is still in its infancy and is yet to come of age. We well know that cellular therapies may bring variable results in few patients, with marginal or no effect in others, unlike the mainstream pharmaceutical drugs. An explicit understanding of whether and how much the patient could benefit out of these therapies is very critical. In this context, attempt is to identify the right customers, and put an effort to inform customers of what can be expected from cellular therapies, and exceed their expectations if possible. This will be the key for maintaining sustainability.

In India, the ICMR (Indian Council of Medical Research) in its most recent draft guidelines on stem cell research in 2017 [7] has chalked out various obligations to be fulfilled by those in stem cell research. Similarly, the starting point of research is from the donors of biological tissues. An appropriate and ethically designed informed consent will give every information that the donor needs to know, including information on their rights and responsibilities. Researchers must obtain consent from the ethics' committees on the content, wording and language of the document; taking utmost care that no one is threatened or coerced to part with a tissue even if it is a biological waste. Similarly, one must share every information that will generate confidence and ensure safety of patients at all times. The objective of the quality group is to ensure all activities are consistent with the laid down policies, and in compliance with the national guidelines [8]. There should be adequate measures to help patient and donor privacy, which is more nuanced than it may appear. Cells or tissues with a certain specific disease can be collected from donors for research on that particular disease, but an explicit understanding of the donor benefits, if and when it gets translated clinically, and commercialized, is very important. A lot of emphasis on customer care, donor care, and dissemination of information to them in a timely manner is explained in the guidance document of the AABB and the Netcord [1]. Donation of Limbal tissues, Placental membranes, Cord tissue, Cord blood stem cells etc. for public use, will need this additional attention. Leveraging the power of patient community can be enormously helpful when we plan research using donated cells for a specific disease.

Leadership

Leaders at all levels must align their efforts and resources to the mission and vision statement of the organization to achieve the optimum environment that will propel growth. Leadership must be given realistic time lines so that there is no pressure on researchers to hide unfavorable findings or share unsuccessful experiences. Involvement of the top management should be very tangible, encouraging and consistent, if continuous quality improvement is to be expected. Responsibilities of the management are to be made known to the team and are written clearly in the quality document and distributed to all team members [2]. Regular interactions with the management teams and ability to pass on information to the employees at the lower most level helps confidence building especially in the Stem cells' research area where we have experienced surprises, delays, disappointments and challenges!!

People Engagement

Skilled, competent, engaged, accountable, responsible and empowered employees are an asset to the organization and will serve a significant role in achieving quality and mission statement goals. The organization's quality document thus defines the training goals, job responsibilities, annual performance review plans and employee growth very meticulously. This is a great way to have inclusive growth [1, 2]. Stem cells' based conferences, workshops and tutorials are very relevant and must be encouraged, otherwise people tend to apply their traditional 'Pharma' based understanding on to stem cells, which is not acceptable. Specific training is to be imparted as per needs, on a continuous basis. If the organization does not care for continuous growth of its employees, we well know that the competitor benefits. Satisfied and motivated staff alone contributes to innovation, learning, and cost reduction.

Various Approaches for Systematic Quality

A 'process based' quality approach in cellular based medicine has been very useful. The processes and activities and their inter-relatedness needs to be understood well to effectively and efficiently achieve reproducibility and consistency [2]. While we learnt that a system based approach is more desired by senior management to do a root cause analysis, a mix of both, has been more rewarding in a day-to-day practice. These usually lead to corrective actions and preventive actions which is one of the important objectives of the exercise. More recently, like us, increasing number of cellular therapists are embracing the 'risk based approach' to rule out possible un-desirable outcomes. A 'Risk based approach,' for products and services therefore forms a significant part of the QMS. Addressing contamination, batch failures, product mix up, issues with parallel processing versus serial processing, have all been well analyzed using this approach [9]. In cellular therapies, there are innumerable variables as mentioned earlier; variations exist in practically every aspect beginning from raw materials, processes, testing methods, excipients and other enablers, reagents etc. Hence, understanding the risk due to variability is paramount. We felt that this approach is most suitable if applied right from the time of product development. This has been a point of challenge and debate, and has been the subject matter of several conferences. The aim is always to keep as many factors constant as possible so as to be able to predict a reasonable outcome, batch after batch or case after case. No two similar looking cell types behave similarly in growth or in differentiation; as an example, Mesenchymal stem cells (MSCs) from different sources like adipose tissue and bone marrow look similar, but do show mild variations in phenotypes, growth potentials and differentiation capabilities [10]. Thus, one process or derivation protocol does not fit all cell types.

Continuous Improvement

Improvement and innovation is growth and is a 'must' for all industries. As organizations needs to keep abreast with the dynamically changing and evolving business approaches and yet remain competitive, one must factor this in the policy statement [2]. Thus, change is the only 'constant.' An improvement also opens new ways for new opportunities and is an ongoing process. Several examples of this approach can be cited; for instance although there are well tried and tested cord blood stem cell enrichment methods used by most cord blood banks, mild modification of the cord blood sedimentation and centrifugation speed often yielded better stem cell counts. Similarly, minimization of incubation times of cell culture showed no adverse effects on the outcome, decreasing amount and frequency of cytokine helped cell propagation, usage of newer cell culture media, use of cost effective culture options, etc. are examples of meaningful process improvements. This, in our own experience, is an ongoing process.

Evidence-Based Decision Making

A regular analysis and evaluation of accrued data and dissemination of the information to those concerned, is very important. A periodic review will throw light on good and bad trends. All accepted improvements, however incremental are generally for a business, technical, cost or time reasons. Whether or not to establish and run all a newer transfusion transmissible test on donor samples will depend on the epidemiological significance. For example, Chaga's disease which is prevalent in Latin America and United States of America, will necessitate their biological donors' blood be tested for its freedom, but may not be applicable for donors in India [11]; whereas, testing for malarial parasites in donor blood is more significant in endemic zones like India. There are innumerable such situations in the cell therapy context that one can easily spot. QMS of the organization must value this finding and implement only what is contributory to quality. This has to be, however, supported by adequate data.

Relationship Management

To achieve sustained progress, an organization must be proactive in maintaining relationship with all its stakeholders like vendors, suppliers, service partners etc. [2]. Guidelines outlined by the AABB and the FACT attach high significance to vendor audits, vendor rating, vendor behavior and punctuality and integrity of partners. This may sound qualitative, but a semi qualitative rating scale can be developed by organizations employing commonly used criteria. This removes subjectivity and bias, to a large extent. AABB and other guidelines also emphasize the need to have a 'MoU' (Memorandum of Understanding) between the service seekers and providers, be it the waste management agencies, or the HLA laboratories and the likes, to explicitly document all terms of the contract. Going further, in the 'MoU' for outsourcing of tests to a laboratory, one must specify the use of the methods and kits [1]. Every Quality Policy should focus on this aspect, as these external 'customers' and service providers, left to themselves, could shake the balance resulting in deviation.

Quality Control (QC)

It involves activities pertaining to testing and evaluation of product against the set specifications and reporting the results to appropriate stakeholders. QC is specifically focused on finding the defects and reporting them to production, administration and quality assurance (QA) [12]. The pursuit of quality in disciplines such as stem cell science is no easy task. Working with living cells, establishing their functionality and quantifying it continue to be very daunting. It has been described as something close to the quest for the Holy Grail. The industry has been facing a great challenge in setting up standards, however minimum, followed by method development, consistency, repeatability, and acceptable range of results.

It is of utmost importance to assure stakeholders for the accuracy and precision of QC, especially in multicenter clinical trials involving stem cells by minimizing risk to participants and setting stringent statistical power. The current guidelines in this area are focused on precision of testing methods at individual research/clinical trial centers thereby minimizing testing/re-testing variability. Moreover, today, increasingly, the services and products of health care industry are constantly on the lookout for innovative and newer performance standards as a process to measure quality. While there are many aspects that are measurable and reasonably comparable, there are some that are yet subjective. The QC testing outcomes and harmonization of global methods requires strong coordination and agreement between QC laboratories. The accuracy, reproducibility and precision of QMS depends on the interactions among people involved in different functions like testing equipment variability, calibration, and maintenance; biomechanical factors; participants' cooperation, motivation, and effort during procedure; and knowledge, skills, and training of laboratory personnel.

Cellular therapists have to address this challenge well, if they have to show effectiveness, and quantify them. While performing quality checks, there are challenges peculiar to stem cells. In the allogenic mode, cells from one donor can be given to many patients. The input variability can be reduced by careful selection of comparable input material to the process. In parallel, the variability due to in-process conditions can be dealt with by incorporating manufacturing processes altogether in automated mode which is expected to confer better control and reproducibility [13]. This 'off-the-shelf' so called non-self-source strategy in cell therapeutics mimics closely to biopharmaceutical practices [13, 14] in terms of establishment of product stability in advance.

On the contrary, cell therapeutics based on autologous model poses its own set of challenges. It includes optimizing custom made processes right from logistics through processing, quality control checks and delivering cost-effective cellular therapy. To add to the complexity, the short term preservation methods create further challenges [15]. There might be cases where allogenic cells are transported to the clinic in non-cryopreserved state which demands on-the-production-day delivery to avoid any cell losses. This may result in inevitable cost elevation. This puts pressure on researchers to find effective production strategies directed for allogenic cell therapy to achieve reproducibility. It is nearly impossible to have common production process since cell therapy products are governed by varied indications, different mode of delivery and cell type diversity which demands custom made process development and validation [13]. A close QA oversight on this is an imperative, and will minimize deviations.

Need for Reference Centers (Centers of Excellence), Benchmarks and International Standards

Non availability of controls and standards for quality control experiments is a matter of great concern. We know that use of controls during the manufacturing and testing of clinical-grade cells is very critical. There are a few companies that are making efforts in this direction, but challenges associated with transportation of live cells exist. It is desirable that affordable standards be made by a centralized domestic agency, on a regular basis, to avoid problems due to expiry, cell death, poor preservation, loss in transit etc. Exchange or purchase of specific cell lines across borders must also be made easy by governmental agencies, to support research. Each country can have some 'centers of excellence' that can handle this responsibility. Several hurdles can be overcome effectively by participating in accreditation programs. These accreditation agencies have developed certain quality standards, and provide guidance on policies, processes and procedures prevalent in the cellular therapy business. The accreditation serves as a preparatory training that helps comply with regulatory guidelines as and when they come into force. It is necessary to harmonize various aspects and parameters put out by regulatory as well as accreditation agencies and follow them with stringency. It has been a great learning experience for our teams, while complying with standards from accreditation agencies. A well planned quality program successfully helps clearing the assessments.

Cost Considerations

The 'cost of quality' concept is debatable and many authors establish the same in many different ways. Some authors assume it as the costs due to the lack of quality. Others maintain it as a cost of producing 'good quality' products. Another group of authors refer cost of quality to the term 'cost of poor quality.' Crosby, in his book, refers cost of quality as 'price of conformance' as the prevention and appraisal cost and 'price of nonconformance' as the failure costs [16, 17]. Both views are acceptable. Thus, the cost of quality is the sum of all the expenditures incurred by an organization in prevention of poor quality. In other words, the cost incurred to ensure and evaluate that the quality requirements are being met as specified in the quality policy. The cost of quality can be subdivided in to three sub-parts as, costs incurred for prevention, appraisal and failure. This essentially refers to prevention of errors, elevation of quality and bringing down incidences of failure. This is an ideal strategy to increase profit margins at the same time as gaining valuable customer confidence. Since this is an ongoing expenditure, managements must make provision for this in their annual budgetary plans.

Validation Plans

Validation outlines the principles involved in the qualification of a facility, defining the areas and systems to be validated, and provides a written program for achieving and maintaining a qualified facility. Process validation consists of data gathered throughout the design and manufacturing of a stem cell product in order to confirm that the process can reliably put out products of a determined standard. This is to ensure that varied inputs lead to consistent and high quality outputs. Process validation is an ongoing process that must be tweaked if and when relevant manufacturing feedback is gathered. End-to-end validation of production processes is essential in determining product quality because quality cannot always be determined by finished product inspection alone. Process validation can be broken down into three steps as process design qualification, process qualification, and continued process verification [18]. This will indirectly define the quality of reagents, incubation periods, equipment status, raw material specifications, storage specifications, transport requirements etc. and also define the permissible limits of variation.

Challenges with Scalability, and Animal -Free Reagents

In cellular therapy practice, safety of final products remains one of the main concerns. Use of animal-source free reagents and media is a pre-requisite for the production of clinical grade stem cells [19]. Most researchers currently use reagents and media derived from animal source, which is a huge compromise with safety [20]. For example fetal bovine serum (FBS) which is routinely used for stem cell growth, may harbor potential xenogenic antigens, infectious agents and bovine viral agents which could transmit the infection to the recipient [21, 22, 23]. It is also to be noted here that the precise composition of FBS is not clear and there are batch to batch variations due to pooling of plasma from multiple donors [24]. To avoid these risks, it is customary to validate the use of media and reagents prior to use, and set some safety steps ahead of time. A better scenario is when you can make products without bovine derivatives.

The processes should also ensure microbiological safety and avoid genomic instability driving transformation and senescence or decrease of cell functions (immune-regulation, differentiation potential). This gets demonstrated through validation of processes and long term studies. The culture processes therefore should be reproducible, robust, and efficient.

Another pertinent issue is to address the 'up-scaling' of clinical grade stem cells. Most therapies need cells in millions, and that must come from as minimum passages as possible. Thus, a single batch of stem cells at one passage must yield as many as possible stem cells. This requires that the systems, culture vessels and related processes must be up-scaled to suite bulk production. This bulk production, apart from meeting the set specifications, must be reproducible in nature. There are numerous approaches for up-scaling of stem cell production including use of multi-stacked flasks, use of disposable bioreactors, bead-based bioreactors, and traditional stainless steel bioreactors etc. [25]. The up-scaling continues to be challenging since there are different factors involved like paracrine signaling, phenotypic stability of the cells and the media and reagents used for the same. Defining quality in up-scaling is thus continues to be a great challenge!

QMS Implementation

A well-articulated document is job half done. That the devil is in the details becomes obvious during implementation and training. Typically, QMS discharges its functions through many tools and techniques that include but not limited to periodic review of documentation, document control, policy framework, quality manual (QM), QA review reports, standard operating procedures (SOPs) reviews, contingency operations, disaster management plan (DMP), master formula records (MFR), testing specifications, final product specifications, certificate of analysis (CoA), validations, charts, log sheets, algorithms check lists etc. The most critical part of implementation is a review of internal and external audits against the set of standards, review of stability studies, review of proficiency studies, review of testing procedures, review of raw material testing results, review of purchase procedure for critical raw materials etc. [1, 26]. This is a long journey, and one can make constructive changes all along. Companies who have reported lesser batch failures and lesser non compliances are those with very tough internal QA teams who keep a close check on overall compliance at all the times. We well know that most of the existing regulations were based on those developed for drugs or proteins or genetic therapies, whereas the use of cells as therapeutic agents requires a better understanding of processes and quality. Cell based therapies could be structural or systemic or both, which regulations committee must keep in mind. This is not trivial as one has to define quality for the two components and also the combinatorial product!

Conclusion

Industry standards are not always easy to follow and copy, and many times are unenforceable in the 'stem cell' field, for reasons that are understandable. It results into general perception that this is a poorly regulated industry. Both lack of proper regulatory standards and introduction of very stringent guidelines could restrict the successful translation of the tremendous advances taking place in this field into the clinic. While it is imperative that this industry needs to be regulated like the 'Pharma' & the 'Biologics' industry, we cannot lose sight of the fact that throttling it by introducing impossible, unjustified and repetitive experiments could be unproductive or counterproductive.

Though, the definition of QMS above has touched upon all the aspects, there are many factors which cannot be controlled merely by system calibration which emphasizes the role of optimization, standardization and QC. The clinical study investigators rely on trial results provided the interpretations are not biased by error in the measurement or test method. This highlights the importance of precision and accuracy based QC assurance since it has direct impact on the design, conduct and outcome analysis of multicenter trials. This can be achieved by implementation of stringent statistical power and minimizing participant risk and laboratory burden.

Summarizing, it is evident that in any new activity pertaining to biologic manufacturing, a methodical and strategic quality approach with a business process improvement in mind, can greatly eliminate the causes of defects and errors in complex manufacturing and service processes such as 'Cellular therapies.' The long-term outlook for regenerative medicine predicts an increased need for scalable cell expansion technologies that utilizes non-animal derived materials and is compatible with limited number of downstream processing steps.

While the importance of tools and techniques of QMS cannot be ignored by the stem cell industry, quality management has to be given paramount importance by organizations, to be able to deliver to the client's promises and regulatory compliances. QMS is a constantly evolving process and is a continuous learning for all involved directly and indirectly. Thus, quality is a journey not a destination. Some guidelines are stringent to the extent of spreading unnecessary fear among the investors in this industry, which can be avoided. The stem cell industry with pre-set QMS is the need of an hour and we strongly feel that it will ease some burden. Needless to say therefore, a sound QMS will have an unparalleled role in shaping up this novel cell based medicinal industry.

Conflict of Interest

The authors would like to state that there is no conflict of interest whatsoever.

Acknowledgments

We thank the management of ReeLabs Pvt. Ltd. for extending the necessary support for this project.

References

[1] *AABB standards for cellular Therapy Products & Services* (8th edition).

[2] '*Quality Management Principles*' from ISO website www.iso.org accessed on 07 Jul 2017.

[3] Rose KH. *Project Quality Management: Why, What and How.* Fort Lauderdale, Florida: J. Ross Publishing. 2005.

[4] Dr. Chandra V, Dr. Shabari S, Sushilkumar R. Accreditation and Standardization of Cellular Therapies in India. *Telegraft (ISCT).* Apr 2015; 21:2.

[5] Cade Hildreth. *Major Milestone for Stem Cells – 300K Scientific Publications.* 23 August 2017. Online article published on https://www.bioinformant.com/major-mile stone-stem-cells-300k-scientific-publications/. Accessed 26 September 2017.

[6] *Gazette Notification G.S.R.899 (E),* by Ministry of Health and Family Welfare, Government of India (27th December, 2011).

[7] *National Guidelines for Stem Cell Research – 2017.* Jointly released by ICMR and DBT on website www.icmr.nic.in.

[8] Medcalf N, Hourd P, Chandra A and Williams DJ. Quality assurance and GMP in the manufacture of cell-based therapeutics. (June 30, 2014). *StemBook* ed. *The Stem Cell Research Community*, StemBook, doi: 10.38 24/stembook.1.99.1.

[9] Lopez F, Di Bartolo C, Piazza T, Passannanti A, Gerlach JC, Gridelli B et al. Quality risk management model approach for cell therapy manufacturing. *Risk Anal* 2010 Dec; 30(12):1857-71.

[10] Shetty P, Cooper K, Viswanathan C. Comparison of proliferative and multilineage differentiation potentials of cord matrix, cord blood, and bone marrow mesen-chymal stem cells. *Asian J Transfus Sci* 2010 Jan; 4(1): 14-24.

[11] *Chagas disease (American trypanosomiasis) Fact sheet* Updated March 2017. website http://www.who.int/mediacentre/factsheets/fs340/en/. Accessed on 31 Aug.

[12] *Wikipedia.* Quality Control. Website https://en.wikipedia.org/wiki/Quality_control. Accessed on 07 July 2017.

[13] Thomas RJ, Alvin WN, Mark JM, Karen C, Bo K & Christopher JH. The translation of cell-based therapies: clinical landscape and manufacturing challenges. *Regenerative Medicine* 6 Jan 2015; 10:1.

[14] Mason C, Hoare M. Regenerative medicine bio-processing: the need to learn from the experience of other fields. *Regen Med* 2006; 1(5):615–623.

[15] European Medicines Agency. *European Public Assessment Report: Provenge* (2013, updated in 2015). www.ema.europa.eu/ema/index.jsp?curl=pages/medicines/ human/medicines/002513/human_med_001680.jsp&mid=WC0b01ac058001d124 accessed on 15 September 2017.

[16] Crosby P. *Quality improvement through defect prevention.* Philip Crosby Associates. 1985.

[17] Dennis GB. *Cost of Quality, Quality Planning and The Bottom Line.* Oct 20, 2000. http://intra.itiltd-india.com/ quality/QulandRelTools/RR-01-08%20Quality%20Cost .pdf. Accessed on 15 September 2017.

[18] Hummer C, Poppe C, Bunos M et al. Automation of cellular therapy product manufacturing: results of a split validation comparing CD34 selection of peripheral blood stem cell apheresis product with a semi-manual vs. an automatic procedure. *Journal of Translational Medicine*. 2016; 14:76.

[19] Oikonomopoulos A, van Deen WK, Manansala A-R et al. Optimization of human mesenchymal stem cell manufacturing: the effects of animal/xeno-free media. *Scientific Reports* 2015; 5:16570.

[20] Halme DG, Kessler DA. FDA regulation of stem-cell-based therapies. *N Engl J Med* 2006 Oct 19; 355(16): 1730-5.

[21] Ivan A, Indrei LL. Emergence of transmissible disorders, a continuous process--a new type of viral meningoencephalitis. *Rev Med Chir Soc Med Nat Iasi* 2000 Apr-Jun; 104(2):51-5.

[22] Mackensen A, Drager R, Schlesier M, Mertelsmann R, Lindemann A. Presence of IgE antibodies to bovine serum albumin in a patient developing anaphylaxis after vaccination with human peptide-pulsed dendritic cells. *Cancer Immunol Immunother* 2000 Jun; 49(3):152-6.

[23] Spees JL, Gregory CA, Singh H, Tucker HA, Peister A, Lynch PJ, Hsu SC, et al. Internalized antigens must be removed to prepare hypoimmunogenic mesenchymal stem cells for cell and gene therapy. *Mol Ther* 2004 May; 9(5):747-56.

[24] Shih DT, Burnouf T. Preparation, quality criteria, and properties of human blood platelet lysate supplements for ex vivo stem cell expansion. *N Biotechnol* 2015 Jan 25; 32(1):199-211.

[25] Krishna MP, Sunghoon J, Lawrence R and Leo AB. Bioprocessing strategies for the large-scale production of human mesenchymal stem cells: a review. *Stem Cell Research & Therapy* 2015; 6:225.

[26] Schedule M, *Good Manufacturing Practices and Requirements of Premises, Plant and Equipment for Pharmaceutical Products.* The Drugs & Cosmetics Act, 1940 (As Amended Up To 30 June 2005), Ministry Of Health and Family Welfare, Government of India.

INDEX

C

D

E

F

G

H

I

J

K

L

M

N

O

P

Q

R

S

T

U

V

W

Y

β

Adult Stem Cell Released Molecules: A Paradigm Shift to Systems Therapeutics

Author: Greg Maguire, Ph.D.

Series: Stem Cells – Laboratory and Clinical Research

Book Description: Dr. Greg Maguire's thought-provoking book offers a radically new approach to the standard means of developing drugs through the use of the molecules that stem cells release.

Hardcover ISBN: 978-1-53612-623-5
Retail Price: $160

Stem Cells from Culture Dish to Clinic

Editors: Vinod Verma, M.P. Singh, and Manoj Kumar

Series: Stem Cells – Laboratory and Clinical Research

Book Description: The book will help to increase the awareness vis-à-vis stem cell research in developing countries in particular India. The book pronounces the ways which may help us to extract maximum benefits from state of the art technology.

Hardcover ISBN: 978-1-53612-732-4
Retail Price: $230